Oh, *Evolve!*

(Good Luck With That. . .)

THOUGHTS ON. . .

Oh, *Evolve!*

(Good Luck With That. . .)

Book 14 in the "THOUGHTS ON" Series
by David M. Arns

First Edition
Copyright © 2020 David M. Arns.
Eighth Printing, March 2024

All rights reserved.
Paperback ISBN: 979-8-67665-689-8
E-book ISBN: 978-1-64669-229-3

Copyright © 2020 David M. Arns.

All rights reserved. No part of this publication may be reproduced, distributed, or transmitted in any form or by any means, including photocopying, recording, or other electronic or mechanical methods, without the prior written permission of the author, except in the case of brief quotations embodied in critical reviews and certain other noncommercial uses permitted by copyright law.

Books in the "THOUGHTS ON" Series

All the books below are available both in electronic form and in paperback, and are available from the sources mentioned on the website Bible-Author.DaveArns.com.

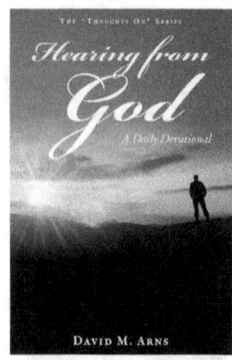 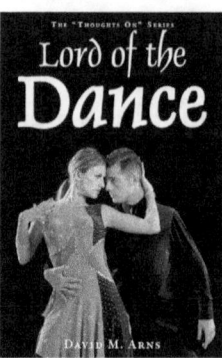 

  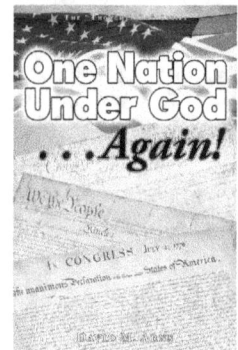

Music in the "Worship On" Series

All the music below is available both as downloadable electronic files and as physical CDs, and are available from the sources mentioned on the website Music.DaveArns.com.

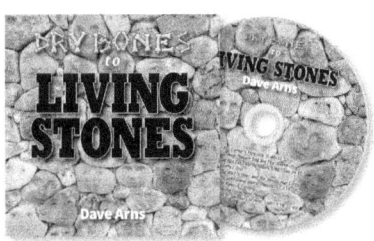

For descriptions and other details of all these books or albums, see the back of this book, or the websites BibleAuthor.DaveArns.com or Music.DaveArns.com, respectively.

Table of Contents

Preface 17
   Typographical Conventions 18
   Images 20

Chapter 1: The $64,000 Question 21
   Catastrophism vs. Uniformitarianism 24
      The Parable of the Candlestick 25
      More on Uniformitarianism 27
   Occam's Razor 27
   But Isn't Evolution a Proven Fact? 29
   The Question of Religion 33
   Evolution as a Social Theory 47
   Evolution and Racism 50
      Did Darwin Espouse This? 52
   The Appeal of Evolution 55
   The Politics of Evolution 56
   Could Both Be True? 58
      The Gap Theory 58
      The Day-Age Theory 60
      Framework Theory 61
      The Implicit Conflict With Rationality 61
      The Implicit Conflict With Scripture 62
   Recognizing Intelligence 65
   Explaining vs. Predicting 67
   Chapter 1 Summary 69

Chapter 2: The Theory of Evolution 71
   The Title of This Book 72
   What Is Evolution? 73
      Muddying the Waters 76
      Microevolution vs. Macroevolution 79
   Sacred Cows of Evolution 81
      Embryonic Recapitulation 81
      The Miller-Urey Experiment 83
      The Phylogenetic Tree 86
      Vestigial Organs 90
      The "March of Progress" 92
      Slaughtering the Sacred Cows 93

    The *Theory* of Evolution 94
        Darwin's Misgivings 94
        The Scientific Method 97
    The Second *Law* of Thermodynamics 98
    The "God of the Gaps" 104
    Division in the Camp 106
    Agreement in the Other Camp 110
    Chapter 2 Summary 113

Chapter 3: A Young Universe 115
    The Big Bang 115
        Creation *Ex Nihilo* 117
        Some Problems with the Big Bang Theory 118
    The Cosmological Principle 123
    Cosmic Microwave Background 125
    The Expanding Universe 127
        Redshift and Blueshift 127
    Blue Stars 130
    Supernova Remnants 135
    Spiral Galaxies 137
    Chapter 3 Summary 140

Chapter 4: A Young Solar System 141
    Where Did the Solar System Come From? 142
        The Roche Limit 146
        Orbital Considerations 147
    The Sun 149
        The "Young Faint Sun" Paradox 149
    Mercury 151
        The Unexpected Magnetic Field 152
        Outgassing of Volatile Compounds 154
    Venus 156
        Retrograde Axial Rotation 157
        Surface Features 157
    Earth 158
    Mars 159
        Liquid Water on Mars 159
        The "Unbalanced" Atmosphere 160
    Jupiter 161
        Retrograde Moons 163
        Jupiter's Moon Io 164
        Jupiter's Moon Europa 168

OH, EVOLVE! (GOOD LUCK WITH THAT...)

- Jupiter's Moon Ganymede........................170
- Saturn...171
  - Saturn's Rings................................171
    - How Old Are the Rings?..................172
  - Saturn's Moon Enceladus....................177
  - Saturn's Moon Titan..........................180
    - Titan's Atmosphere......................180
  - Saturn's Moon Iapetus.......................182
- Uranus..185
- Neptune..186
  - Neptune's Moon Triton......................187
- Pluto...188
  - Pluto's Moon Charon.........................191
- Comets..192
- Chapter 4 Summary.............................195

Chapter 5: A Young Earth............................197
- The "Goldilocks" Planet........................197
  - The Precision of the Universe..............198
  - The Strength of Gravity....................199
  - The Cosmological Constant.................200
  - Protons and Neutrons.......................201
  - Phase-Space Volume.........................202
  - The Precision of the Solar System.........203
  - The Precision of Earth.....................204
- Dating Methods.................................204
  - The Bathtub Illustration....................204
  - Back to Dating Methods.....................205
  - Carbon-14...................................206
  - Potassium-Argon.............................212
  - The "Sweet Spot"............................215
- Earth's Magnetic Field.........................216
- That Pesky Erosion.............................217
- Earth's Oceans.................................219
  - Ocean-Floor Sediment.......................219
  - Chemical Imbalances........................220
    - Sodium...................................220
    - Nickel...................................220
    - Other Chemical Clocks..................225
- The Moon..226
  - The Moon's Origin...........................226
  - The Moon's Construction....................228

11

    The Moon's Magnetic Field. 229
    The Moon's Recession. 231
Chapter 5 Summary 233

Chapter 6: Mutations 235
    Spontaneous Generation 236
        Wikipedia. 236
        Dictionary.com 238
        Encyclopedia Britannica 238
    Fascination with Mutations 240
    What Are Mutations? 241
        A Linguistic Illustration of Evolution 242
    Beneficial Mutations 247
        Equivocally Beneficial Mutations 247
            Antibiotic Resistance 248
        Unequivocally Beneficial Mutations 251
            "Hello, World!" 252
            Back to Mutations. 253
        Lip Service 255
    Fixing Typos 255
    Do Mutations Happen in People? 257
    The Effects of Mutation 262
    How Likely is Evolution? 268
        Doing the Math 269
        Our Hypothetical Organism. 270
        The Surface of the Earth 271
        How Much Time? 272
        Back to Our Hypothetical Organism 272
        Haldane's Dilemma. 276
    "Primitive" Man 279
        Neanderthal Man 280
        Cro-Magnon Man 284
        Java Man. 285
        Piltdown Man. 288
        Nebraska Man. 290
        Australopithecus 291
        Peking Man. 292
        Zinjanthropus. 293
        And On and On. 294
    Survival of the Fittest. 297
        Natural Selection as a Causal Agent 297
        Natural Selection as a Filter. 299

Arrival of the Fittest. 302
   Life "Arising" from Non-Life. 303
   Another Showstopper: Racemization. 305
   Where Did the Information Come From?. 307
   The Elephant in the Living Room. 309
      "Junk DNA" 311
   Irreducible Complexity. 316
   The "Light-Sensitive Spot" 321
   The Clot Thickens. 323
   Reproduction 326
      Kinesin: Life's Delivery Service 329
  Chapter 6 Summary 333

Chapter 7: The Fossil Record. 335
  The Geologic Column. 336
  Fast Geological Processes 339
   Polystrate Fossils 339
   Folded Rock Strata 341
   Bioturbation 343
   Original-Tissue Fossils 344
   Surtsey, Iceland. 346
   Mount St. Helens, United States. 348
      Fast Deposition 349
      Rapid Erosion 350
      Restoration of Flora and Fauna. 351
   Arches National Park 353
   Other Fast Processes 353
      Rock Formation 354
      Coal and Oil 355
      Cave Formations 357
      Gemstones. 359
      Petrification 361
      Glaciers 364
      Coral Reefs. 365
  The Grand Canyon. 366
   The Volume of the Grand Canyon 368
   Colorado River Flow Rate. 369
   Colorado River Sediment Content 369
   Doing the Math 370
   Other Fast Canyons 371
      Canyon Lake Gorge 372
      Mount St. Helens' Little Grand Canyon. 373

- Burlingame Canyon 374
- Providence Canyon 374
- The Cambrian Explosion............................... 375
  - Punctuated Equilibrium 377
- From Invertebrates to Fish............................ 380
- From Fish to Amphibians 382
- From Amphibians to Reptiles 384
  - Walking Reptiles to Flying Reptiles................. 384
- From Reptiles to Birds................................ 387
- From Reptiles to Mammals 390
- From Walking Mammals to Swimming Mammals........ 391
- What Can We Learn from This? 392
- The Worldwide Flood 395
  - The Timeline of Noah's Flood....................... 395
    - The "Fountains of the Great Deep" 397
  - The *Mountains* were Covered?...................... 399
  - How Could the Ark Hold All Those Animals? 400
    - The Ark Was Big 400
    - It Wasn't *That* Many Animals..................... 402
    - The Animals Used Little Floor Space 404
    - For More Detail................................. 405
- The Lack of Transitional Fossils....................... 406
- Chapter 7 Summary 410

Chapter 8: The Next Step 413
- Retaining Your Brain................................. 413
- The Parable of the Car Lot 417
- A Prophetic Message 421
- Requiem for Deep Time 422
- Unreal Scientists 435
  - Leonardo da Vinci (1452–1519)...................... 435
  - Sir Francis Bacon (1561–1626)....................... 437
  - Galileo Galilei (1564–1642)......................... 438
  - Johannes Kepler (1571–1630)........................ 439
  - Blaise Pascal (1623–1662).......................... 440
  - Robert Boyle (1627–1691) 441
  - Isaac Newton (1642–1727).......................... 443
  - Leonhard Euler (1707–1783) 444
  - Carl Linnaeus (1707–1778) 447
  - Louis Pasteur (1822–1895).......................... 448
  - Lord Kelvin, William Thomson (1824–1907)........... 449
  - James Clerk Maxwell (1831–1879).................... 450

 . . .And So Many Others! 452
 But How Important Is It, Really? 452

Appendix A: Bibliography 461
 Websites 461
 Books 462

Glossary 465

About the Author 471
 Books in the "Thoughts On" Series 471
 Music in the "Worship On" Series 478

Oh, Evolve! (Good Luck With That...)

Preface

All Scripture references are from the public-domain King James Version (KJV) of the Bible unless otherwise noted. Other versions of the Bible that may be quoted are as follows:

- AMP: Amplified Bible: Copyright © 1954, 1958, 1962, 1964, 1965, and 1987 by the Lockman Foundation, La Habra, CA, 90631. All rights reserved. www.lockman.org.
- ASV: American Standard Version of 1901: Public Domain.
- BBE: Bible in Basic English: This text is in the public domain and has no copyright. The Bible In Basic English was printed in 1965 by Cambridge Press in England. Published without any copyright notice and distributed in America, this work fell immediately and irretrievably into the Public Domain in the United States according to the UCC convention of that time.
- CEB: Common English Bible: All rights reserved.
- CEV: Contemporary English Version: Copyright © 1995 by American Bible Society. All rights reserved.
- DARBY: Darby Translation: Public domain. First published in 1890 by John Nelson Darby, an Anglo-Irish Bible teacher associated with the early years of the Plymouth Brethren.
- ERV: Easy-to-Read Version: Copyright © 2006 by World Bible Translation Center.
- ESV: The Holy Bible, English Standard Version: Copyright © 2001, 2006, 2011 by Crossway Bibles, a division of Good News Publishers. All rights reserved.
- GWORD: God's Word Translation: Copyright © 2010 by Baker Publishing Group, © 1995 by God's Words to the Nations. All Rights reserved.
- HCSB: Holman Christian Standard Bible: Copyright © 1999, 2000, 2002, 2003 by Holman Bible Publishers. Holman Christian Standard Bible®, Holman CSB®, and HCSB® are federally registered trademarks of Holman Bible Publishers. Used by permission.
- MSG: The Message: Scripture taken from The Message. Copyright © 1993, 1994, 1995, 1996, 2000, 2001, 2002. Used by permission of NavPress Publishing Group.
- NABRE: New American Bible, Revised Edition. © 2010, 1991, 1986, 1970 Confraternity of Christian Doctrine, Inc., Washington, DC. All Rights Reserved.
- NASB: New American Standard Bible: Copyright ©1960, 1962, 1963, 1968, 1971, 1972, 1973, 1975, 1977, 1995 by The Lockman Foundation. All rights reserved.
- NET: New English Translation. The NET Bible®, First Edition (NET); New English Translation, The Translation That Explains Itself™; Copyright © 1996-2005 by Biblical Studies Press, L.L.C. All rights reserved.
- NIV: New International Version: Scripture quoted by permission. Quotations designated (NIV) are from The Holy Bible: New International Version (NIV). Copyright © 1973,

1978, 1984 by International Bible Society. Used by permission of Zondervan Publishing House. All rights reserved.

- NKJV: New King James Version. Scripture taken from the New King James Version®. Copyright © 1982 by Thomas Nelson. Used by permission. All rights reserved.
- NLT: New Living Translation: Holy Bible, New Living Translation, Copyright © 1996, 2004 by Tyndale Charitable Trust. Used by permission of Tyndale House Publishers, Inc., Wheaton Illinois 60189. All rights reserved.
- NLV: New Life Version: Copyright © 1969 Christian Literature International.
- TEV: Today's English Version: Today's English Version Bible. Copyright © American Bible Society, 1966, 1971, 1976, 1992. Used by permission.
- TLB: The Living Bible: Copyright © 1971 by Tyndale House Foundation. Used by permission of Tyndale House Publishers Inc., Carol Stream, Illinois 60188. All rights reserved.
- TPT: The Passion Translation®. Copyright © 2017 by BroadStreet Publishing® Group, LLC. Used by permission. All rights reserved. thePassionTranslation.com
- WEYMTH: Weymouth New Testament in Modern Speech: Third Edition, 1913. Public Domain—Copy Freely.

Typographical Conventions

In Scriptural quotes in this book, emphasis (indicated by **boldface** type, and occasionally ***italic* within the boldface**) may be added by the author to draw attention to the portions of the passage that pertain to the topic currently under discussion. This applies throughout, so "emphasis added by author" doesn't need to be stated in every single instance.

In this book, the generic pronouns "he," "him," and "his" are used whenever explicit inclusion of both gender-specific pronouns would result in grammatical cumbersomeness. We know that in Christ, there is no difference between male and female (Galatians 3:28), so the pronouns used in this way should be read as generic, not masculine.

When you see a number prefixed by a "H" or a "G", it represents the word number Hebrew or Greek dictionaries of *Strong's Exhaustive Concordance,* one of the standard tools for Biblical study: *Strong's Hebrew and Chaldee Dictionary of the Old Testament* (Hebrew Strong's) and *Strong's Greek Dictionary of the New Testament* (Greek Strong's), both public domain. So, for example, "G256" indicates that English word being discussed was translated from the word defined in entry 256 in Strong's Greek Dictionary.

In Scripture quotations, the letter case of the English word "Lord" indicates the standard meanings when quoting from the Old Testament. Mixed Case, as in the word "Lord," indicates the Hebrew name אֲדֹנָי (*Adonay*, H136), while SMALL CAPS, as in the word "LORD" indicates the Hebrew name יְהוָֹה (*Yahweh*, H3068), also known as the Tetragrammaton, which literally means "four letters." And finally, when the original Hebrew uses the name יְהוָֹה אֱלֹהִים (*Yahweh Elohim*, H3068 H430), it is translated and lettercased as "Lord GOD."

In order to retain accents and diacritical marks of the original languages—which are very meaningful—Hebrew and Greek words are rendered as small inline images, as in the previous paragraph. This works fine for e-readers such as iBooks and nook, but doesn't work so well on some Kindle e-readers. Because of an inability of some Kindle devices to scale inline images to a size proportional to the currently selected text size, Hebrew and Greek words will likely be larger than the surrounding text. Also, some Kindle devices can't adjust the vertical alignment of images, so the Hebrew and Greek descenders don't actually descend below the text baseline. Although this looks bad on such Kindle devices, no information is lost. If typographical aesthetics are important to you, you may want to get the iBooks version of this book instead.

Images

Images in this book have come from several sources:

- Some images are from Wikimedia Commons, are licensed in accordance with their requirements, and will be labelled with the appropriate Creative Commons license tag, as well as a hyperlink to the copyright information on Wikimedia Commons. All images from Creative Commons are licensed under one of the following licenses:
    - CC Public Domain: https://commons.wikimedia.org/wiki/Template:PD-US-expired or https://CreativeCommons.org/public-domain/zero/1.0/deed.en
    - CC-BY-SA-2: https://CreativeCommons.org/licenses/by-sa/2.0
    - CC-BY-SA-3: https://CreativeCommons.org/licenses/by-sa/3.0
    - CC-BY-SA-4: https://CreativeCommons.org/licenses/by-sa/4.0
    - CC-BY-2: https://CreativeCommons.org/licenses/by/2.0/deed.en
    - CC-BY-2.5: https://CreativeCommons.org/licenses/by/2.5/deed.en
    - CC-BY-3: https://CreativeCommons.org/licenses/by/3.0/deed.en
    - CC-BY-4: https://CreativeCommons.org/licenses/by/4.0/deed.en
- Some images, primarily the space pictures, are made available by NASA National Aeronautics and Space Administration) and JPL (Jet Propulsion Laboratory), or one of their affiliates. These images, having been paid for by tax dollars, are in the public domain.
- Photo- or illustration captions containing "Wellcome *nnnnnnnn*" indicate that the associated photo or illustration comes from Wellcome Images, a website operated by Wellcome Trust, a global charitable foundation based in the United Kingdom. Refer to Wellcome blog post (archive).
- Some images were created by the author, and are labelled as such.
- Illustrations or photos without associated licensing notification are in the public domain.

Chapter 1:

The $64,000 Question

In any discussion of the relative merits of creation versus evolution, one of the main points of contention—perhaps *the* main point of contention—is the timeframe in which various things happened, or the amount of time it took for some process to take place.

There are many other questions, of course, but the majority of them boil down to the timeframe. So in this chapter, I will introduce the timeframes that the two camps proffer, and then in the remainder of the book, will address only a few Biblical statements or references because, on the whole, those who espouse evolution do not recognize the authority of the Bible. Therefore, most of this book will be looking at the evidence, or lack thereof, for evolution, and compare how the observations fit the creation model. But creationists will appreciate the occasional Biblical reference among the plethora of science-based content.

Terminology-wise, I will use the word "creationist" to refer to those who believe the Bible is accurate when it says a certain thing happened at a particular point in time—even if the specified point in time was not the main emphasis of the passage. And, I will use the word "evolutionist" to describe those who believe that natural, undirected, random events are responsible for all that we observe in the universe today.

As such, "evolutionist" refers to the person who believes one or more of the following things:

- **Cosmological evolution:** That the universe (galaxies, stars, nebulae, etc.) and solar system (planets, moons, asteroids, comets, etc.) came about by random means,
- **Geological evolution:** That the current state of the Earth (rock layers, fossils, tectonic plates, etc.) came about by random means,
- **Chemical Evolution:** That chemicals in a primordial ocean spontaneously collected into organisms that began to live (respond to stimulus, take in nutrients, eliminate waste, and reproduce)—life "arising" by itself.
- **Biological evolution:** That simple life on Earth (viruses, bacteria, etc.) transformed, by random genetic mutations, into ever more complex organisms (e.g., plants, insects, fish, birds, mammals, humans).

To think that "evolution" encompasses all of the above areas is not an exaggeration or a sweeping generality espoused by a rabid creationist; look at these quotes from leading evolutionists:

> Evolution comprises **all** the stages of the development of the universe: the **cosmic, biological, and human** or cultural developments.[1]

> Most enlightened persons now accept the fact that **everything in the cosmos**—from heavenly bodies to human beings—has developed and continues to develop **through evolutionary processes**.[2]

> Our present knowledge indeed forces us to the view that **the whole of reality *is* evolution**—a single process of self-transformation.[3]

Note the not-so-subtle jab at creationists in the second quote above: it's the "enlightened persons" who accept the "fact" that everything evolved. Clearly, if you don't believe that, you're *not* enlightened.

---

[1] Theodosius Dobzhansky, "Changing Man," *Science,* Vol. 155, January 27, 1967, p. 409. Emphasis added.
[2] Rene Dubos, "Humanistic Biology," *American Scientist,* Vol. 53, March, 1965, p. 6. Emphasis added.
[3] Julian Huxley, "Evolution and Genetics," in *What Is Man?* (Ed. by J. R. Newman, New York, Simon and Schuster, 1955), p. 278. Bold emphasis added; italic emphasis in original.

Chapter 1: The $64,000 Question

Let's think a little bit more about the quotes above, which imply, or even bluntly state, that "the whole of reality *is* evolution." C.S. Lewis noted in the 1940s[4] that there is a very large disconnect between what evolution says and what we observe when we look at how nature actually works:

> By universal evolutionism I mean the belief that the very formula of universal process is from imperfect to perfect, from small beginnings to great endings, from the rudimentary to the elaborate, the belief which makes people find it natural to think that morality springs from savage taboos, adult sentiment from infantile sexual maladjustments, thought from instinct, mind from matter, organic from inorganic, cosmos from chaos. This is perhaps the deepest habit of mind in the contemporary world. **It seems to me immensely unplausible, because it makes the general course of nature so very unlike those parts of nature we can observe.**

In other words, "If all of nature acts as the theory of evolution claims, why can't we ever see it acting that way?" Lewis continues:

> **The obviousness or naturalness** which most people seem to find in the idea of emergent evolution thus **seems to be a pure hallucination.**

In the context of a creation/evolution discussion, the main anchor points in time are the beginning of the universe, the formation of the Earth and its oceans, and the appearance of life:

	Creationist Chronology	**Evolutionist Chronology**
Beginning of the Universe	≈6,000 years ago	≈13.8 billion years ago
Formation of Earth	≈6,000 years ago	≈4.5 billion years ago
Formation of Oceans	≈6,000 years ago	≈4.4 billion years ago
Appearance of Life	≈6,000 years ago	≈4.3 billion years ago
Appearance of Mankind	≈6,000 years ago	≈2.4 million years ago
Worldwide Flood	≈4,500 years ago	(dismissed entirely)

Note: The evolutionist chronology is revised from time to time, so various sources may give different opinions.

---

[4] C.S. Lewis, *The Weight of Glory*, chapter "Is Theology Poetry?" (HarperCollins e-books, 1949, 1976).

Note that there is a point in the creationist's timeline that evolutionists typically dismiss entirely, and that is the timing of Noah's Flood. However, this is a very significant event, because it dramatically influenced the geologic column (rock strata and the fossils therein) as well as various attributes of human DNA (since the Biblical record has all subsequent humanity being descended from Noah). We will cover these topics in more detail in later chapters.

There have been some attempts to "improve" the Biblical account of creationism, all of which try to combine the Biblical narrative with various aspects of current evolutionary thinking, in spite of the fact that such combining of mutually exclusive views is inherently self-contradictory. (Some of these proposed hybrids will be covered below in "Could Both be True?".) So in this book, my usage of the word "creationism" refers to the version that follows the Biblical narrative: creation was completed in six days, Adam and Eve were real people, and there was a worldwide flood. If this sounds a bit too religious for you, please stick with me; I think you will come across some very thought-provoking discoveries.

The enormous timeframes espoused by evolutionists are often referred to as **deep time**, or a **long-age** viewpoint, as in the "millions" of years ostensibly required for the evolution of man, or the "billions" of years required for the universe, or the Earth, to arrive at the state they are in now. These terms will be used later in this book when referring to the enormous timespans required by the evolution model.

By the way, the title for this chapter, "The $64,000 Question," comes from an old TV game show shown during the 1950s, in which progressively harder questions were asked and the prize money awarded to the contestants doubled (roughly) with each new question, up to a maximum of $64,000. Thus, "the $64,000 question" has become a figure of speech meaning "the most important question," or at least "an exceedingly important question." And indeed, a correct answer to the question of whether God is actually the Creator of the universe, the solar system, the Earth, and all life—whether God is *your* Creator Who loves you dearly and created you for a purpose—is an exceedingly important question. Truly, it is *the* most important question.

Catastrophism vs. Uniformitarianism

These two rather long words represent one of the important distinctions between the creation viewpoint and the evolution viewpoint. Catastrophism is a term that denotes occurrences that happen because of sudden and unusual events; for example, creation itself, or Noah's Flood. Clearly, creationists believe that catastrophism is the significant factor in the story of history.

Evolutionary theories absolutely insist on uniformitarianism, because if non-uniform events happened at particular times, it would point to an overseeing Mind causing the events and choosing the timing. And that insistently points to a Creator, which is unacceptable. But uniformitarianism continuously causes problems for evolutionists, although it is rarely admitted.

The Parable of the Candlestick

Here is a typical kind of problem caused by uniformitarianism; I call it "The Parable of the Candlestick."

A scientist walks into a room and sees a candle burning on the table. He measures it and finds that it is 13 inches tall. He sits down and observes it for an hour and then measures it again: it is now 12 inches tall. He sits down and observes it for another hour and then measures it a third time: it is now 11 inches tall. An evolutionist walks into the room and they strike up a conversation.

Evolutionist: See that candle? It's been burning for a year.

Scientist: How do you know? Have you been watching it for a year?

Evolutionist: No, but processes such as candles burning take a really long time to happen.

Scientist: Well, it doesn't take that long; I've just measured how fast it takes for this candle to burn down, and it certainly hasn't been burning for a year!

Evolutionist: Actually, it has, I'm afraid. It takes a long time for these things to happen.

Scientist: I just measured this candle a little while ago, and after a known amount of time, I measured it again, and then did the same thing again. It burns down at a rate of one inch per hour.

Evolutionist: That may be, but it has been burning for a whole year.

Scientist: That makes no sense at all! If it burns down at a rate of one inch per hour—and it does, I just observed it doing so—that means that a year ago, it would have been 730 feet tall!

Evolutionist: Your calculations must be wrong.

Scientist: This candle is about an inch in diameter, and since candle wax is about 1g/cc, that candle would have weighed almost 250 pounds a year ago! It wouldn't have had the structural strength to stand up; its own weight would have crushed the base! Which, as you can see, it hasn't.

Evolutionist: It must be an exceptionally strong candle.

Scientist: And if it was 730 feet tall a year ago, there would be a hole in the ceiling and roof above it. There isn't!

Evolutionist: Well, maybe the hole in the ceiling was patched by a reservoir of plaster contained above it. And you can't actually see the roof, so maybe the hole is still there.

Scientist: But a year ago, a 730-foot candle would have been high in the air, and the flame would have been quickly snuffed out by wind or rain!

Evolutionist: It's a special kind of wax that stays lit even under the most extreme circumstances.

Scientist: What? Do you have any evidence for that?

Evolutionist: Well, it's obvious that it has to be, to survive the wind and rain up there. And besides, the flame could have been re-lit by lightning strikes.

Scientist: Are you serious? But here's another thing: if this candle had been burning for a year, there would be a huge cone of candle drippings at the base. There isn't! Only two or three drips have reached the bottom of the candle!

Evolutionist: Yes. This special kind of wax doesn't drip much.

Scientist: That's certainly convenient, isn't it? But you also have to consider the wind gusts several hundred feet in the air: they would have broken the candle off the very first day!

Evolutionist: It's always possible that wind gusts didn't happen in the vicinity of the candle. That's possible, you know.

Scientist: You're proposing that wind gusts wouldn't happen in this particular place, even though they happen everywhere else on the planet? Do you have any actual reason to say that?

Evolutionist: It's self-evident that the wind gusts didn't happen because the candle obviously didn't break off...

In the above parable, as in real life, evolutionists are loath to admit that things don't take enormous amounts of time, because their entire doctrinal system depends on it. Rather than acknowledge that things could happen because of an overseeing Mind causing events for a purpose, evolutionists would rather add to the main theory one or more supplemental theories, for which there is no evidence and which fulfill no purpose except to lend credibility to the uniformitarian tenet of faith. Such supplemental theories are called "rescue devices;" we'll see many of them in the following pages.

More on Uniformitarianism

So creationists believe that catastrophism plays a significant part in the history of the universe, and evolutionists say uniformitarianism is the way of things. Again, uniformitarianism is the belief that things happen because of continuous and uniform processes. Evolutionists would say that star formation, planetary formation, and the development of life are all uniformitarian processes; they came about slowly and gradually, over enormous spans of time. Clearly, evolutionists believe that uniformitarianism is the significant factor in the story of history.

Closely related to uniformitarianism is **naturalism**. Like evolution itself, naturalism too is a religious conviction; specifically, that there is no God.[5] According to naturalism, everything that happens is a result of natural forces and processes, because there is no such thing as a *supernatural* force or process.

Both creationists and evolutionists have existing worldviews and presuppositions at the outset: Creationists have the presupposition that there is a God who created the universe, and evolutionists have a presupposition that there is no God, so therefore he couldn't have created the universe. We all have access to the same data, and the same experiments; what is different, because of our presuppositions, is how we *interpret* the data. We all *see* the same evidence, but in order to draw conclusions about what the data *means,* we employ our worldview—our fundamental beliefs about the nature of reality.

So, the approach taken in this book will be to compare the two interpretations—evolution and creation—and notice which interpretation is more consistent with observational science. In other words, we'll notice which interpretation better matches the universe as we observe it, and the laws of physics that we can test and measure.

Occam's Razor

Occam's Razor is a problem-solving principle that it is used in many different areas of science, and the creation/evolution question is no exception.

---

[5] Psalm 14:1 (NET): Fools say to themselves, "There is no God." They sin and commit evil deeds; none of them does what is right.

It was named after an English Franciscan friar by the name of William of Ockham (c. 1287–1347), a scholastic philosopher and theologian.[6]

In essence, Occam's Razor (also known as "the law of parsimony") states that when trying to determine the validity of competing hypotheses, the one that makes the fewest assumptions is more likely to be correct. In other words, "the simplest solution is most likely the right one."

For example, suppose you are awakened one night by the sound of a violent, howling windstorm. But, soon you go back to sleep anyway. The next morning when you go outside, you discover that a large tree branch has broken off and landed on your car, causing severe damage. What happened? Actually, you don't know. You didn't see what happened, and neither you nor anyone else on your block has security cameras aiming this direction. Not only that, but after talking to your neighbors, none of them saw what happened either.

So, are you just stuck, never knowing what happened to make the tree branch fall on your car? Yes and no. You will never know for *absolute certain* what happened, but you do know your car was fine when you parked it there yesterday. You also know that there was a violent windstorm last night, so you'd probably assume that the wind broke the tree branch, whereupon it fell onto your car and damaged it. True, it is an assumption, but it's very likely to be true, based on your experience with wind in the past. This conclusion is made even more likely to be true if you see other branches and twigs lying on the ground up and down your street.

However, there is another possibility as well. It's possible that Stevie, that kid that you used to make fun of in grade school, used the windstorm to help him finally take revenge on you for being so mean to him all those years ago. He only lives a couple hours away now, so it would have been easy for him to check the weather forecast and see that a windstorm was coming into your area, and then he could have bought an electric winch at an all-night hardware store on his way here. He could have easily gotten here from his city in time, and parked half a block away to watch until the storm was at its peak. Then, getting out the winch, he could have thrown a strap over the branch, anchored the other end to the base of a different tree, and turned on the winch. The winch put ever more force on the branch until it broke, falling onto your car, causing thousands of dollars' worth of damage. Stevie, with grim satisfaction, extricates the strap from the fallen branch, unties it from the base of the other tree, throws it in his trunk, and drives home, chuckling all the way.

---

[6] en.wikipedia.org/wiki/Occam's_razor; accessed May 26, 2020.

So which scenario is more likely to have actually happened? After all, in the second scenario, *every bit of it* was doable. Could Stevie have remembered your making fun of him? Yes. Could he have waited all these years to take revenge? Yes. Could he have checked the weather forecast, seen the incoming windstorm, driven here, and bought an electric winch on the way? Yes. Could he have gotten here before the storm, sat in his car, picked out the perfect branch to pull down, thrown a strap over it, secured his winch to another tree, and turned it on long enough to break the branch, sending it onto your car? Yes.

All of those things are possible. But you would probably never conclude that that is how the tree branch fell on your car. Why? It's self-evident that an explanation that makes *one* assumption (the wind broke the branch) is enormously more likely to be correct than an explanation that makes *twelve* assumptions (the revenge of Stevie).

And if you insisted on concluding that the revenge of Stevie was the cause of the car damage, people would wonder why in the world you are thinking that. "Is he paranoid?" "Is he trying to hide something?" "Is there an ulterior motive?" It's intuitive to everyone that the more assumptions you have to make to explain some historical event, the less likely your explanation is to be true.

So, Occam's Razor is the principle that would conclude that, of the two options presented, the tree branch fell because of the wind broke the branch, rather than the long-delayed revenge of some kid you went to school with years ago. Occam's Razor would conclude that precisely *because* there were far fewer assumptions that were necessary to make the option possible.

Occam's Razor is very useful in deciding creation/evolution questions, because, as we saw above: the more assumptions required, the less likely the actuality.

But Isn't Evolution a Proven Fact?

Far from it. However, to listen to the movies, TV shows, science documentaries, and school textbooks, you'd swear that nothing has been more conclusively proven. But you'd be wrong. Ponder the following comments on the subject of **paleoanthropology** (the study of human fossils), made mostly by non-creationists.

Sherwood L. Washburn of the University of California, Berkeley, commented:

> It may be more useful to regard the study of evolution as **a game rather than as a science**.[7]

After Sir Peter Medawar, who won the Nobel Prize in medicine, noted that paleoanthropology was a "comparatively humble and **unexacting kind of science**," Andrew Hill, from Yale University, agreed:

> It has certainly been possible to get away with being **an unexacting practitioner**.[8]

Note that both of the above quotes included the word "unexacting." What does this word mean? Basically, it means to "fudge." Not the chocolate confection but, as the Oxford English dictionary defines it, "to present or deal with something in a vague or inadequate way, especially so as to conceal the truth or mislead," and to "adjust or manipulate facts or figures to present a desired picture."[9] The more you study the evolutionary belief system, the more you will see the aforementioned "unexactingness;" that is, widespread fudging.

Elsewhere, Hill commented:

> Compared to other sciences, **the mythic element is greatest in paleoanthropology**.[10]

Interesting word, that. What does "mythic" mean? It means "relating to or resembling myth." But what does "myth" mean? It means "a widely held but false belief or idea."[11] So yes, characterizing as "mythic" the evolutionary tales of apes gradually turning into humans is entirely appropriate.

---

[7] Sherwood L. Washburn, *Abstracts of the 71st Annual Meeting, American Anthropological Association*, Washington, DC, 1972, p. 121. Emphasis added.

[8] Andrew Hill, "The gift of Taungs," review of *Hominid Evolution*, edited by D. V. Tobias, *Nature* 323 (18 September 1986): p. 209. Emphasis added.

[9] Oxford Dictionary of English, online version 14.4.72, Copyright 2022 Mobisystems, Inc.

[10] Andrew Hill, "The Myths of Human Evolution," review of *The Myths of Human Evolution*, by Niles Eldredge and Ian Tattersall, *American Scientist* 72 (March–April, 1984): p. 189. Emphasis added.

[11] *New Oxford American Dictionary*, Copyright © 2010, 2019 by Oxford University Press, Inc. All rights reserved.

Commenting on the rarity of paleoanthropologists actually getting to study fossils they themselves didn't find, Milford Wolpoff, of the University of Michigan, states:

> When the only people who can comment are the discoverers or friends of the discoverers, there is no sense of independent observer. **We're not practicing science. We're practicing opera.**[12]

Norman Macbeth, a Harvard-trained lawyer, studied evolution for years and was unimpressed by the scientific rigor. He wrote a book[13] describing his findings, and he noted that there were serious gaps in the evidence for evolution and errors in the reasoning of evolutionists. He claimed that evolution itself had become a religion (which, as we'll see below, is indeed the case). The alleged evidence for evolution, he charged, was not of the quality that would stand up in a court of law.

In the previous paragraph, the phrase "the reasoning of evolutionists" was used. Let's take a closer look at that phrase, because there is an inherent self-contradiction in the concept itself. Evolution, by its very definition, believes that all organisms, along with all their capabilities, arose by random, aimless, accidental processes. That would necessarily include our brains, with which we think. If our brains' functioning is simply a result of random, accidental, aimless processes, what is the basis for trusting its conclusions? *There is none.* If the thinking processes in our brains are the results of aimless and random occurrences, we can have no confidence in any conclusions that our brains come up with.

C.S. Lewis noticed this show-stopping problem back in the 1940s.[14] Referring to the claim put forth by evolutionists that their worldview is supported by reason, he says:

> ...those who ask me to believe this world picture also ask me to believe that Reason is simply the unforeseen and unintended by-product of mindless matter at one stage of its endless and aimless becoming. Here is flat contradiction. They ask me at the same moment to accept a conclusion and

---

[12] Gibbons, "Glasnost for Hominids," p. 1467. Emphasis added.
[13] Norman Macbeth, *Darwin Retried: An Appeal to Reason* (Boston: Gambit Incorporated, 1971).
[14] C.S. Lewis, *The Weight of Glory*, chapter "Is Theology Poetry?" (HarperCollins e-books, 1949, 1976).

to discredit the only testimony on which that conclusion can be based. The difficulty is to me a fatal one; and the fact that when you put it to many scientists, far from having an answer, they seem not even to understand what the difficulty is. . .

Lewis continues, "The man who has once understood the situation is compelled henceforth to regard the scientific cosmology as being, in principle, a myth; though no doubt a great many true particulars have been worked into it."

Another book along the same lines as the *Darwin Retried* book mentioned above, entitled *Darwin on Trial,* was written by law professor Phillip E. Johnson, from the University of California at Berkeley. Johnson, the only creationist quoted in this section, makes the following observations:

- Evolution is grounded on philosophical belief known as naturalism, rather than on scientific facts.
- The belief that a large amount of empirical evidence supports the theory of evolution is a false impression.
- Evolution is itself a religion, even though it denies a supernatural God.
- If the hypothesis of evolution were subjected to rigorous examination of the evidence, and without an initial presumption of its validity, it would have been abandoned long ago.
- Since atheism is a basic supposition in the theory of evolution, one can't also use evolution to "prove" there is no God because of its inherently circular reasoning. As in, "I can prove that God doesn't exist by first assuming that God doesn't exist."

Johnson also astutely notes:

> The story of human descent from apes is not merely a scientific hypothesis; it is the secular equivalent of the story of Adam and Eve, and a matter of immense cultural importance. . . The needs of the public and the [secular scientific] profession **ensure that confirming evidence will be found,** but **only an audit performed by persons not committed in advance** to the hypothesis under investigation can tell us whether the evidence has any value as confirmation.[15]

---

[15] Phillip E. Johnson, *Darwin on Trial* (Washington, DC: Regnery Gateway, 1991), p. 83. Emphasis added.

David van Reybrouck has studied the artwork created to describe paleoanthropological discoveries, and after examining 355 different paintings, drawings, and sculptures of supposed **hominins** (evolutionary ancestors of humans), wrote an article for the journal *Antiquity*. In the article, he notes that the pictures, drawings, and reconstructions:

- **Always** go beyond the archaeological data.
- **Always** involve the **speculations and prejudices** of the fossil discoverers, who advise the artists.
- **Always** involve interpretations that are heavy on unsubstantiated theory.
- **Always** are **nonobjective** but are trusted as being accurate.
- Are used so extensively because **they sell evolution so effectively.**[16]

Van Reybrouck's conclusion is that "A good drawing is like a Trojan horse: to be rhetorically effective, its interpretation must be hidden inside."[17]

So, in answer to the question posed in the title of this section, "But isn't evolution a proven fact?", the answer, as confirmed by the evolutionists quoted above, is a resounding "Not at all!"

For a much more thorough treatment of the sloppy and shady handling of fossils, along with biased and unsubstantiated—though well publicized—conclusions, see *Bones of Contention,* by Marvin Lubenow.[18]

The Question of Religion

It's not uncommon for evolutionists to dismiss the claims of creation science out of hand, simply because they feel that creationists are just religious nuts. (If that statement sounds exaggerated, I invite you to research it yourself.) On the other hand, evolutionists claim that their viewpoints are solidly backed by science.

Creationists are often characterized as being "unscientific" or "pseudoscientific;" for example, Wikipedia describes creationism as "a religious belief" that follows the "creation myths" in the Bible book of Genesis, and teaches

---

[16] David Van Reybrouck, "Imaging and imagining the Neanderthal: the role of 355 technical drawings in archaeology," *Antiquity* 72 (March 1998): p. 62. Emphasis added.

[17] Ibid., p. 63.

[18] Marvin L. Lubenow, *Bones of Contention: a Creationist Assessment of Human Fossils* (Baker Books, Grand Rapids, Michigan), 2004. BakerBooks.com.

"pseudoscientific creation science."[19] This book will show in many places that young-Earth creationism is solidly scientific, and in the process, we will also discover some "skeletons in the closet" of the theory of evolution.

It's interesting to note that when **plate tectonics** (the study of continents and sea floors on separate plates that can move independently) was first proposed in a book written by the creationist Antonio Snyder,[20] he had gotten the idea from Genesis 1:9–10, which talks about the seas having been gathered into "one place." Unfortunately, his book was virtually ignored, probably because Darwin's book *Origin of Species* came out the same year, and took the world by storm.

Then in the early 1900s, the German meteorologist Alfred Wegener supported Snyder's concept of plate tectonics in his own book.[21] But for almost the next 50 years after that, the huge majority of scientists dismissed the idea out of hand; they accused the few scientists who promoted the tectonic-plate concept of dabbling in pseudo-science that violated basic principles of physics.[22]

Sound familiar? They accused the tectonic-platists of engaging in pseudoscience, just like evolutionists accuse creationists of the same thing today. Yet today, plate tectonics is an accepted, confirmed, and repeatedly measured fact. Could that happen with creationism? I think so.

What is science? The *New Oxford American Dictionary* defines it as "the intellectual and practical activity encompassing the systematic study of the structure and behaviour of the physical and natural world through observation and experiment."[23] Notice that science is study *through observation and experiment.* That is of paramount importance.

But neither creationists nor evolutionists can go back in time to *observe* how the universe, Earth, and life began, so how can we approach it scientifically? We can observe the laws of physics in action today and, assuming that the laws of physics have not changed, conclude how they must have worked in the past. Remember, evolution depends on uniformitarianism, so there is

---

[19] en.wikipedia.org/wiki/Creationism; accessed May 26, 2020.
[20] Antonio Snider, *Le Création et ses Mystères Devoilés,* Franck and Dentu, Paris, 1859.
[21] Alfred Wegener, *Die Entstehung der Kontinente und Ozeane,* 1915.
[22] Ken Ham and Bodie Hodge, editors, *The New Answers Book 1,* Answers in Genesis (Master Books, Green Forest, Arkansas, 2006), p. 188.
[23] *New Oxford American Dictionary,* Copyright © 2010, 2019 by Oxford University Press, Inc. All rights reserved.

rarely an objection to assuming that laws of physics still behave today like they used to.

But the past itself is not observable, so both schools of thought require assumptions. But since we're discussing science, we would approach things scientifically, with as much observation and repeatable experimentation as possible, right? Well, maybe not, with evolution.

Here's a quote from Ernst Mayr, one of the 20th century's most distinguished evolutionists:

> Evolutionary biology, in contrast with physics and chemistry, is an historical science—the evolutionist attempts to explain events and processes that have already taken place. **Laws and experiments are inappropriate techniques** for the explication of such events and processes. Instead **one constructs** a historical narrative, consisting of **a tentative reconstruction** of the particular scenario that led to the events one is trying to explain.[24]

Did you see that? Laws and experiments are *inappropriate* techniques for explaining evolution? Science, by its very definition, requires experimentation, and experimentation itself requires an understanding of the physical laws involved! And the whole point of experimentation is to observe the results and see if they bear any resemblance to theoretical predictions.

And then Mayr says that "instead" of experimentation, one "constructs" a "reconstruction?" Basically, you make stuff up? This would not even be a hypothesis, because you can test a hypothesis and determine whether its predictions correlate with experimental results; i.e., with reality. But with evolution, that is apparently unnecessary.

The problem with "constructing a reconstruction" is of course that, since there is no way to disprove events in the past, all sorts of fanciful ridiculosities are tolerated. James Conant, Harvard chemist and former president of Harvard, said it well:

> The sciences dealing with the past, stand before the bar of common sense on a different footing. Therefore, a grotesque account of a period some thousands of years ago is taken seriously, though it be built by piling special assumptions on

---

[24] Ernst Mayr, "Darwin's Influence on Modern Thought," *Scientific American,* July 2000, p. 80, 83. Emphasis added.

special assumptions, *ad hoc* hypothesis on *ad hoc* hypothesis, and tearing apart the fabric of science whenever it appears convenient. **The result is a fantasia which is neither history nor science.**[25]

In Mayr's article quoted above, he said, "No educated person any longer questions the validity of the so-called theory of evolution, which **we know now to be a simple fact**" (emphasis added). Notice the not-so-subtle implication here: "No *educated* person any longer questions" evolution. Therefore, clearly, creationists are uneducated and simple folk—gullible, superstitious, naïve, and misguided. Poor things. Pat them on the head and maybe they'll go away.

Richard B. Goldschmidt, while he was a professor at the University of California, stated something very similar to Mayr's comment, in its dismissiveness of anyone who disagreed with him:

> Evolution of the animal and plant world is considered by **all those entitled to judgment** to be effective for which no further proof is needed.[26]

Note the condescending attitude: if you don't believe in evolution, you are not "entitled to judgment." Wow. In other words, "We will reject, out of hand, any evidence that disagrees with our position." Is that good science?

So why do evolutionists believe in evolution without rigorous scientific proof? It's simple: *because they want to believe in it.* They are very sincere in their beliefs, but their beliefs are based on a worldview that the universe originated by itself, using nothing more than naturalistic processes. What do you call that type of belief system? It's called "a religion." "Religion" is defined as "a cause, a principle, or an activity pursued with zeal or conscientious devotion."[27] Clearly, a religion need not include a concept of a supernatural God.

But seriously, how do I have the gall to call evolutionism a religion? Isn't that just a tit-for-tat reaction to evolutionists claiming that creationism is a pseudoscience? Actually, no. Read this quote from Dr. Michael Ruse, a

---

[25] James Conant, Ph.D., *Science and Common Sense*, Yale University Press, 1951, Chapter 10, "The Study of the Past," p. 278.
[26] Richard B. Goldschmidt, *American Scientist* 40:84 (1952). Emphasis added.
[27] *The American Heritage Dictionary of the English Language.* 1996.

philosopher of science—especially evolution—who has written several books on evolutionary theory and Darwinism:

> Evolution is promoted by its practitioners as more than mere science. **Evolution is** promulgated as an ideology, **a secular religion**—a full-fledged alternative to Christianity, with meaning and morality. . . **Evolution is a religion.** This was true of evolution in the beginning, and it is still true of evolution today.[28]

This candid admission is very eye-opening: note that he acknowledges that evolution is "a secular religion." So, the investigation of how life could have evolved on Earth—with a prerequisite assumption that life *did* evolve on Earth—is a religion. Fascinating. But not surprising, once you think about it.

And Ruse's statement was not just a sudden knee-jerk reaction to some unexpected event; he had been thinking about this for a long time. In fact, *years* before making the above comment, he stated, at the annual meeting of the American Association for the Advancement of Science in 1993:

> . . .at some very basic level, evolution as a scientific theory makes a commitment to a kind of naturalism, namely that at some level **one is going to exclude miracles** and these sorts of things, **come what may.**[29]

What does the phrase "come what may" mean? According to the Oxford English Dictionary, it means "no matter what happens."[30] In the context in which the above statement was made, it means, "We will continue to believe that there is no God, regardless of contrary evidence, logic, or repeatable observations and experiments." It's nice to see that secular scientists are so open and unbiased in their quest to discover truth.

---

[28] Michael Ruse, "Saving Darwinism from the Darwinians," *National Post*, May 13, 2000, p. B-3. Emphasis added.

[29] Carl Wieland, "The Religious Nature of Evolution" *Journal of Creation* 8(1):3–4, April 1994. [Online at creation.com/the-religious-nature-of-evolution; accessed May 7, 2022.] Emphasis added.

[30] *Oxford Dictionary of English,* online version 14.4.72, Copyright 2022 Mobisystems, Inc.

Ruse goes on to say:

> . . .**evolution,** akin to religion, **involves making certain a priori or metaphysical assumptions,** which at some level cannot be proven empirically.

The implication of secular scientists in general, and evolutionists in particular, is that if anything supernatural (like creation) is tolerated, it will somehow hinder our ability to learn about the universe. So anything supernatural is summarily dismissed, regardless of whether there is anything credible to replace it with. Naturalism—attempting to explain the universe without acknowledging God's existence—is absolutely essential to believe in evolution.

So hostile to all things spiritual is the secular definition of science, that this hostility is openly acknowledged, as in this quote from Professor Richard Lewontin, an ardent evolutionist:

> We take the side of science *in spite* of the patent absurdity of some of its constructs, *in spite* of its failure to fulfil many of its extravagant promises of health and life, *in spite* of the tolerance of the scientific community for unsubstantiated just-so stories, because **we have a prior commitment, a commitment to materialism.** It is not that the methods and institutions of science somehow compel us to accept a material explanation of the phenomenal world, but, on the contrary, that we are forced by our *a priori* adherence to material causes to create an apparatus of investigation and a set of concepts that produce material explanations, no matter how counter-intuitive, no matter how mystifying to the uninitiated. Moreover, that materialism is absolute, for **we cannot allow a Divine Foot in the door.**[31]

Here's another very enlightening admission, by evolutionist and paleontologist Stephen Jay Gould:

> **Our ways of learning about the world are strongly influenced by the social preconceptions and biased modes of thinking** that each scientist must apply to any problem. The stereotype of a fully rational and objective "scientific

---

[31] Lewontin, R., "Billions and Billions of Demons," *The New York Review,* January 9, 1997, p. 31. Italic emphasis in original; bold emphasis added.

method," with individual scientists as logical (and interchangeable) robots is self-serving mythology.³²

With such candid admissions that science, when confined to materialism, is impotent to deliver rational explanations of the universe, and that materialistic scientists are biased against all things supernatural, it's astonishing to see how many of them *remain* evolutionists, deliberately closing their eyes to evidence that points in a direction they don't like.

Dr. Colin Patterson is the Senior Principal Scientific Officer in the Paleontology Department of the British Museum of Natural History. In a talk to a roomful of prominent evolutionists at the American Museum of Natural History, he compared amino acid sequences in several proteins of different animals. The evolutionary relatedness of these animals had been taught for decades in classrooms all around the world. Dr. Patterson shocked the audience by stating that this new information contradicted the theory of evolution.

Dr. Patterson said, in part, "**The theory makes a prediction**; we've tested it, and **the prediction is falsified precisely. . . . Evolution was a faith.**" He stated that he had been "duped into taking evolutionism as revealed truth in some way," and that "**evolution not only conveys no knowledge but seems somehow to convey anti-knowledge,** apparent knowledge which is harmful to systematics *[the science of classifying forms of life]*."³³

And they're not the only ones. William Provine, of Cornell University, was the Charles A. Alexander Professor of Biological Sciences at the Department of Ecology and Evolutionary Biology there. In an article he wrote in *Origins Research,* he stated, "Let me summarize my views on what modern evolutionary biology tells us loud and clear." What might you expect a professor in a university's Department of Evolutionary Biology to say? Perhaps something about how he believes life evolved?

Actually, no. Here's what he said that evolutionary biology tells us, "loud and clear:"

> **There are no gods,** no purposes, no goal-directed forces of any kind. **There is no life after death.** When I die, I am

---

³² Gould, Stephen J., *Natural History* 103(2):14, 1994.
³³ "Prominent British Scientist Challenges Evolution Theory," Audio Tape Transcription and Summary by Luther D. Sunderland, personal communication. Speech given on November 5, 1981. Emphasis added.

absolutely certain that I am going to be dead. That's the end for me. There is no ultimate foundation for ethics, no ultimate meaning to life, and no free will for humans, either.[34]

This is tragic. I am not being facetious in the least here: this poor man must have been so miserable and empty inside. With that worldview, which is an inevitable result of the atheistic religion of evolutionism, he had no hope. His life had no purpose. No meaning. No comprehension of right and wrong. Hopefully, he had a change of heart on his deathbed, because if not, he was in for a very rude awakening once he breathed his last—Provine died in 2015.

Kenneth Hsu, who is not a creationist, writes the following in response to another scientist who was pointing out the failures of evolution:

> . . .nevertheless, I agree with him that **Darwinism contains "wicked lies;"** it is not a "natural law" formulated on the basis of factual evidence, but **a dogma, reflecting the dominating social philosophy of the last century.**[35]

So evolution is a "dogma" and a "social philosophy." Could this be why evolution is defended so "religiously?" I think so. As we saw above, evolution itself *is* a religion.

But, the religion of evolution is a subset—one aspect—of a larger religion, and that is **naturalism,** also known as **atheism.** Naturalism, as its name implies, is the philosophical belief that everything arises exclusively from natural causes. Therefore, no supernatural or spiritual explanations are even considered; such explanations for anything are dismissed out of hand. Although naturalism is very similar to evolutionism, naturalism encompasses evolutionism because one could be totally uninterested in how the universe, the Earth, or living organisms came about, but still want to exclude God from his life.

Very similar to both of these religions is **secular humanism** (or just plain "humanism"), but humanism has a more pronounced point of emphasis: not only is there no supernatural God, but *we* are God. So yes, humanism is a religion, and its believers are so zealous in devotion to their faith that it puts most Christians to shame.

---

[34] William B. Provine, *Origins Research* 16, no. 1 (1994): 9. Emphasis added.
[35] K. J. Hsu, *Journal of Sedimentary Petrology* 56(5): 725–730 (1986). Emphasis added.

Chapter 1: The $64,000 Question

For example, John Dunphy, a outspoken humanist, stated it this way:

> I am convinced that the battle for humankind's future must be waged and won in the public school classroom by teachers who correctly perceive their role as the **proselytizers of a new faith: a religion of humanity** that recognizes and respects the spark of what theologians call **divinity in every human being.** These teachers must embody the same selfless dedication as the most rabid fundamentalist preachers, for they will be **ministers of another sort, utilizing a classroom instead of a pulpit to convey humanist values in whatever subject they teach, regardless of the educational level**—preschool day care or large state university. The classroom must and will become an arena of conflict between the old and the new—the rotting corpse of Christianity, together with all its adjacent evils and misery, and **the new faith of humanism.**[36]

As you can see, the much-promoted "separation of church and state" doesn't exist in America. However, the "separation of Christianity and state" certainly does: nearly every classroom in the country tirelessly sermonizes humanism every school day. Humanism, with its two manifestations of naturalism and evolutionism (and others as well), is the state-sponsored religion of the USA, and it is preached vociferously in virtually every classroom across the country.

Some of my readers may consider me to be exaggerating when I claim that humanism (and its manifestation, evolution, which is the subject of this book) is a religion—a godless religion, but a religion nonetheless. However, this is not just my opinion. In the 1961 legal case *Torcaso v. Watkins* (which addressed the legality of requiring a religious test for public office), the rationale includes:

> . . .religions in this country which do not teach what would generally be considered a belief in the existence of God, are Buddhism, Taoism, Ethical Culture, **Secular Humanism,** and others.[37]

---

[36] J. Dunphy, "A Religion for a New Age," *The Humanist,* Jan.–Feb. 1983, p. 23, 26; as cited by Wendell R. Bird, *Origin of the Species—Revisited,* Vol. II (New York, NY: Philosophical Library, c. 1989), p. 257. Emphasis added.
[37] Torcaso v. Watkins, 81 S. Ct. 1681 (1961). Emphasis added.

Because of the belief system of humanism, and the tireless effort of its devotees, the Pledge of Allegiance has been removed from schools, because it acknowledges God. Likewise, prayer was removed from the classroom, along with Bibles, any mention of the Ten Commandments, and prayer before sporting events. We've made it illegal to teach morality in schools, and now the majority of our students (as well as the adults they've grown into) are acting immorally or amorally. Seriously, what did we expect?

On the other hand, humanism/naturalism/evolutionism is ubiquitous in state-run schools, textbooks, media, museums, and so forth. Many Jesus-followers are unaware of, or indifferent to, the intensity of the effort to remove God from society.

So, back to evolution itself: Dr. Phillip Johnson is the Jefferson E. Peyser Professor of Law at the University of California, Berkeley. In his book, called *Darwin on Trial,* Johnson quotes a statement from George Gaylord Simpson that ends with "Man is the result of purposeless and natural process that did not have him in mind."

Here is Johnson's response to Simpson's statement:

> Because the scientific establishment has found it **prudent to encourage a degree of confusion** on this point, I should emphasize that **Simpson's view** was not some personal opinion extraneous to his scientific discipline. On the contrary, he **was merely stating explicitly what Darwinists mean by "evolution."** The same understanding is expressed in countless books and articles, and where it is not expressed, it is pervasively implied. Make no mistake about it. **In the Darwinist view, which is the official view of mainstream science, God had nothing to do with evolution.**
>
> Naturalism is not something about which Darwinists can afford to be tentative, because their science is based upon it. As we have seen, **the positive evidence that Darwinian evolution either can produce or has produced important biological innovations is nonexistent.** Darwinists know the mutation-selection mechanism can produce wings, eyes, and brains not because the mechanism can be observed to do anything of the kind, but because their guiding philosophy assures them that no other power is available to do the job.

> The absence from the cosmos of any Creator is therefore the essential starting point for Darwinism.[38]

In other words, one can claim that evolution is "scientific" and creationism is not, *only* if "science" is defined in such away to preemptively disallow creation before it is even considered as a possibility. Which is exactly what evolutionists do. Carl Sagan's comment in his *Cosmos* series is a perfect example of this; he said: "We're here. That *proves* we evolved!"

So if the absence of any Creator is the initial assumption that is taken for granted, then evolution begins to sound reasonable. This means, then, that all observations will be interpreted in such a way as to be compatible with the presupposition that evolution is a fact (and this practice is ubiquitous in evolutionary circles). Then, this "evidence" is, in turn, used to "prove" evolution, and we'll hope that people don't notice the circular reasoning.

Case in point—as Marvin Lubenow points out:

> Further, many important fossils are the subject of intense controversy among evolutionists regarding the date, the category, or both. The two matters are sometimes related. **The category to which a fossil is assigned sometimes determines the date assigned to it. Or the date of the fossil sometimes determines the category to which it is assigned.** In our college, that would be called cheating.[39]

But why would evolutionists do that? The answer, related to the statement made above, about the denial of God being "the essential starting point for Darwinism," is this: "In the evolutionist mind, the **fossils exist to serve evolution—not objective science.**"[40]

Another example: Dr. Richard Dawkins, the Charles Simonyi Professor of the Public Understanding of Science at Oxford University was asked the question "Is atheism the logical extension of believing evolution?" He answered, "My personal feeling is that **understanding evolution led me to atheism.**"[41] In another place, he said, "The more you understand the significance

---

[38] Phillip E. Johnson, *Darwin on Trial* (Washington, DC: Regency Gateway, 1991). Emphasis added.

[39] Marvin L. Lubenow, *Bones of Contention* (BakerBooks, Grand Rapids, Michigan), 2004, p. 18. Emphasis added.

[40] Ibid, p. 19. Emphasis added.

[41] Laura Sheahen and Dr. Richard Dawkins, "The Problem with God: Interview with Richard Dawkins," BeliefNet.com/story/178/story_17889.html. Emphasis added.

of evolution, **the more you are pushed** away from an agnostic position and **towards atheism.**"[42]

And Dawkins is not just a "casual" evolutionist. I hesitate to use the word "rabid," but read the following quote and decide for yourself. Phillip Johnson, in his book *Darwin on Trial*, describes and quotes Richard Dawkins:

> When he contemplates the perfidy of those who refuse to believe, Dawkins can scarcely restrain his fury. "It is absolutely safe to say that, if you meet somebody who claims not to believe in evolution, **that person is ignorant, stupid or insane (or wicked**, but I'd rather not consider that)." Dawkins went on to explain, by the way, that **what he dislikes particularly about creationists is that they are intolerant.**[43]

"Kettle, meet Pot."

Why are people drawn to evolution? As we saw above, it often leads to atheism. But why would people be drawn to atheism? Jeremy Rifkin, a well-known atheist, says it this way:

> We no longer feel ourselves to be guests in someone else's home and therefore obliged to make our behavior conform with a set of preexisting cosmic rules. It is **our** creation now. **We** make the rules. **We** establish the parameters of reality. **We** create the world, and because **we do, we** no longer feel beholden to outside forces. **We** no longer have to justify our behavior, for **we** are now the architects of the universe. **We are responsible to nothing** outside ourselves, for **we are the kingdom, the power, and the glory forever and ever.**[44]

Does the quote above remind you of a similar quote by someone else?[45] So evolution, which presumes naturalism (the belief that there is no supernat-

---

[42] Richard Dawkins, "On Debating Religion," the Nullifidian Vidian (December 1994).

[43] Phillip E. Johnson, *Darwin on Trial* (Regnery Publishing, Washington DC, 1991), p. 7. Emphasis added.

[44] Jeremy Rifkin, *Algeny* (New York: Viking Press, 1983), p. 244. Emphasis added.

[45] Isaiah 14:12–17 (NKJV): How you are fallen from heaven, O Lucifer, son of the morning! How you are cut down to the ground, You who weakened the nations [13]For you have said in your heart: 'I will ascend into heaven, I will exalt my throne above the stars of God; I will also sit on the mount of the congregation on the farthest sides of the north [14]I will ascend above the heights of the clouds, **I will be like the Most High.** [15]Yet you shall be brought

ural God), leads directly to a religion of self-worship, where people can do anything they please, because they don't have to answer to God—in their own minds, they *are* God.

But think about it: If you don't ultimately have to give account to God, *you don't count.* You have no value. You have no worth. You have no purpose. You have no significance. You are irrelevant. You are an accident. Your life doesn't matter, regardless of who you are.

For Jesus-followers who actively fellowship with Christ every day, such an attitude as Rifkin's—such a worldview as that—is almost incomprehensible. *But some people actually live like that.* Such loneliness! Such emptiness! We, as Jesus-followers, need to pray and intercede diligently for such people, so they will discover the reason for which they were put on this planet.

Author Tom Bethell demonstrates an accurate understanding of the reasons behind the hostility toward challenges to the evolutionary faith when he says:

> Evolution is perhaps the most **jealously guarded dogma** of the American public philosophy. **Any signs of serious resistance to it has encountered fierce hostility** in the past and it will not be abandoned without a tremendous fight. The gold standard could go (glad to be rid of that!), Saigon abandoned, the Constitution itself slyly junked. But **Darwinism will be defended to the bitter end.**[46]

Evolutionism is indeed a case of defending a religious belief system. Indeed, evolutionism has become the unofficial state-sanctioned religion of the United States.

As an example of that last statement, the famous Russian-American evolutionist Theodosius Dobzhansky quotes the geologist and paleontologist

---

down to Sheol, to the lowest depths of the Pit [16]Those who see you will gaze at you, and consider you, saying: 'Is this the man who made the earth tremble, who shook kingdoms [17]who made the world as a wilderness and destroyed its cities, who did not open the house of his prisoners?'

[46] Tom Bethell, *The American Spectator* (July 1994), p. 17. Emphasis added.

Pierre Teilhard de Chardin (who was involved with the discovery of the fossils dubbed the Peking Man):

> **Evolution is a light** which illuminates all facts, a trajectory which **all lines of thought must follow.**[47]

Note: "all lines of thought *must* follow" evolution. That is nothing less that a statement of faith, which disregards, out of hand, anything that contradicts it. As we'll see below, this includes repeatable experiments and observations. The above quote was written for *The American Biology Teacher*, so it is no exaggeration to say that evolutionism has become the unofficial state-sanctioned religion of the United States, taught in public schools everywhere.

Clearly, the above quote from Dobzhansky is simply a secular version of what Jesus said:

> John 8:12 (CEB): Jesus spoke to the people again, saying, "**I am the light** of the world. Whoever follows me won't walk in darkness but will have the light of life."

Because evolution is indeed a religion, as amply attested to by the evolutionists above, there are many places in this book that refer to the "belief system" of evolution, the "doctrine" of evolution, the "philosophy" of evolution, the "faith" of evolution, or even the "dogma" of evolution. If you are an evolutionist, please don't take this as a personal affront; it is not intended to be so. I just feel it is important for people to realize that evolution is no less a religion than Christianity.[48]

Why is it significant that evolution is a religion? Because it sets the presupposition through which evidence is interpreted. Christians have a presupposition that the Bible is true, and evolutionists have a presupposition that naturalism—a universe *without* a supernatural God—is true. Because evolutionists have a primary presupposition that there cannot be a God, all observations are interpreted in that context, whether they fit or not, and any evidence that might militate against evolutionism is summarily dismissed.

---

[47] Theodosius Dobzhansky, "Nothing in Biology Makes Sense Except in the Light of Evolution," *The American Biology Teacher* (March 1973): p. 129. Emphasis added.

[48] Indeed, evolution is *more* of a religion than Christianity, because Christianity, by its very nature, is a personal relationship with the intelligent Creator, and not simply a belief system, a mindset, a philosophy, a set of doctrines, or a worldview. Christianity can be *reduced* to a mere religion by removing the relationship at its core, but then what is left is simply a system of dos and don'ts and duties that have been eviscerated of their power to transform people's lives.

So the creation/evolution question is *not* a debate between science and religion, as is often claimed; it is a debate between whether to interpret data with *this* presupposition or *that* presupposition. Again, evolutionists and creationists have access to the same information; it's all in how the information is interpreted. Both creationism and evolutionism have worldview presuppositions, and both have to deal with **historical science** (theories about occurrences in the past, which no human currently alive has seen) and **observational science** (theories pertaining to the operation of math and physics, which can be tested with repeatable experiments).

There is much data, much evidence, in this book. If you have been brought up to believe in evolution, I would encourage you to consider the statements in this book as if Creation were true. In other words, have a mindset of "If Creation were actually true, what would that imply?" Looking at the data in this book, while keeping an open mind to the possibility that maybe, just maybe, there is a God who interacts with his Creation, may be an enlightening experience indeed.

Evolution as a Social Theory

What does evolution have to say about making "progress?"[49] Basically, it is survival of the fittest (we'll cover this in more detail in a later chapter). In other words, those organisms that adapt, survive, and those that don't, don't. An evolutionist would say—as Carl Sagan did in his *Cosmos* series—"We're here. That *proves* we evolved!"

But what are the results, the effects, of a worldview of evolution? As shown above, it leads to despair, because if evolution is true, none of us have any value, importance, purpose, or reason to go on living. We're all just biological accidents anyway.

Beyond that, the "survival of the fittest" mindset claims that the survivors are *better* than those that don't survive. After all, if they weren't "fit" enough to survive, just go ahead and let them die off. Why not? We have no value or purpose, remember, being biological accidents and all.

And the fact that we are here now just shows that we are better than those who came before us, because, plainly, *we survived,* so therefore we must have been the fittest. Those who are inferior don't last. By the inexorable logic of the religion of evolution, what exists nowadays *must* be superior to what used

---

[49] My appreciation to Skip Moen (SkipMoen.com/2009/08/devolving-evolution-as-social-theory) for starting me thinking along this line.

to exist. There is no other possibility, assuming the belief system of evolution.

This means that we could not possibly learn from history, because we're more advanced than they are. What could we possibly learn from people from the past? Previous generations were not as informed as we are, not as capable, not as equipped to deal with the universe as we are—*not as evolved.* And the morals that they had back then? Unenlightened nonsense. Every generation is better, smarter, and more capable than the previous one, which is why popular TV shows nowadays usually portray the kids as smarter than their parents, and the parents as hopelessly out of touch, basically obsolete. Grandparents even more so.

And because we are so much further advanced than those "unevolved" people way back when, we have outgrown their antiquated ideas of morality. We are well on our way to utopia, you see.

Or maybe not. What are some of the likely—and yes, *inevitable*—results of the religion of evolution? Because there is no standard of right and wrong, it is basically the law of the jungle: whoever is strongest, rules.

This plays out in our modern world, so enthralled with the lack of accountability that evolution offers, in the following ways:

- **Abortion.** Because babies are not strong, they have no power. And because we're all just biological accidents, it's not any more objectionable to kill a human baby than it is to put down a dog or step on an ant. And since existence is meaningless and people have no value, go ahead and kill the kid if he's inconvenient.
- **Selfishness.** Anything I do for others just decreases my own supply of resources and reduces my ability to compete in this dog-eat-dog world. Generosity and self-sacrifice are antithetical to natural selection, so I always need to take care of myself first.
- **Addictions.** Since my life has no meaning, *why not* get hooked on drugs, alcohol, or sex? After all, "If it feels good, do it," right?
- **Bullying.** When one person bullies another, it is an example of how people desire strength and power. And with an evolutionary worldview, it's simply natural selection, and the bullied one is showing his lack of "fitness" to survive.
- **Child Abuse.** Why not? If the kid gets on your nerves, teach him who's boss. If he survives, he'll be the stronger for it, and if he doesn't, well, apparently he wasn't fit to. And if he doesn't survive, it's no big

loss; you can always make another one.
- **Human Trafficking.** Since the people being trafficked are simply inconsequential objects to be used, go ahead and use them, deriving as much profit from them as possible before they are used up. Then just throw them away and go get a replacement.
- **Euthanasia.** When someone becomes infirm and no longer productive, he's just a drain on resources. Besides, old people are from a previous generation, and we've evolved past them—they are obsolete. Since there's no meaning in life or value in people, sure, let's off them if it's more convenient for us.
- **Genocide.** To justify the extermination of any people group, just appeal to the fact that they're "inferior," or "less evolved" than you. Then, just start killing them. Evolution conveniently dispenses with the concepts of right and wrong, so if you wipe out a people group, that just proves that, according to evolution's notion of "survival of the fittest," they just weren't fit to survive, right? (If you think it's an exaggeration to say that evolution can cause and justify genocide, please read the stories about the Tierra del Fuegians and the Tasmanians.[50])
- **Amorality.** Since we all are just biological accidents, anyone's idea of "right" and "wrong" is simply an opinion, and it is no more valid than someone else's opinion. So no one can object to my behavior on the basis of morality; after all, morality is just a desperate construct of those who don't have enough brute strength to survive.
- **Suicide.** I have no purpose and no value, being a biological accident and all, so when I get tired of my empty existence, *why not* blow my brains out? Since there's nothing after physical death, it's a good way to stop all the suffering.
- **Anarchy.** The next logical step after bullying is groups of bullies trying to destroy whomever stands in their way. Since law enforcement agencies are simply institutionalized attempts to legislate morality, which is itself baseless, let's get rid of anyone who stands in our way. In other words, "When I have the gun, I'm the boss."

This list could get much longer, but I'm sure you see the common thread: people have no value, so do with them as you please. I apologize if my bluntness in the above list was somewhat jarring, but we need to realize that there

---

[50] Marvin L. Lubenow, *Bones of Contention* (BakerBooks, Grand Rapids, Michigan), 2004, Chapter 13: "The Voyage."

will inevitably be unpleasant consequences if we remove God from our collective worldview.

And spiritually speaking, evolution, since it discounts the existence of God, also necessarily discounts the reality of a Judgment Day, in which we will all stand before God and give an account for how we conducted our lives and what we accomplished in them. A belief in evolution would unavoidably convince people that there was no need to prepare for what happens after physical death, because with the evolutionary belief system, there *is* nothing after physical death.

And those evolution-endorsing people in the world today who are working so diligently to help others, and to support the underprivileged and downtrodden: they are behaving honorably. But wait—no, "honor" is a word that has no meaning if evolution is true. The dignity they are endeavoring to give to others only exists because we have the inherent dignity of being created by a loving God, in His image, and for a glorious purpose. But if we disallow God in our worldviews, dignity and honor vanish, being only imaginary constructs of the "less fit." Dignity and honor have no basis and no meaning for the bunch of purposeless, valueless, accidents of nature that evolution says we are.

We can't have it both ways: if dignity and honor are real, that fact militates against evolution. And if evolution is real, that militates against dignity and honor. I applaud those people who claim to believe in evolution, but who still are selflessly and tirelessly working on behalf of underprivileged and downtrodden. They don't believe in evolution as much as they claim to, because they *know* down deep that dignity and honor are real things, and thus, humanity cannot be just meaningless, purposeless accidents of biology.

Evolution and Racism

You may have noticed that in the bulleted list above, all the items have a common factor: they are all reasonable, natural, predictable, and very logical results of a worldview where there are different classes of people, and some are better—superior, more valuable—(more "evolved") than others.

At this point, let me clarify something: I am not saying here that evolutionists are racists; in fact, I would say that very few of them are overtly racist. However, *the theory itself* is inherently and unavoidably racist, and evolutionists need to walk a very fine line to promote a racist theory but not succumb to racism themselves.

But am I really saying that racism (or its superset, classism), is related to the theory of evolution? Actually, yes. Although racism has been around for thousands of years, there has been a dramatic uptick in the last century or two. Why is that? Could it be that a worldview in which "newer is better," and in which "survival of the fittest" is the operative principle, actually *promotes* a feeling of superiority in some people? Could the worldview of evolutionism actually whip not only eloquent speakers, but *entire nations* into a frenzy against a different people-group, as Hitler did in Germany? I think so.

And I'm not the only one. Sir Arthur Keith was a passionate evolutionist, and here is his take on it:

> The German Fuhrer. . . consciously sought to make the practice of Germany conform to the theory of evolution.[51]

In the same book, Keith makes another statement about Hitler:

> The leader of Germany is an evolutionist, not only in theory, but, as millions know to their cost, in the rigor of its practice. For him, the national "front" of Europe is also the evolutionary "front;" he regards himself, and is regarded, as the incarnation of the will of Germany, the purpose of that will being to guide the evolutionary destiny of its people.[52]

Where did Sir Arthur Keith get this idea that Hitler was actively promoting and "assisting" evolution? Perhaps from Hitler's book, *Mein Kampf*. Take a look at some of these statements from *Mein Kampf*:

- Hitler referred to some people as "lower human types."
- He berated the Jews for bringing "Negroes into the Rhineland" with the aim of "ruining the white race by the necessarily resulting bastardization."
- He referred to certain people as "monstrosities halfway between man and ape."
- He bemoaned those evangelical Christians who journeyed to "Central Africa" to establish "Negro missions," because it would inevitably result in turning "healthy. . . human beings into a rotten brood of bastards."
- In the chapter "Nation and Race," Hitler wrote, "The stronger must dominate and not blend with the weaker, thus sacrificing his own great-

---

[51] Sir Arthur Keith, *Evolution and Ethics* (New York: G. P. Putnam's Sons, 1947), p. 230.
[52] Ibid., p. 10.

ness. Only the born weakling can view this as cruel, but he, after all, is only a weak and limited man; for if this law did not prevail, any conceivable higher development *(Hoherentwicklung)* of organic living beings would be unthinkable."

- He also wrote, "Those who want to live, let them fight, and those who do not want to fight in this world of eternal struggle do not deserve to live."[53]

Well, *that's* certainly enlightening: according to this, the Jewish Holocaust of World War II was not evil; it was just that the Jews who died didn't deserve to live.

Did Darwin Espouse This?

Although Darwin spoke eloquently and humbly in many cases, his ideas betray an undeniably racist worldview.

For example, take a look at this quote, from Darwin's book *The Descent of Man:*

> At some future period, not very distant as measured by centuries, **the civilized races of man** will almost certainly exterminate, and replace, **the savage races** throughout the world. At the same time, the anthropomorphous apes. . . will no doubt be exterminated. The break between man and his nearest allies will then be wider, for it will intervene between man in a more civilized state, as we may hope, even than **the Caucasian,** and some ape as low as a baboon, instead of **as now between the negro or Australian and the gorilla.**[54]

Is Darwin saying the "negro" and the "Australian" aborigine are somewhere between the "civilized races" (the Caucasians, of course) and the baboon? It certainly sounds like it.

Fifty pages later, he writes:

> It has often been said. . . that man can resist with impunity the greatest diversities of climate and other changes; but this is true only of **the civilized races.** Man in his wild condition

---

[53] Adolf Hitler, *Mein Kampf* (Boston: Houghton Mifflin Co., 1943), pp. 286, 295, 325, 402, 403, 285, 289 respectively.

[54] Charles Darwin, *The Descent of Man* (London: John Murray, 1901), pp. 241–242. Emphasis added.

seems to be in this respect almost as susceptible as his nearest allies, **the anthropoid apes,** which have never yet survived long, when removed from their native country.[55]

Interesting: he refers to "man in his wild condition." What did Darwin mean by "man in his wild condition?" He meant the ones who were the most "primitive" and "uncivilized;" that is, the ones least like Caucasians.

Ardent evolutionist Stephen Jay Gould, of Harvard University, commented on Darwin's book *The Origin of Species* and said:

> Biological arguments for racism may have been common before 1859, but they **increased by orders of magnitude** following the acceptance of evolutionary theory.[56]

In Gould's book, he cites several sources to support his statement about how much racism increased after Darwin's *Origin of Species* came out. But here are a couple more Gould didn't mention.

About 60 years after *Origin of Species,* Edwin G. Conklin wrote some statements that revealed his evolutionary worldview, replete with racism. From 1908 to 1933, Conklin was a biology professor at Princeton University, and in 1936—the year that the Olympic games were held in Berlin, under Hitler's watchful eye—Conklin was the president of the American Association for the Advancement of Science.

Conklin wrote:

> Comparison of any modern race with the Neanderthal or Heidelberg types shows that all have changed, but probably **the negroid races more closely resemble the original stock** than the white or yellow races.[57] . . . Every consideration should lead those who believe in the **superiority of the white race** to strive to preserve its purity and to establish and maintain the segregation of the races, for the longer this is maintained, the greater the preponderance of the white race will be.[58]

---

[55] Ibid, pp. 291–292. Emphasis added.
[56] Stephen Jay Gould, *Ontogeny and Phylogeny* (Cambridge, Mass: Harvard University Press, 1977), p. 127. Emphasis added.
[57] Edwin G. Conklin, *The Direction of Human Evolution* (New York: Scribner's, 1921), p. 34. Emphasis added.
[58] Ibid., p. 53. Emphasis added.

About the same time, Henry Fairfield Osborn was writing his impressions of evolution's accomplishments. Osborn was for 25 years the president of the Board of Trustees of the American Museum of Natural History, and he also was a professor of biology and zoology at Columbia University.

Here are Osborn's thoughts:

> The Negroid stock is even more ancient than the Caucasian and Mongolians, as may be proved by an examination not only of the brain, of the hair, of the bodily characteristics. . . but of the instincts, the intelligence. The standard of **intelligence of the average adult Negro is similar to that of the eleven-year-old youth of the species** *Homo Sapiens*.[59]

Wow. So Negroes are not even *Homo sapiens*. In a book honoring the evolutionist school teacher John T. Scopes (of the famous "Scopes Monkey Trial"),[60,61] Osborn stated:

> The ethical principle **inherent in evolution** is that only the best has a right to survive. . .[62]

Make you wonder who is qualified to determine who the "best" people are, doesn't it? And doesn't that sound disturbingly similar to some of Hitler's quotes cited above? Conklin's quotes are from 1921, Hitler's from 1923, and Osborn's from 1926.

One thing that is amazing about the fact that racism still exists and is rampant in the world today, is that genetically, we are all the same race, and this is experimentally demonstrable. Through DNA analysis, geneticists have discovered that those factors determining appearance—things like skin color, eye color and shape, hair color and texture, body build, and so forth—are determined by 0.01%–0.012% of our DNA.[63,64]

---

[59] Henry Fairfield Osborn, "The Evolution of Human Races," *Natural History*, April 1980, p. 129—reprinted from January/February 1926 issue. Emphasis added.

[60] John Morris, "Did the Evolutionists Present a Good Case at the Scopes Trial?" *Acts & Facts* August 1, 1995. [Online at ICR.org/article/did-evolutionists-present-good-case-at-scopes-tria; accessed June 16, 2020.]

[61] William Jennings Bryan, "Mr. Bryan on Evolution" *Acts & Facts,* March 1, 1991, reprinted from August 1925. [Online at ICR.org/article/mr-bryan-evolution; accessed June 16, 2020.]

[62] Henry Fairfield Osborn, *Evolution and Religion in Education* (London: Charles Scribner's Sons, 1926), p. 48. Emphasis added.

[63] N. Angier, "Do races differ? Not really, DNA shows," *New York Times* web, August 22, 2000.

Do you see the significance to that? If what we considered differentiators of race is from 0.01% of our DNA, that means we have 99.99% in common with each other! The genetic differences that cause variations in external appearance are 16 to 20 times *smaller* than the genetic differences between two random people, *even from the same people group!* So the "race" differences that we so often hear about on the news, where one group is attacking another, are genetically insignificant and unsupported by reality.

All humans are the same race, as shown by the fact that humans are **interfertile:** that is, a healthy male and a healthy female from *any* people-group can mate and have babies. That couldn't happen if we were actually different races.

The Appeal of Evolution

Why does the theory of evolution even exist? Isn't it self-evident that if there is a house, somebody built it? (Hebrews 3:4.) That if there is a smartphone, somebody manufactured it? That if there is a design, there is a designer? That if there is a garden, there is a gardener? That if there is a building, there is a builder?

Yes, all those are self-evident. *But*—and this is a big "but"—those self-evident answers also come with another implication that is very disconcerting, even alarming, to those who want to run their own lives as if they were in charge. That other implication is this: If there is a creation, *there is a Creator.* And if there is a Creator, there is intelligent purpose and intent behind all there is.

Even more unsettling is the obvious corollary implication: If I was intentionally created for a purpose, I will be held accountable for how much of my purpose—or how little—I accomplished during my lifetime.

And here is where the Body of Christ—the Church, at large—is at fault. For the most part, the Church has represented God so poorly that people actually *don't know* that He is the best, most loving, most faithful, most caring, good, and kind Person they could ever imagine. Because the Church has represented God so poorly, people think God is like us, only more so, and therefore people want to avoid God as much as possible.

---

[64] S. C. Cameron and S. M. Wycoff, "The destructive nature of the term race: growing beyond a false paradigm," *Journal of Counseling & Development,* 76:277–285, 1998.

One of the results of the Church being so embarrassingly un-Christlike is that people in the sciences and elsewhere want to find reasons, processes, and facts that omit any need for God to be in the picture. And the theory of evolution fills the bill nicely. So even though there is a creation, we don't need a Creator, because everything in the universe created itself (and never mind the intrinsic self-contradiction in that statement).

Plus, all relevant processes have to be random, because intentionality implies a Mind, which in turn implies a Creator, and if one wants to keep God out of the picture, the only alternative to intentionality is randomness. So even though random processes are by definition undirected, we'll assume that we all just got lucky enough, often enough, for long enough, that the finely tuned, stunningly complex, interconnected, and interdependent systems we observe *everywhere* just fell together by accident (and never mind the implications of those pesky observations).

So for everyone reading this book, proponents of evolution are not the enemy. They are people who are acting in a very reasonable way, considering how God's people have fallen down on the job of accurately representing Christ in the earth. So even though the theory of evolution is scientifically untenable, those who believe in it are still God's handiwork, created in His image and likeness, and are people for whom Jesus died.

As such, our responsibility is to treat them with love, honor, and respect, praying that they will encounter God for themselves and submit their lives to Him. Because when they do, entire new vistas of scientific inquiry, all of which glorify the infinitely loving and powerful Creator, will open up to them, and they will see as never before that "The fear of the Lord is the beginning of knowledge" (Proverbs 1:7).

The Politics of Evolution

One of the most common rebuffs to anyone who would dare to question the doctrine of evolution is, "You can't argue with it. It's *science*." Almost as if, "If science says something, it's indisputable." The problem is that *science* doesn't say anything; *scientists* say things. And since they are human, scientists, in whatever field of study, have to deal with hopes, fears, insecurities, desires, pride, employment concerns, politics, and so forth, just like everyone else.

Cosmologists, astronomers, geologists, paleoanthropologists, and all the rest, are fallible human beings. So again, "science" doesn't say anything, but people often interpret scientific evidence in a way that reinforces what they

CHAPTER 1: THE $64,000 QUESTION

want it to say. No one, creationist nor evolutionist, is immune to this temptation. But not everyone *succumbs* to this temptation, as evidenced by people who set out to prove a viewpoint wrong, and upon actual research end up adopting the viewpoint they set out to discredit (e.g., Josh McDowell, Lee Strobel, and others). So as you read through this book, and as you do further study on this topic, let the evidence itself speak to you.

So is there politics in the fields of study related to evolution? Look at this statement, from *Nature:*

> I don't need to tell you how things are, Miss Franklin. Non-scientists think of science as universal. Celestial, even. But science is terrestrial. Territorial. Political.[65]

Most anthropologists have rarely if ever dealt with the actual fossils they write about. Because the fossils are so fragile, and irreplaceable, they are kept under extremely high security, including—and I'm not exaggerating here—bank vaults, rooms with three-foot-thick concrete walls, and more. So what do they study? Plaster casts, which are of varying quality and accuracy. More on this in "Chapter 7: The Fossil Record." But *rarely* do they see the fossils themselves, so it's a great privilege to be able to see the real deal.

Author Roger Lewin, in his book *Bones of Contention* (not to be confused with Marvin Lubenow's book by the same name), quoted Donald Johanson, the discoverer of the "Lucy" fossil:

> . . .only those in the inner circle get to see the fossils; **only those who agree with the particular interpretation of a particular investigator** are allowed to see the fossils.[66]

Apparently Johanson himself operates the same way. Adrienne Zihlman is a paleoanthropologist from the University of California at Santa Cruz. When Zihlman was working at the Cleveland Natural History Museum, she contacted Johanson, asking for permission to see the fossils he had discovered in Ethiopia. He stated that he would give permission only if he could review any article she might write before sending it to any journal. She thought that sounded suspicious, so she declined.

Another blatant example of the "good ol' boy" nature of evolutionist politicking is how Wikipedia continuously blocks or removes articles that dare

---

[65] William Nicholson, *Nature* 422 (March 20, 2003): 259.
[66] Roger Lewin, *Bones of Contention* (New York: Simon and Schuster, 1987), 24.

to imply that evolution isn't as open-and-shut as it's purported to be; see the article "Here's How We Get Around the Wikipedia Roadblock."[67]

So is there objective science in the field of evolution? Not as much as you might think. Is there politicking? More than you might imagine.

Could Both Be True?

Since the theory of evolution has gained popularity in scholarly circles, some Christians, perhaps intimidated by the intelligence of evolution's proponents, have suggested that "maybe they're both true." After all, perhaps God created the universe and all life by using the process of evolution. Couldn't God have done that?

Actually, no. It's not that God is somehow lacking in power, but if you actually think about what the proposed hybrid theories entail (and there are several; see below), it quickly becomes clear that the question itself is a self-contradiction. Let's take a look.

All the attempts to hybridize the Biblical account of creation with the atheistic idea of deep time, include a desire—at least a small one—to preserve the integrity of the Bible. Therefore, this section will include more Scripture passages than most others in this book.

The Gap Theory

The Gap Theory is one approach to add the ideas of evolution to the Biblical account of creation. The Gap Theory proposes that there is an enormous gap of time between Genesis 1:1 and Genesis 1:2, during which all sorts of imaginative things happen.[68]

The Gap Theory goes like this:

Genesis 1:1–2: In the beginning God created the heavens and the earth. *(Then there was a gap of millions or billions of years, during which the Earth was ruled by Lucifer and populated by plants, animals, and soulless people. When Lucifer rebelled, God executed judgment by flooding the earth, forming the fossils. The people died too, but they were soulless, so it's okay.)* ²The earth was without form, and void; and darkness was on

---

[67] EvolutionNews.org/2020/06/heres-how-we-get-around-the-wikipedia-roadblock/; accessed July 5, 2020.

[68] W. W. Fields, *Unformed and Unfilled*, Burgeners Enterprises, Collinsville, Illinois, 1976, p. 7.

the face of the deep. And the Spirit of God was hovering over the face of the waters. . .

It was initially proposed by Episcopius, a Dutchman, in the early 1600s, and then popularized by Thomas Chalmers, a Scottish theologian in the early 1800s. Then G. H. Pember wrote a book entitled *Earth's Earliest Ages* in 1884, in which he promoted the idea, and finally Arthur Custance wrote his book *Without Form and Void*, in which he gave the idea an aura of credibility.

Unfortunately, even Bible commentators got swept up in the idea of science "correcting" the Bible. In the Scofield Reference Bible, Scofield states:

> Relegate fossils to the primitive creation, and no conflict of science with the Genesis cosmogony remains.[69]

Even Dake's Annotated Reference Bible succumbs to "validating" the Bible through the ever-changing speculations of naturalistic geologists:

> When men finally agree on the age of the earth, then place the many years (over the historical 6,000) between Genesis 1:1 and 1:2, there will be no conflict between the Book of Genesis and science.[70]

So why does the Gap Theory, an attempt to retain the integrity of Bible while also accommodating deep time, fail?

- The Gap Theory *doesn't* retain the integrity of the Bible: the Bible clearly states that God did all of creation in six days (Exodus 20:11, 31:17), not "billions of years plus six days."
- God said of His creation that "It was good" (Genesis 1:4, 10, 12, 18, 21, 25) and "It was very good" (Genesis 1:31). How could that have been true if there was already suffering, death, and judgment?
- The Gap Theory says that death, disease, and destruction all existed *before* the fall of man, but the Bible says those things were *because* of the fall of man (Romans 5:12, II Corinthians 15:21).
- The Gap Theory is logically inconsistent with itself: in an effort to accommodate "geological time," the Gap Theory was invented to avoid

---

[69] C. I. Scofield, Ed., The Scofield Study Bible, Oxford University Press, New York, 1945. (Originally published as The Scofield Reference Bible; this edition is unaltered from the original of 1909.)

[70] F. H. Dake, Dake's Annotated Reference Bible, Dake Bible Sales, Lawrenceville, Georgia, 1961, p. 51.

catastrophistic events like Noah's Flood as an explanation of where fossils came from. This was because "everybody knew" that fossils need millions and millions of years to form. But the catastrophistic event of Lucifer's Flood makes the theory moot: replacing one catastrophistic event (Noah's Flood) with another (Lucifer's Flood) removes any need for the theory itself.

- If the fossil record was created by Lucifer's Flood, what did Noah's Flood do, geologically speaking? Apparently nothing. At this point, some Gap Theorists relegate Noah's Flood to a local event, which further contradicts the Bible's description of a *global* flood (Genesis 6:17, 7:18–24)
- Proponents of the Gap Theory ignore the evidence for a young Earth (and there is a lot; see "Chapter 5: A Young Earth," "Chapter 6: Mutations," and "Chapter 7: The Fossil Record" for details).
- Not only does the Gap Theory disagree with the Bible, it disagrees with uniformitarianism as well, because of the posited globalness and catastrophic nature of "Lucifer's Flood."

The Day-Age Theory

In the Day-Age Theory, sometimes called "theistic evolution," the "days" that Genesis 1 refers to are not 24-hour days, but are really "age-long" days, indefinitely and arbitrarily long periods of time, which might overlap. This supposedly reconciles the Bible to the "fact" that geologic ages were "millions and millions of years" or even "billions and billions of years" long.

As we will see in most of the rest of this book, deep time (the idea of the universe and the Earth being billions of years old) is *far* from settled. There are numerous serious problems with deep time and uniformitarianism. So there's really no reason to even consider a compromise such as the Day-Age Theory, because there are few reasons to believe that the universe is that old, and many reasons *not* to believe it.

The Day-Age Theory was first put forth in 1823 by the respected Anglican theologian George Stanley Faber (1773–1854), and the respect that he had among church people made it easier for him to popularized it. The problem is—well, *one* of the problems is—that the wording of Genesis disallows age-long days: each day was described as singular "evening" and singular "morning," not "billions of evenings and billions of mornings."

The Day-Age Theory, like all the other attempts to combine the Bible's account with incompatible theories, is made much more palatable by a worldview in which we reduce God's power to the level of our power, God's wisdom to the level of our wisdom, and God's abilities to the level of our abilities. There seems to be an underlying tacit assumption that, since *we* can't do it, neither can God. This way of thinking only betrays the fact that we don't know God very well.

Framework Theory

The Framework Theory dismisses the Genesis account altogether, in the sense of being accurate. Instead, it assumes that the story of creation in Genesis is merely a literary framework—an allegory, a parable, a metaphor. But certainly not anything that can be taken literally.

In other words, "We know what the Bible says, but it doesn't really mean it."

The Implicit Conflict With Rationality

All of the above theories fail for the same reason: they conflict with God's words in the Bible. If you consider the Bible to be authoritative, you cannot remain intellectually honest while simultaneously believing in evolution or any attempt to combine the Biblical record and evolution.

However, if you don't yet consider the Bible to be authoritative, I invite you to carefully examine the content of the remainder of this book, following up on footnote references as much as you like, and then honestly ask yourself: "Which model—the creation model or the evolution model—makes predictions that more closely match repeatable observations and experimentation?" If you are honestly seeking the answer to that question, and you aren't going into it with your mind already made up, I think you'll be surprised at how well the creation model fares in the comparison.

Here's where all attempts to blend deep time with the Biblical account of creation run into trouble. If God's intention was to create the universe and everything in it, His actions were clearly directed toward that outcome. But the theory of evolution, by its very definition, states that everything came about through random and undirected changes. So the question "Couldn't God have created life by using evolution?" boils down to this: "Couldn't God intentionally guide a process and still keep it random?" Boiling it down even further, it becomes, "Couldn't God direct something without directing it?" Do you see the self-contradiction?

The most fundamental law of rational thought is the Law of Non-Contradiction, which states that no aspect of reality can be both true and false at the same time and in the same sense. If we reject this concept, *all* knowledge of *every* kind becomes meaningless, including what God says, because He is the source of knowledge.[71]

But that is what results from the attempt to claim that "God created the universe through evolution," because in order to do so, God would have to direct the process (for Him to create), and simultaneously *not* direct the process (for it to be evolution). And, God would be using a process that inherently doesn't need Him to exist. Why would He do that?

So it becomes clear, by the very definitions of the words used, no answer is possible, because the question itself—"Couldn't God create the universe through evolution?"—is nonsensical. It is intrinsically a logical absurdity. It cannot be answered in any way, no matter how smart or powerful you are, because the question itself doesn't even make sense.

The Implicit Conflict With Scripture

There are other problems with attempting to combine creation (intelligent, purposeful design) with evolution (random, undirected changes)—I mean, besides the ones discussed above. Suppose God actually authored evolution and therefore, through countless generations of small adjustments, mankind finally arrived on the scene. That scenario implies countless generations of organisms that *died* long before Adam and Eve arrived, which contradicts what Paul said about what brought death into the world.

Here's how Paul describes it:

Romans 5:12a (CEV): Adam sinned, and **that sin brought death** into the world.

TLB: When Adam sinned, sin entered the entire human race. **His sin spread death throughout all the world,** so everything began to grow old and die. . .

---

[71] I Samuel 2:3 (NASB): Boast no more so very proudly, do not let arrogance come out of your mouth; for **the Lord is a God of knowledge,** and with Him actions are weighed.

Proverbs 1:7 (NIV): **The fear of the Lord is the beginning of knowledge,** but fools despise wisdom and discipline.

Proverbs 2:6 (ESV): For the Lord gives wisdom; **from his mouth come knowledge and understanding.** . .

NLV: This is what happened: Sin came into the world by one man, Adam. **Sin brought death with it.**

NLT: When Adam sinned, sin entered the world. **Adam's sin brought death. . .**

TPT: When Adam sinned, the entire world was affected. **Sin entered human experience, and death was the result.**

So did death show up when Adam and Eve sinned, or had death been around for countless eons, as required by evolution? The proposed evolution-creation hybrid theory states that death was here before Adam was, but the Bible states that Adam was here before death was. They can't both be right, since they say mutually exclusive things. If you put any stock at all in the Bible, all proposed evolution-creation hybrid theories go down in flames.

Related to that, evolutionists would agree that plant life arose before mankind, and therefore thorns and thistles were around long before people were. But God said, *after* Adam and Eve sinned:

Genesis 3:18 (CEV): Your food will be plants, but **the ground will produce thorns and thistles.**

So again, you can't combine evolution and creation into a single theory, because they say opposite things. Are you going to believe that thorns and thistles preceded man, or the other way around? It depends on where you put your faith; you can't have it both ways.

Dr. James Barr is the Regius Professor of Hebrew at Oxford University. Although he himself does not believe Genesis is true history, he admitted that the language of Genesis 1 does present it that way:

> So far as I know, there is no professor of Hebrew or Old Testament at any world-class university who does not believe that the writer(s) of Gen. 1–11 intended to convey to their readers the ideas that (a) creation took place in a series of six days which were the same as the days of 24 hours we now experience, (b) the figures contained in the Genesis genealogies provided by simple addition a chronology from the beginning of the world up to later stages in the biblical story, (c) Noah's Flood was understood to be world-wide and ex-

tinguish all human and animal life except for those in the ark.[72]

Both creation and evolution require faith to believe in them. Which one is more tenable will be indicated by seeing which one is more supported by repeatable observations and experimentation, and which one is the model whose predictions are more accurately borne out in the universe.

So, in short, it is impossible to believe in creation and evolution at the same time, because of the unavoidable contradictions between them, a few of which are shown in this table:

Biblical Creation Account	Evolutionary Conjectures
The Earth existed before the sun and the stars	The sun and the stars existed before the Earth
Initially, the Earth was covered with water	Initially, the Earth was a ball of magma
The oceans existed before dry land	Dry land existed before the oceans
Life was first created on land	Life first "arose" in the oceans
Plants were created before the sun	Plants "arose" long after the sun
Birds existed before land animals	Land animals existed before birds
Whales existed before land animals	Land animals existed before whales

So let's think about the spiritual implications when we try to insert evolution, either via the Gap Theory, the Day-Age Theory, or any of their variations, into the Biblical record:

- Eons of death, decay, and sickness (diseased dinosaur fossils have been found) existed before Man was ever created.
- Since Jesus was the Creator (John 1:3), when Jesus saw *everything* that He had made, and pronounced it "very good" (Genesis 1:31), He must have included death, decay, and sickness in that pronouncement.
- Moses claimed that the Lord made heaven and earth in six days (Exodus 20:11), but he was apparently mistaken, because it should have been "six days plus billions and billions of years."

---

[72] Letter to David C. C. Watson, April 23, 1984.

- Why did Jesus, through the stripes on His back, secure our healing (Isaiah 53:5, I Peter 2:24), if He had already declared sickness to be "very good?"
- When the Bible says that death entered the world through Adam's sin (Romans 5:12), apparently that sin wasn't much of a big deal, because death was already in the world. And it was "very good."
- When Jesus considered death to be an enemy, and destroyed its power (I Corinthians 15:26), what's the big deal, if death was "very good?"

. . .and so forth. As you can see, attempting to mix evolution with Scripture immediately corrupts Scripture to the point of meaninglessness. In adhering to man-made traditions (like evolution), we become, like the Pharisees, those who make the Word of God of no effect (Mark 7:13). We *don't* want to go there.

So without further ado, let's get into the science of comparing creationism and evolutionism.

Recognizing Intelligence

When paleoanthropologists discover human bones and stone tools together, they typically get all excited. Why is that? Because the stone tools are undeniably the work of intelligent beings who created those tools to accomplish some deliberate purpose.

And the above response is perfectly reasonable: if there are stone tools—for example, sharpened bits of flint with notches on the sides so they can be secured on to the shafts of arrows or spears—it is obvious that those stone tools didn't build themselves. They didn't just "happen" as a result of rain and lightning and sun beating down on some rocks for centuries. No, it is obvious that they were intentionally fashioned.

Then those same paleoanthropologists turn around and say, with a straight face, that meticulously assembled protein molecules, which are *orders of magnitude*[73] more complex than stone tools, fell together accidentally, with no

---

[73] An "order of magnitude" is a difference of a factor of ten. So:
- If a is 1 order of magnitude greater than b, then $a = 10b$, or $a = 10^1 \times b$.
- If a is 2 orders of magnitude greater than b, then $a = 100b$, or $a = 10^2 \times b$.
- If a is 3 orders of magnitude greater than b, then $a = 1000b$, or $a = 10^3 \times b$.
- If a is 4 orders of magnitude greater than b, then $a = 10,000b$, or $a = 10^4 \times b$.

. . .and so forth. As you can see, it doesn't take many orders of magnitude to get really big numbers.

need for intelligence to accomplish it. Now how in the world does *that* work? Intelligence, doing intentional work, is required for simple things, but not for enormously complex things? That makes about as much sense as saying, "It'll take muscles to lift this toothpick, but not to lift that redwood tree."

Something about "straining at a gnat and swallowing a camel" comes to mind. . .

Dr. Michael Denton, a molecular biologist who is not a Christian, acknowledges that chance, as the cause of life, doesn't stand a chance:

> It is the sheer universality of perfection, the fact that everywhere we look, to whatever depth we look, **we find an elegance and ingenuity of an absolutely transcending quality, which so mitigates** *[sic; read "militates"]* **against the idea of chance.**
>
> Alongside the level of ingenuity and complexity exhibited by the molecular machinery of life, **even our most advanced artifacts appear clumsy.** We feel humbled, as neolithic man would in the presence of twentieth-century technology.
>
> It would be an illusion to think that what we are aware of at present is any more than a fraction of the full extent of biological design. In practically every field of fundamental biological research **ever-increasing levels of design and complexity are being revealed at an ever-accelerating rate.**[74]

Notice Denton's words: the "elegance and ingenuity of an absolutely transcending quality" that is overwhelmingly strong evidence "against the idea of chance." And "ever-increasing levels of design and complexity are being revealed" faster and faster, the more we learn. Those are amazing statements from such a high-caliber scientist.

What's even more astonishing is that Richard Dawkins, a *very* zealous and passionate spokesman for evolution, states:

> We have seen that **living things are too improbable and too beautifully 'designed' to have come into existence by chance.**[75]

---

[74] Michael Denton, *Evolution: A Theory in Crisis*, Adler & Adler Publishers, Bethesda, Maryland, 1986, p. 32.

[75] Richard Dawkins, *The Blind Watchmaker*, W. W. Norton & Co., New York, 1987, p. 43.

That is astonishing! In his book, *The Blind Watchmaker* (quoted above), Dawkins admits that life is "too improbable and too beautifully designed to have come into existence by chance." Yet the book was promoting evolution! If life did not come about by nondirected (purposeless, unintentional) causes, it must have come about by directed (purposeful, intentional) causes. *There is no other option.* Dawkins admits that it couldn't have been done by chance, but then states that it was done by chance. That is both puzzling and sad.

Explaining vs. Predicting

In any context in which two competing viewpoints are mutually exclusive, like creation and evolution, it is logically impossible to adhere to both, since they say opposite things: if one is true, the other can't be. In such a context, adherents of both sides will present arguments that predict certain things. As part of the scientific method, those arguments are then tested by observation, to see if the predictions turned out as expected.

The party whose prediction turned out to be correct, based on repeatable observation and experimentation, has every right—and even the responsibility, for the furtherance of science—to ask the party whose prediction did *not* match observations, "How do you explain that?" The party whose prediction was wrong will then need to do one of two things. First, he could present a plausible explanation of why his viewpoint is still correct, in spite of his prediction not being confirmed by observation and experimentation. Or, second, he could adjust his viewpoint so that future predictions are more likely to be borne out by real-world observation and experimentation.

So if a theory states that a particular experimental result or real-world observation will occur, and that result/observation actually does occur, that is good for the theory: results and observations support the validity of the hypothesis. Since the prediction was accurate, the hypothesis is more trusted to make further predictions.

On the other hand, if a theory states that a particular experimental result or real-world observation will occur, and that result/observation does *not* occur, that is bad for the theory: results and observations do not support the validity of the hypothesis. Since the prediction was *not* accurate, the hypothesis is less trusted to make further predictions and, as stated above, the proponents of the hypothesis are obliged to explain why, if their theory is true, it does not

---

Emphasis added.

match real-world results or observations. Or, revise their hypothesis and try again.

One thing you often see after examining much evidence related to the question of evolution versus creation is this: there are numerous verifiable, observable, undeniable processes and conditions that exist in nature, whose very existence are a significant deterrent to the process of evolution taking place. In other words, real-world observations fly in the face of the theory of evolution; they contradict each other. In all such places, a creationist would be very justified in asking an evolutionist, "Given your theory, how do you explain that?"

On the other hand, if the evolutionist asked the creationist the same question, the creationist would be very justified in replying, "We don't need to *explain* it; we *predicted* it!" In other words, real-world observations support the Creation model far more often than they support the evolution model. We don't need to explain away inconsistencies between the predictions of the model and the observations of nature; the observations support creation much better than they do evolution.

For people steeped in public education in America, where evolution is taught exclusively, that statement might sound ludicrous at first. However, please continue reading, and follow the numerous footnotes whenever you want to delve into greater detail. There are *so* many repeatable experiments and observations that support the creation model better than the evolution model, that the hardest thing about writing this book was deciding which information *not* to include. I didn't want this book to be a thousand pages long, but it easily could have been.

Chapter 1 Summary

Here are the most important points in this chapter:

- Evolution requires **deep time** or **long ages** (millions or billions of years) for its processes to take place.
- **Catastrophism** implies sudden and/or unusual events could be responsible for what we observe, and **uniformitarianism** states that what we observe is the result of only continuous, uniform processes.
- **Occam's Razor** is the problem-solving principle that states "The theory that requires fewer assumptions is more likely to be correct."
- **Laws and experimentation** are allegedly "inappropriate techniques" to explain evolution.
- Evolution is no less a **religion** than Christianity. Evolution is just a religion in which we are our own gods.
- According to the evolutionary worldview, **we cannot, or should not, learn from history,** because we have evolved since then, and our forefathers' ideas are hopelessly outdated.
- Evolution provides a justification for **abortion, euthanasia, genocide,** and other behaviors that devalue certain classes of people. Specifically, evolutionism promotes **racism.**
- Evolution provides a convenient **way to remove God** from our lives, because buildings don't imply builders anymore.
- Hybrid theories that claim that God could have created everything through evolution, are **inherently self-contradictory,** and thus cannot possibly be valid, because they claim mutually exclusive conditions simultaneously.
- Any theory makes **predictions,** and if experimentation and observation don't match the predictions, proponents of the theory are obligated to **explain** why, if the theory is true, observations don't concur.

OH, EVOLVE! (GOOD LUCK WITH THAT. . .)

Chapter 2:

The Theory of Evolution

The theory of evolution has caused untold confusion, consternation, debate, and resentment over the decades and especially the last century. Most of the hard feelings are between people who, on the one side, believe that the universe, the solar system, and life itself came to be via the process of evolution and, on the other side, those who believe it was all created through an act of God.

But why are there so often hard feelings between these two groups of people? Religious people—even Christians[76]—tend to feel threatened by anyone challenging their belief systems, whereas those in the sciences tend to look down on those who believe in God as deluded, but usually harmless people who need religion as a "crutch" in order to deal with reality.

But ignorance on either side tends to make the discussion descend into a shouting match, complete with aspersions thrown both directions. But this is completely unnecessary for the Jesus-follower who is willing to do his home-

---

[76] The distinction here between religious people and Christians is necessary, because there are many kinds of religions and religious people, but only one kind of Christian: a person who has submitted himself to Jesus and patterns his life after that of Jesus. The various Christian denominations are caused by differences in worship styles, modes of ministry, and so forth, but whatever the denomination, Jesus remains the only way to peace with God. Religions tend to have one thing in common: attempting to reach (and/or appease) God through our own efforts. This is fundamentally different from Christianity, in which God, knowing that we are incapable of bridging the gap between us, came to us in the person of Jesus to bridge that gap for us. Therefore, Christianity is a personal relationship with loving, providential God.

work. There is *so* much evidence for the Creation model, and *so* many problems with the theory of evolution, that it isn't a problem at all for a diligent Jesus-follower to confront the assertions of evolution head-on and come out on top, sticking strictly to scientific evidence.

That is the purpose of this book. By and large, Christians are far too often woefully misinformed when it comes to scientific topics. Such a state often results from being intimidated by scientific discussions. But since the entire universe was created by God, He understands that scientific basis for everything He created. Indeed, He *invented* the scientific basis for everything He created.

Think about it: if the heavens do indeed declare His handiwork, as Psalm 19 states, what we observe in the universe will indeed point to God—to an intelligent, thoughtful, caring Person. And that is exactly what we find. What we *don't* find is evidence of random, undirected, chaotic processes that mindlessly came up with the exquisitely interconnected and mutually dependent systems time after time after time.

So let's launch into it. In some cases, discussions will include math or physics content, but it will not be unnecessarily complex. It will be presented in a way that a high-school graduate who paid attention in class should be able to follow after going over it a time or two. For many of the specialized scientific terms used in this book, definitions are available in the Glossary; refer to that for clarification whenever necessary. Also, in case a particular term is not in the Glossary, have a dictionary handy in case you come across a word you're not familiar with. There are free and low-cost dictionary apps for your smartphone and tablet; grab one of those if you don't have one already. If you find some concept tricky to understand, please don't give up and remain in the dark about the idea under discussion—Christians remaining in the dark about scientific topics is the very problem this book is intended to overcome.

I haven't met any high-school graduate yet who couldn't understand the concepts contained in this book if he was motivated to learn it, but I have seen people give up when they realized it would take some work. So please don't allow yourself to fall into the latter category.

The Title of This Book

The title of this book, *Oh, Evolve!*, comes from a bumper sticker I saw years ago. The idea that comes across from the bumper sticker is that the owner of the car disapproves of someone—perhaps people in general—and wants them to become better in some unspecified area, and the "obvious" way to do that is to evolve.

It's similar to telling someone, "Oh, grow up!" with the implied request that the recipient stop being childish and act his age. But telling someone "Oh, *Evolve!*" includes a subtle insinuation that evolution is actually a thing. That assertion is what we will examine in detail in this book.

The book's subtitle, "Good Luck With That. . .", is also very intentionally chosen, because evolution depends on luck. *Loads* of luck. *Ridiculous* amounts of luck. Absolutely *impossible* quantities of luck.

What is luck? It is the outcome resulting from random factors. "Good luck," or "being lucky," is when random factors operate in your favor, or to your benefit or advantage. Conversely, "bad luck," or "being unlucky," is when random factors operate to your detriment or to your disadvantage.

The theory of evolution, in order to actually happen, requires trillions upon trillions of "lucky breaks," each with great impact, and vanishingly few "unlucky breaks," and those few are relatively minor in impact. So yes, you would certainly need "good luck" for evolution to occur. So much, in fact, that if we applied those same probabilities to any other area of science, or life in general, those knowledgeable in the field would immediately label it an "impossibly small" likelihood of occurring.

What Is Evolution?

Before we get into too much detail, we need to define exactly what it is we're talking about here to avoid misunderstanding and confusion. Exactly what *is* evolution?

The definition of evolution, in the biological sense, is as follows:

> Evolution is gradual genetic improvement in heritable characteristics in successive generations of living organisms. Changes in genetic material are caused by random and undirected mutations, and the gradual improvement of a species, or the change of one species into another, is brought about by natural selection—also known as "survival of the fittest"— the continuance of beneficial mutations and the "weeding out" of harmful ones.

At first glance, the above definition sounds reasonable; the process of evolution sounds plausible. After all, we know that mutations exist, right? And we know that all mutations are either beneficial or not, right? And if mutations are beneficial, they would help the organism survive, and if they're not, the organism would die out, right?

As in most other fields of study, there is *so* much more to it than a "first glance" communicates. And those other factors—the ones that lie beneath the surface-level façade of plausibility—are the killers for the theory of evolution.

Biological evolution was actually postulated by other people before Charles Darwin, but his book *The Origin of Species* seemed to take the world by storm, because at that time there was a philosophical preference toward removing God from everyday life. And although Darwin was not the originator of this philosophical preference to eliminate God, he certainly subscribed to it.

As a young man, Darwin attended seminary to become a parson, and was intrigued by William Paley's book *Natural Theology*,[77] which eloquently points out how created things have meticulously designed features, which plainly point to an intelligent and stunningly skilled Designer. Darwin was so impressed with Paley's book that in a letter to his friend John Lubbock, Darwin stated that he hardly ever admired a book more than that one.

But like so many people do, Darwin suffered a tragedy in his life, and he blamed God for it. Darwin's daughter Annie died, and in anger, Darwin turned away from God:

> Annie's cruel death destroyed Charles's tatters of beliefs in a moral, just universe. Later he would say that this period chimed the final death-knell for his Christianity. . . . Charles now took his stand as an unbeliever.[78]

Apparently, in his desire to "punish" God for letting his daughter die, Darwin first rejected an unbiblical, philosophical version of creationism, and then,

---

[77] William Paley (1743–1805) wrote several books showing evidences of creation. Perhaps the most well-known is his 1802 work entitled *Natural Theology*, subtitled *Evidences of the Existence and Attributes of the Deity*. The books discuss **natural theology** (seeing evidence of God from what He has made), *à la* Romans 1:19–20, and makes use of the **teleological argument**: the argument for the existence of God, or at least for an intelligent Creator, based on the overwhelming evidence of intelligent design in the natural world.

[78] A. Desmond and J. Moore, *Darwin: The Life of a Tormented Evolutionist*, W. W. Norton & Company, New York, 1991, p. 387.

throwing the baby out with the bathwater, he rejected Biblical creationism as well. His new philosophy, which he heavily promoted through his book *Origin of Species,* was that it was "unscientific" to acknowledge a Creator, and Darwin made it his aim to "ungod" the universe.[79]

So Darwin's book was just what his readers, eager to rid their lives of God, were looking for. But something that most people are unaware of is that Darwin himself questioned his theory, because of observations that not only didn't support his theory, but militated against it.

Darwin asked:

> "Why, if species have descended from other species by insensibly fine gradations, do we not everywhere see innumerable transitional forms?"[80]

In another place he admitted:

> "To the question why we do not find a rich fossiliferous deposits belonging to these assumed earliest periods I can give no satisfactory answers."[81]

Indeed, these are excellent questions. With one species slowly and gradually changing into another over many generations, there should be fossilized transitional forms all over the place. *But there aren't.* If such transitions actually happened, intermediate forms should be found both in fossils and in living organisms. Existing classes should overlap. Distinct boundaries ought to be the exception rather than the rule. Darwin's own observations showed a troubling lack of evidence for his own theory. And, in the more than 150 years since his book was published, we're *still* looking for partly-this-and-partly-that transitional fossils.

The "missing links" are still missing. True, there are occasional "discoveries" that are lauded as missing links, but they are *so* vague and *so* ambiguous that they are not at all convincing, even to many evolutionists. Or, they are shown to be wrong. Or, they are shown to be hoaxes, Or, they are shown to be equivalent to modern man. More about these examples in a later chapter.

---

[79] Marvin L. Lubenow, *Bones of Contention* (BakerBooks, Grand Rapids, Michigan), 2004, p. 94.

[80] Darwin, Charles, *On the Origin of Species by Means of Natural Selection, or the Preservation of Favoured Races in the Struggle for Life.* 1859, London: John Murray, p. 158.

[81] Geoffrey Simmons, M.D., *What Darwin Didn't Know,* 2004 (Harvest House Publishers, Eugene, Oregon), p. 298.

Here in the real world, there exists the Law of Cause and Effect. In a nutshell, it states that "Every effect must have an adequate cause." This is merely the formalization of two definitions: "A **cause** is that which produces an effect," and "An **effect** is that which is produced by a cause." A very normal and healthy response to perceiving anything that is unexpected or unknown is, "Why did *that* happen?" Indeed, most of science is devoted to answering that question in a wide variety of contexts.

However, a problem with answering the question "Why did that happen?" can arise when people have a pre-existing bias against some answer: even if evidence points in that direction, they *will not* consider it, because they have already decided that such an answer could not possibly be correct. In many cases, this kind of bias may not even be intentional, but may be a result of previous training or philosophical preference. We'll see many examples of that in this book, where significant evidence that points to a young universe and solar system, a young Earth or a worldwide flood, or recent appearance of life, is ignored or dismissed out of hand, because it points to something of which the average evolutionist is terrified: a Creator.

So in the real world, every effect must have a cause, but in the world of evolutionary credulity, the laws of physics can be set aside if they become inconvenient. In this magical land, plans draw themselves, mechanisms create themselves, the Second Law of Thermodynamics conveniently doesn't apply, chaos spontaneously falls into order, information writes itself, and life spontaneously "arises." Yet, far too often, evolutionists accuse creationists of being "pseudoscientific" because they propose an adequate cause—a Creator—for the effects we see everywhere.

As you keep reading, ponder these things, while intentionally resisting any pre-existing biases, and then make up your own mind which scenario is more plausible.

Muddying the Waters

One of the reasons that there is so much confusion and misunderstanding about the viability of evolution as a theory is the inconsistent way the word is bandied about in popular culture.

For example, using your favorite search engine, look up the phrase "evolution of computers." Or "evolution of the telephone." Or "evolution of the car." Or any other piece of technology. There are numerous articles on the web that purport to show you the evolution of whatever invention you just looked up. Look at the images below; do you see the problem?

CHAPTER 2: THE THEORY OF EVOLUTION

> Note: In this book are many photos and illustrations with greater detail than can be shown here, even with images as wide as the whole page. Also, many of them are in full color, and lose much impact when printed in black and white. Therefore, every image has a QR code at its lower corner: just scan this with your phone to see the full-res, full-color, zoomable image.

Image in the Public Domain

Image courtesy Premeditated, Creative Commons CC BY-SA 4.0

Does this illustrate the evolution of computers?

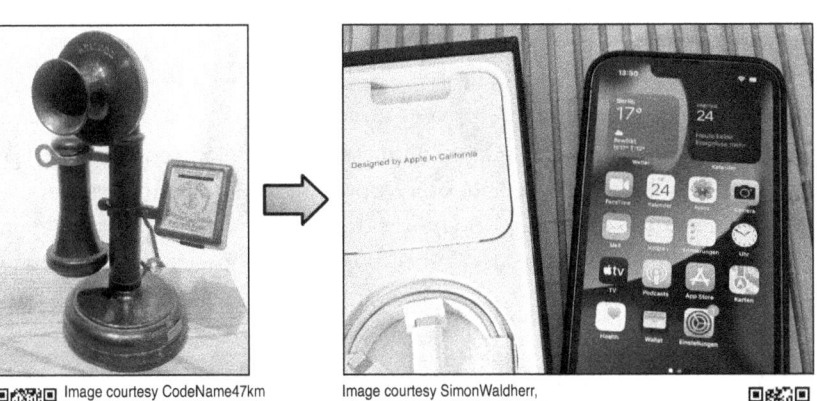

Image courtesy CodeName47km, Creative Commons CC BY-SA 4.0

Image courtesy SimonWaldherr, Creative Commons CC BY-SA 4.0

Does this illustrate the evolution of the telephone?

77

Oh, Evolve! (Good Luck With That...)

Does this illustrate the evolution of the car?

Evolution, as shown in the definition shown earlier, is improvement brought about by *random and undirected* changes. Does anyone actually want to allege that the changes and improvement in computers, telephones, cars, or anything else was brought about by random, undirected changes? Would anyone really claim that the improvements to the three technologies shown above were accomplished with no intention, goals, decisions, intelligence, foresight, planning, or awareness? Of course not; that is patently ridiculous, and you'd be laughed out of the room if you presented that statement as anything other than a joke.

It has taken *millions* of man-hours of labor by very smart people, using very sophisticated machines, and with very specific goals toward which they were working, in order to come up with the technological tools that we take for granted and use every day. In other words, it is not *at all* the evolution of those devices that people talk about; it is the intentional development, by intelligent designers, of those devices, through carefully selected and meticulously implemented changes. That is a far cry from the "evolution" of those devices!

As another example of such misleading usage, think of the name of one of the standards in broadband wireless communication for mobile phones. Although sometimes marketed as 4G, it's not quite there (technically, 3.95G), but it is very useful regardless. Do you recognize what technology this is? It's LTE, which you may have seen as a communication protocol on your cell phone. What does LTE stand for? It stands for "Long-Term Evolution." Again, did it arise from earlier technologies by unintentionally introducing random and undirected changes of an indiscriminate nature, in arbitrary locations in the hardware and software? Of course not.

So the salient point here is that evolution, by its very definition, cannot have anything done on purpose. *All* changes must be random, undirected, haphazard, accidental, and without method, foresight, goals, intention, or conscious decisions. Seriously, could you even build a doghouse that way? Could you even toast a slice of bread? Not a chance.

Microevolution vs. Macroevolution

There are two "flavors" of evolution that are commonly talked about: microevolution and macroevolution. What's the difference, and why is it important?

Microevolution is not disputed, although it's misleading to call it "evolution." This is simply variation within a kind, using already-existing genetic information. Even though a particular manifestation of a particular characteristic may not have been seen before, it was always latent within the DNA. Typical examples of this are the different breeds of dogs, or the "zonkey" shown at right, that are possible by intentional crossbreeding.

Image in the Public Domain (PD-US-expired)

All dog breeds came from selectively breeding members of the dog kind.

Microevolution is simply the expression of various characteristics already latent in the organism's DNA. Offspring of the same parents can result in dramatically different appearances, personalities, body builds, preferences, behaviors, and so forth, as anyone who has a brother or sister or kids knows.

So microevolution can be illustrated by a deck of playing cards: it is simply like the varying combinations you can get with a standard deck of 52 cards. Those 52 cards can you give numerous combinations, but no matter how many times you shuffle them, you still have only 52 cards.

So technically, microevolution is not really evolution, in the sense that the theory of evolution requires, because no genetic information is added to the organism's DNA; it's simply that different combinations of attributes are expressed. Attributes can be emphasized by selective breeding, but all breeding must happen within the organism's "kind," as described in Genesis 1:21–25. Organisms within a kind are interfertile; they can produce offspring. Attempts to crossbreed animals from different kinds (for example, dogs and cats) fail, because their genetic material is incompatible.

Image courtesy sannse (CC BY-SA 3.0)
A "zonkey" (a cross between a zebra and a donkey) comes from selective breeding of members of the horse kind.

Macroevolution, on the other hand, is the kind that the theory of evolution demands, because that is the version in which one kind of organism changes into another kind. This is the kind of evolution for which no transitional forms have been found in the fossil record, even though they would be everywhere if the process had actually happened. This is also the kind of evolution that supposedly adds functional genetic information, which conveniently "arises" out of nothing by sheer accident, to the DNA.

Macroevolution can also be illustrated with a deck of cards: it would be like shuffling the 52 cards and then finding more suits than you started with (maybe clubs, diamonds, hearts, spades, triangles, and circles) and brand new pip values of cards (maybe ace through 12 plus jack, queen, king, duke and duchess). Sure, no problem: when you shuffle four suits of 13 cards, don't you usually end up with six suits of 17 cards? That happens all the time, right? This macroevolution is what has never *ever* been observed.

Macroevolution is taken for granted by evolutionists, and observations are massaged as necessary to make them fit into the presupposed sequence. If you are thinking that last statement is being overly critical, I invite you read *Bones of Contention*[82] by Marvin Lubenow, for numerous documented examples of this happening.

The most foundational, important, and far-reaching difference between microevolution and macroevolution is that *microevolution does not require new*

---

[82] Marvin L. Lubenow, *Bones of Contention* (BakerBooks, Grand Rapids, Michigan), 2004.

genetic information to come into being, while macroevolution does. A lack of understanding of this fact has led to the common "micro-versus-macro fallacy," in which people mistakenly think that if microevolution happens for long enough, the result will be macroevolution. It won't. Using a mathematical analogy, the micro-versus-macro fallacy claims that if you start with 1, and add zero enough times, you'll get 2.

In this book, the term "evolution" will refer to macroevolution unless clearly specified otherwise.

Sacred Cows of Evolution

There are many ideas that evolutionists have put forth over the years, that "prove" evolution. They are presented as front-page "headline news"—they are slick and they appeal to the masses, so they are quite effective at selling the whole idea.

They have a habit of being disproven, however, as more testing is done or more information becomes available. Once it becomes inarguable that these ideas were wrong all along, retractions—if they ever appear—are far from headline news. Rather; they are figuratively in Section D, page 18, column 4, toward the bottom of the page, in fine print.

The result is that the public hears only the fanfare of new discoveries, not the sheepish retractions, should they ever actually appear. So unless John Q. Public is paying attention, it seems like there are more confirmations to evolution every week or so, and nothing saying otherwise. Can you say Madison Avenue?

Embryonic Recapitulation

Embryonic recapitulation was a theory put forth by Ernst Haeckel that claimed that the similarities between developing embryos of various animals showed their respective evolutionary journeys.

Haeckel expressed it as "ontogeny recapitulates phylogeny." This twenty-dollar phrase simply means, in laymen's language, that the way an organism comes into being as it develops from a fertilized egg ("ontogeny"), illustrates by visually repeating ("recapitulates") the process of how that type of organism evolved in the first place ("phylogeny").

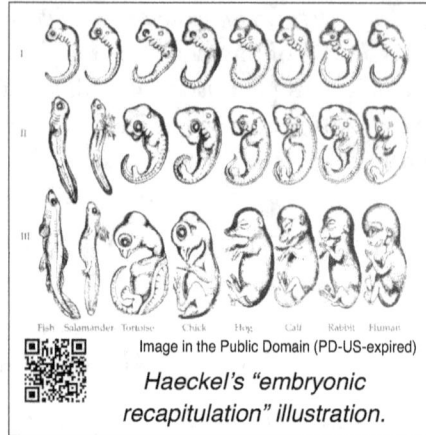

Haeckel's "embryonic recapitulation" illustration.

The image to the right is often touted as strong evidence of evolution. But again, let's go beyond the surface. When Haeckel's illustrations are compared side-by-side with actual photographs of the respective embryos, it becomes obvious that the drawings were faked; they were made to look a lot more similar than they actually were. In some cases, Haeckel used the same woodcut for multiple species, because he was so confident he was right. In other cases, he simply drew them in such a way as to support his theory.[83]

I remember seeing Haeckel's illustrations when I was in school and being indoctrinated into evolution. What makes it even more annoying is that his misrepresentations were known and exposed in the 1860s! It had been *more than a century* since the misrepresentations were exposed, and it was still being taught in schools. Now *that* is lousy science.

Not only did Haeckel doctor the drawings so they would say what he wanted them to say, but he cherry-picked the ones that would best support his theory, rejecting many other examples that didn't indicate what he wanted them to indicate. And, what Haeckel said was the "early" stages of development were actually more "middle" stages, because earlier, they looked *more* different. Since that too militated against his theory, he omitted those images as well.

Jonathan Wells, with a Ph.D. in molecular- and cell biology from Berkeley, compares Haeckel's drawing to the Cambrian Explosion (a burst of life forms, discussed in "Chapter 7: The Fossil Record"):

> It's just false that embryos are the most similar in the earliest development. Of course, some Darwinists try to get around Haeckel's problems by changing their tune. They use evolu-

---

[83] Lee Strobel, *The Case for a Creator* (Zondervan, Grand Rapids, Michigan, 2004), p. 48.

tionary theory to try to explain why the differences in the embryos are there. They can get quite elaborate. But that's doing the same thing that the theory savers were doing with the Cambrian explosion. **What was supposed to be the primary evidence for Darwin's theory**—the fossil or embryo evidence—**turns out to be false, so they immediately say, well, we know the theory is true, so let's use the theory to explain why the evidence doesn't fit.**[84]

The Miller-Urey Experiment

The Miller-Urey experiment in 1952 was widely heralded as strong evidence that life could have spontaneously arisen on Earth through natural means. The goal was to create amino acids, which are the "building blocks" of life.

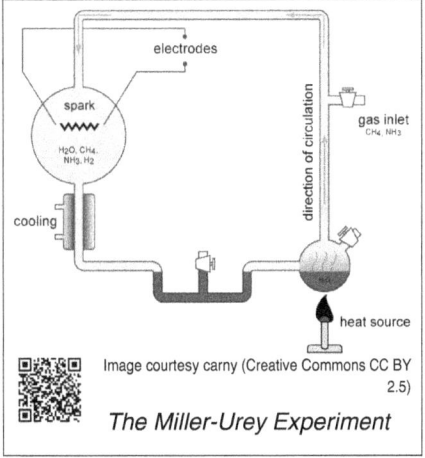

Image courtesy carny (Creative Commons CC BY 2.5)

The Miller-Urey Experiment

Miller introduced into his apparatus what was then thought to be the components Earth's "early atmosphere:" water, methane, ammonia, and hydrogen. He then caused electrical discharges to simulate lightning, and after a while, sure enough, amino acids and some organic compounds formed. That sounds pretty hopeful, doesn't it?

But just looking at the apparatus in the illustration makes you wonder: how much of the experiment was actually left to chance, compared to how much was intelligently designed? And how did Miller know what the "early" atmosphere of Earth was like? And what kind of amino acids and organic compounds were created? The details shown below are left out of the textbooks heralding Miller's "successful" experiment, and the reason will become obvious as you read.

Miller designed his experiment very intentionally, using experience gained from years of research in chemistry. He carefully chose which gases to include, and which gases *not* to include. Then, as soon as amino acids formed, he had

---

[84] Lee Strobel, *The Case for a Creator* (Zondervan, Grand Rapids, Michigan, 2004), p. 50.

to quickly remove them from the system, because the same environment that created them would also destroy them moments later.[85]

Note that Miller left out oxygen from his gas mixture. Why was that? It is well-known that oxygen destroys biological compounds such as amino acids. So amino acids could not have formed with oxygen present in the atmosphere. But if oxygen was *not* present in the atmosphere, we wouldn't have had an ozone layer (ozone is a form of oxygen), so the ultraviolet light from the sun would have destroyed the amino acids. So amino acids can't form *with* oxygen, and they can't survive *without* oxygen. That about covers it, wouldn't you say?

And, by the way, there has always been oxygen in the Earth's atmosphere:

> There is no scientific proof that Earth ever had a non-oxygen atmosphere such as evolutionists require. Earth's oldest rocks contain **evidence of being formed in an oxygen atmosphere.**[86]

> The only trend in the recent literature is the suggestion of **far more oxygen in the early atmosphere than anyone imagined.**[87]

Well, since life couldn't have arisen in the atmosphere, maybe it arose in the oceans. But no, that doesn't work either. Assuming that an amino acid did form in the ocean, the water—through a process called **hydrolysis**—would quickly have split it apart. And according to numerous reports, water is usually present in the oceans.

So the Miller experiment did result in amino acids being formed, even though they had to be quickly extracted so they were not destroyed again. What kind of amino acids were formed? There are about 500 kinds of amino acids, but only 20 of them are used in DNA. How many did the Miller experiment yield? He found three for sure, and maybe two more.[88]

But there's something else too. Three-dimensional molecules can be in one form or its mirror image; these are called "left-handed" and "right-handed" molecules. Experiments such as Miller's create just as many left-handed as

---

[85] Ken Ham and Bodie Hodge, *The New Answers Book 2,* Answers in Genesis (Master Books, Green Forest, Arkansas, 2008), Kindle edition, Location 1193 of 7000.

[86] H. Clemmey and N. Badham, "Oxygen in the Atmosphere: An Evaluation of the Geological Evidence," *Geology* 10 (1982): p. 141. Emphasis added.

[87] Thaxton, Bradley, and Olsen, *The Mystery of Life's Origin,* p. 80. Emphasis added.

[88] en.wikipedia.org/wiki/Miller–Urey_experiment; accessed July 16, 2020.

right-handed molecules; this is called a **racemic** mixture. The problem is that life processes use *only* left-handed amino acids—if even *one* right-handed amino acid were present in a protein, the protein would fail in its function. (More on this below.)

A few paragraphs ago, "organic compounds" were mentioned in passing: Miller's experiment resulted in organic compounds being formed. That sounds encouraging; they would be conducive to life, wouldn't they? What kinds of organic compounds were actually formed? Formaldehyde (embalming fluid) and cyanide, to name a couple. Formaldehyde is a highly toxic compound that destroys proteins,[89] and of course, cyanide is poisonous as well. Maybe those organic compounds are not so conducive to life after all.

Developmental biologist Jonathan Wells states:

> So we remain profoundly ignorant of how life originated. Yet the Miller-Urey experiment continues to be used as an icon of evolution, because nothing better has turned up. **Instead of being told the truth, we are given the misleading impression that scientists have empirically demonstrated the first step in the origin of life.**[90]

And amino acids, as mentioned above, are just the "building blocks." The building blocks of what? Of proteins, which are the workhorses in living cells. Chains of amino acids are assembled to form proteins of very specific shape and function. In a human cell, there are about 200,000 different proteins, all of which need to work properly, or the cell dies. About 2,000 of these proteins are enzymes that facilitate the construction of the other 198,000 proteins. Astronomer Fred Hoyle, who openly says he is not a Christian, says the likelihood of these 2,000 enzymes forming (not even counting the other 198,000 proteins), is 1 chance in $10^{40,000}$.[91]

How much more ridiculous could a belief in evolution get? Claiming that the natural processes in the Miller-Urey experiment shows the ability to create life, on the evidence that a few amino acids were formed, is similar to someone claiming to have the ability to build the Empire State Building, and his evidence is that he made a brick. Or, "I could drink the whole ocean. My

---

[89] Lee Strobel, *The Case for a Creator* (Zondervan, Grand Rapids, Michigan, 2004), p. 38.
[90] Jonathan Wells, *Icons of Evolution* (Washington, DC: Regnery Pub., 2000), p. 24. Emphasis added.
[91] John W. Oller, Jr., "Not According to Hoyle," *Acts & Facts,* December 1, 1984. [Online at ICR.org/article/not-according-hoyle; accessed July 24, 2020.]

evidence is that I drank a glass of water." Or even, "I could create the hardware for the whole internet; I know because I made this piece of wire." How believable is that?

So, although the Miller-Urey experiment was widely heralded as the explanation for how life formed naturalistically, it, along with all of its successors trying to accomplish the same thing, dramatically fail to live up to the marketing:

> The story of the slow paralysis of research on life's origin is quite interesting, but space precludes its retelling here. Suffice it to say that at present **the field of origin-of-life studies has dissolved into a cacophony of conflicting models, each unconvincing, seriously incomplete, and incompatible with competing models.** In private even **most evolutionary biologists will admit that science has no explanations for the beginning of life.**[92]

So what are these intelligent scientists who do such experiments really saying to us? They are saying, "If I can create life in the lab, I will have shown that no intelligence was needed to create life." Ouch.

The Phylogenetic Tree

One of the trains of thought Darwin pondered was how one kind of organism could change into another. He postulated a great number of tiny changes over long ages, and these changes would eventually add up to completely different kinds of animals or plants.

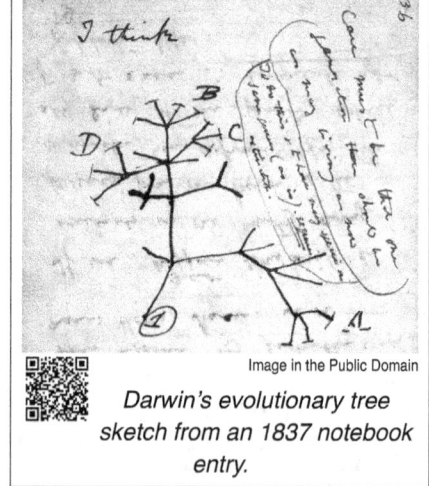

Image in the Public Domain
Darwin's evolutionary tree sketch from an 1837 notebook entry.

Darwin's first sketch of the idea is shown here. This entry in his notebook was dated 1837, and he was apparently not terribly confident in the validity of the idea: see the "I think" in the upper left corner of the dia-

---

[92] Michael J. Behe, "Molecular Machines," *Cosmic Pursuit*, Spring 1998, pp. 30–31. Emphasis added.

gram. But he developed the idea, and discussed and illustrated it in his book *Origin of Species,* published in 1859.

In a "scientific" world in which a belief in God was becoming increasingly unfashionable, people wanted a naturalistic mechanism which could account for the various species of plants and animals that we see today. Such a drawing, which describes the supposed ancestry of all organisms, living and dead (at least to the limits of its level of detail), is called an **evolutionary tree** or a **phylogenetic tree**.

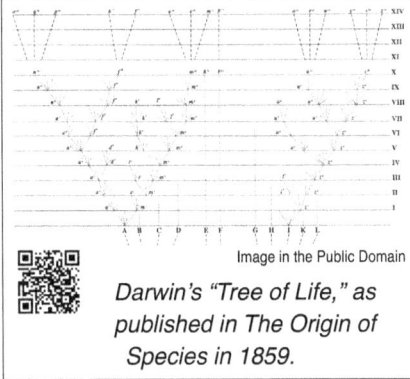

Darwin's "Tree of Life," as published in The Origin of Species in 1859.

The word "phylogenetic" has Greek roots; it comes from "phulon," which means "race" or "tribe," and "gignomai", which means "be born," or "become." So "phylogenetic" implies "how the various races came to be."

So from Darwin's humble sketch in 1937, through a more detailed rendition in *Origin of Species* in 1859, through the much more elaborate and tree-like 1879 representation by Ernst Haeckel shown here (yes, the same one who pushed "ontegeny recapitulates phylogeny," discussed earlier), such images stick in the mind of people the world over, as evidence—even "proof"—that we all came from a common ancestor long ages ago.

The trouble is, it's wrong. It illustrates a long evolutionary history that never happened.

Now I realize that my own statement to that effect carries little weight, so allow me to get a second (and third, etc.) opinion. Here's what Jonathan Wells, a geologist, physicist, molecular- and cell biologist from Berkeley, says about Darwin's "Tree of Life:"

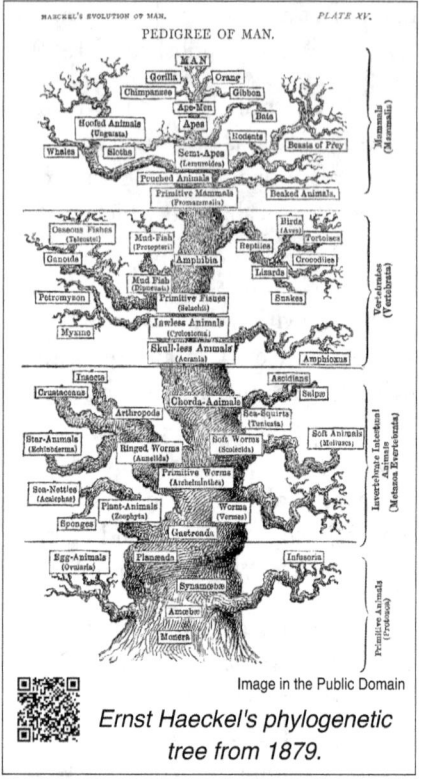
Ernst Haeckel's phylogenetic tree from 1879.
Image in the Public Domain

> As an illustration of the fossil record, **the Tree of Life is a dismal failure.** But it *is* a good representation of Darwin's theory. . . A key aspect of his theory was that natural selection would act, in his own words, "slowly by accumulating slight, successive, favorable variations" and that "no great or sudden modifications" were possible. . . **Darwin knew the fossil record failed to support his tree.** . . . Darwin believed that future fossil discoveries would vindicate his theory—but that hasn't happened. Actually, fossil discoveries over the last 150 years have turned his tree upside down by showing the Cambrian explosion was even more abrupt and extensive than scientists once thought. . . If you consider all of the evidence, **Darwin's tree is false as a description of the history of life.** I'll even go further than that: it's not even a good hypothesis at this point.[93]

We'll discuss the Cambrian Explosion in some detail in "Chapter 7: The Fossil Record," but for now, suffice it to say that it was such an abrupt appearance of numerous and diverse life forms, completely formed and with no ancestry, that it has been called the "Biological Big Bang." Exactly the opposite of what Darwin's theory stated.

---

[93] Lee Strobel, *The Case for a Creator* (Zondervan, Grand Rapids, Michigan, 2004), p. 43. Emphasis added.

Jonathan Wells is certainly not the only scientist who is abandoning the so-called "Tree of Life." Consider these as well:

- **"Biologists need to depart from the preconceived notion** that all genomes are related by a single bifurcating tree."[94]
- The tree of life **"lies in tatters, torn to pieces by an onslaught of negative evidence...** [D]ifferent genes told contradictory evolutionary stories."[95]
- **"We've just annihilated the tree of life.** It's not a tree any more, it's a different topology entirely."[96]

And there are many more.[97,98,99] Another enlightening tidbit: In one study of gene sequence data, predictions based on Darwin's tree were 99% wrong.[100]

---

[94] Dagan, Tal and William Martin. 2006. "The tree of one percent." *Genome Biology.* 7: p. 118. [Online at GenomeBiology.biomedcentral.com/articles/10.1186/gb-2006-7-10-118#B9; accessed July 25, 2020.] Emphasis added.

[95] Lawton, G. 2009. "Why Darwin Was Wrong About the Tree of Life." *New Scientist.* 2692: 34–39. [Online at NewScientist.com/article/mg20126921-600-why-darwin-was-wrong-about-the-tree-of-life/?ignored=irrelevant; accessed July 25, 2020.] Emphasis added.

[96] Michael Syvanen, co-editor of *Horizontal Gene Transfer* (1998) and a medical biochemist at the University of California at Davis, quoted in: Lawton, G. 2009. "Why Darwin Was Wrong About the Tree of Life." *New Scientist.* 2692: 34–39. [Online at NewScientist.com/article/mg20126921-600-why-darwin-was-wrong-about-the-tree-of-life/?ignored=irrelevant; accessed July 25, 2020.] Emphasis added.

[97] Thomas, Brian, "Darwin's Evolutionary Tree 'Annihilated'" February 3, 2009. [Online at ICR.org/article/darwins-evolutionary-tree-annihilated; accessed July 25, 2020.]

[98] Thomas, Brian. "Shared Genes Undercut Evolutionary Tree" February 25, 2011. [Online at ICR.org/article/shared-genes-undercut-evolutionary; accessed July 25, 2020.]

[99] Thomas, Brian. "Can't See the Forest for the Trees" October 28, 2015. [Online at ICR.org/article/cant-see-forest-for-trees; accessed July 25, 2020.]

[100] Ciccarelli, F. D. et al. 2006. "Toward Automatic Reconstruction of a Highly Resolved Tree of Life." *Science.* 311 (5765): 1283–1287.

Vestigial Organs

Another idea that has been touted to the masses over the decades as proof for evolution is that of "vestigial organs"—organs that were formerly useful, but as we've evolved, they were no longer needed. Scores of examples were cited of biological items that had no purpose anymore, and their purposelessness was cited as proof of evolution.

The vermiform appendix, whose obsolescence was used to support evolution until it was discovered to not be obsolete.

But (and I'm sure you see where this is going) the list of vestigial organs has been shrinking for some time now. Organs that once were considered vestigial include the appendix, the coccyx, the small muscles around the ears, the tonsils, the pineal gland, the thymus, and many more. In 1890, the list of vestigial organs in humans had 180 entries, but by 1999, the number of entries had dropped to zero.[101] That's quite a reduction in size, don't you think? What caused such a drastic change?

Before we answer that question, we need to define more precisely what a vestigial organ is. It is typically defined as "a biological structure that no longer has any useful function, and therefore represents the remains of organs that were useful in the evolutionary past." According to one evolutionary reference,[102] a vestigial organ is an organ that "has lost its function in the course of evolution, and is usually much reduced in size." (Note that in both these definitions, evolution is assumed in the definition, and then the resulting definition is used to "prove" evolution. Circular reasoning—you'll see that a lot in arguments for evolution.)

As an example of the circular reasoning noted above, look at this definition:

> *Useless Organs Prove Evolution.* Science has piled up still further evidence for its case. It has found a number of **useless**

---

[101] Jerry Bergman, "Do any vestigial organs exist in humans?" *Journal of Creation* 14(2):95–98, August 2000. [Online at Creation.com/do-any-vestigial-organs-exist-in-humans; accessed January 16, 2022.]

[102] Gamlin, L. and Vines, G., *The Evolution of Life* (New York: Oxford University Press, 1987).

organs among many animals. They have no apparent function and **must therefore be a vestige** of a once useful part of the body. **A long time back** these vestigial organs **must have been** important; now they are just reminders of our common ancestry. One example is the vermiform appendix which not only is **utterly useless** in human beings but which often causes great distress.[103]

Do you see the circularity? First, assume evolution is unquestionable. Second, interpret the data so it corresponds with your assumptions. Third, use your conclusions to confirm evolution. Or, more specifically for this case, assume evolution is unquestionable, then note that these existing organs are now useless (or at least *appear* to be; more on that below), then conclude that they must have been useful back in the evolutionary past (because they wouldn't have evolved in the first place unless they were useful). How far back? A "long time back." Why do we say that? Because evolution, which requires deep time, is unquestionable.

But even more than that, notice the hubris in the above paragraph: *Three times* in one paragraph these organs are stated to be "useless." Note the presumption of omniscience here: "If we don't know what purpose these organs have, they must not have any purpose." Not even a hint of a possibility of "They may do something that we don't yet know about." (See this same mindset in "Junk DNA" in Chapter 6: Mutations.)

But if we think about it for a moment, we realize that it is inherently impossible to prove that an organ is useless. Why? Because a use could be discovered in the future. In *every one* of the 180 cases of the 1890 list of useless organs, they were considered purposeless organs only until someone discovered their purpose.

Moreover, if anyone actually did discover a vestigial organ, and could prove it to be actually useless, it would still not prove evolution. Rather, it would prove *devolution*. Evolution claims that *more* order, *more* genetic information, *more* complexity, *more* functionality results from random processes. Yet in the case of vestigial organs, a *loss* of functionality is touted as evidence of evolution. That's certainly convenient, isn't it? "If functionality increases, it proves evolution. If functionality decreases, it still proves evolution."

---

[103] Perkel, A. and Needleman, M.H., *Biology for All*, (New York: Barnes and Noble, 1950), p. 129.

The "March of Progress"

Perhaps the most recognizable marketing tool for evolution is the famous "March of Progress" illustration, similar to the illustration shown here, and originally shown in the Time-Life Nature Library book *Early Man*, published in 1965, with revisions in 1968 and 1973.

Image by Rudolph Zallinger in the Time-Life book *Early Man* (Howell, 1965)

The "March of Progress" illustration.

This "March of Progress" is a series of fifteen figures, originally published as a 36-inch foldout. It was a masterpiece of propaganda, posted in its entirety on many a schoolroom wall, and many people considered it "proof" of evolution. The proverbial Madison Avenue advertising machine had done its job well.

At this point, you may be thinking, "Now he's going to say that further research has shown that there were errors in the illustration." But no; it's actually much worse: This parade of figures was known to be fake when it first came out: the book itself *admits* the parade is fiction! Furthermore, F. Clark Howell (the author of the book) and the Time-Life editors he was working with were fully aware that this illustration was fabricated—artificial, phony, fraudulent, fake—evidence for evolution.[104]

The visual image was so powerful that few people bothered to look into the book itself, where it admitted that the parade was fictitious. The apes on the left end of the series are all represented as **bipedal** (walking on two feet), but apes aren't; they are **quadrupedal** (walking on all fours). A small explanatory note in the book stated that:

> Although **protoapes and apes were quadrupedal, all are shown here standing** for purposes of comparison.[105]

But as you can see, they weren't just standing; *they were walking.* Just like apes don't. But did any evolutionists raise an uproar about such a glaring

---

[104] Marvin L. Lubenow, *Bones of Contention* (BakerBooks, Grand Rapids, Michigan), 2004, p. 39.
[105] F. Clark Howell, *Early Man,* rev. ed. (New York: Time-Life Books, 1968), p. 41.

error? No, because this was one of the most successful marketing campaigns in the history of evolution. But how about the rest of the "parade?" That's accurate, isn't it?

Actually, no. *Any* collection of items can be placed into a arrangement to make them appear sequential, but there is precious little—even using evolutionary thinking—to nail down when, or even *if,* these various nuances of mankind existed. As we'll see in the "'Primitive' Man" section in "Chapter 6: Mutations," there is much disagreement among evolutionists as to where in the "history" of mankind these specimens belong, and even whether they were pure apes or pure humans.

Sure, the "parade" is not accurate, but why let petty details like accuracy get in the way of a good inculcation?

And it *was* a good one, too. It still is. It was brilliant advertising, scientific fraud notwithstanding. Marvin Lubenow has done creation seminars all over the world, and he has been amazed at how widespread the awareness of the "March of Progress" parade is. It is almost a universal icon of evolution, and Lubenow discovered it was well-known in Japan, Australia, Tasmania, Lithuania, Canada, Barbados, and other countries.[106]

As mentioned above, it was known even before it was published that the parade did not reflect reality. But it took more than 30 years for a more-or-less "official" retraction to be issued. In 2000, J. J. Hublin, of the Max Planck Institute for Evolutionary Anthropology in Leipzig, Germany, wrote:

> The once-popular fresco showing a single file of marching hominids becoming ever more vertical, tall, and hairless **now appears to be a fiction.**[107]

Note: It "now" appears to be a fiction. As in, "Oh, my goodness! We never realized that before!" Seriously?

Slaughtering the Sacred Cows

All of the above "sacred cows" of evolution have been discredited. There are many other highly-touted ideas that have been cited as evidence for evolution as well, and one by one, as more research is done, they too are falling (or have already fallen) by the wayside.

---

[106] Marvin L. Lubenow, *Bones of Contention* (BakerBooks, Grand Rapids, Michigan), 2004, p. 167.
[107] *Nature* 403 [27 January 2000]: p. 363. Emphasis added.

As you continue in this book, you will see many such ideas mentioned, and evidence that threw those ideas into disfavor. Granted, many people still hang onto the ideas tenaciously, despite their absent or vanishing scientific support, but that simply emphasizes the religious nature of naturalism in general, and evolutionism in particular. As various topics come up for the rest of this book, keep close tabs on your own beliefs, scientific and otherwise, and regularly ask yourself, "Do I have a good reason for believing this?" You'll be surprised at how much science supports creationism, and how little supports evolutionism.

The *Theory* of Evolution

It might strike you as odd, in a chapter entitled "The Theory of Evolution," to have a subsection whose title text is also "The Theory of Evolution." But the reasoning becomes more clear when you realize that the subsection title italicizes, and hence emphasizes, the word "Theory."

Although evolution is often presented as indisputable fact in books, magazines, public schools, TV, movies, and statements made by devoted evolutionists, it most definitely is not. It is a *theory*. As such, we need to treat it as possibility, but certainly not a foregone conclusion.

Although evolution is only a theory, and there is much evidence for a young universe, a young Earth, and life having been created, evolutionists repeatedly deny the possibility that their theory could be wrong. There seems to be an attitude of "truth by consensus." In other words, "It must be true because everybody believes it."

However, it is *not* true that "everybody" believes in evolution: as we will see in later chapters, *numerous* people question the validity of the whole evolutionary concept. And these people are not simple, gullible folk dabbling in pseudoscientific superstition, as evolutionists often paint them; they are world-class scholars, researchers, astronomers, geologists, physicists, cosmologists, and more, with advanced degrees in their fields. We will hear from many of these people in later chapters.

Darwin's Misgivings

When Darwin came out with his theory, he admitted that there were significant areas of biology that it completely failed to explain.

He said it like this:

> In the four succeeding chapters, the most apparent and gravest difficulties on the theory will be given: namely, first, the difficulties of transitions, or in understanding how a simple being or a simple organ can be changed and perfected into a highly developed being or elaborately constructed organ; secondly, the subject of instinct, or the mental powers of animals; thirdly, hybridism, or the infertility of species and the fertility of varieties when intercrossed; and fourthly, the imperfection of the Geological Record.[108]

But it was back in 1859 that Darwin said that, so by now, more than 150 years later, surely we have found the evidence he was unaware of. Haven't we? Let's check:

1. **Transitions:** Darwin couldn't understand how a simple organ or organism could spontaneously transform into a complex one. And now, since the development of the field of genetics, we understand that all of life requires enormous amounts of information, which is contained in the DNA. Put that together with the awareness that random processes cannot create information, we see that evolution is farther from addressing this problem now than it was 150 years ago. More on this topic in "Where Did the Information Come From?" in Chapter 6.

2. **Instinct:** Darwin was very understandably flummoxed by the whole idea of animal instinct. Many animals have very complex, almost intelligent-looking behavior in regard to their self-preservation, habitat construction, mating rituals, and so forth. How did such behavior originate? Somewhere along the line, an organism of each type had to be the first to "figure it out." But acquired knowledge is not genetically transmitted to offspring, and the offspring "learning" from the parents' example requires yet more instincts, so instincts had to exist before instincts could be gained. So again, evolution can't explain it.

3. **Hybridism:** Why certain animals can interbreed and other ones cannot is understandable with the knowledge of genetics we have now. But evolution still can't explain how any animals got here in the first place, or how one kind spontaneously transforms into another. More about this in *"Arrival* of the Fittest" in Chapter 6.

---

[108] Charles Darwin, *On the Origin of Species by Means of Natural Selection*, 1859, "Introduction," p. 2.

4. **Imperfection of the Geologic Record:** Darwin assumed that the glaring absence of fossilized transitional forms—upon which evolution theory absolutely depends—was simply because we hadn't found them yet, but surely they would be discovered as more paleontology was done. This too has turned out to be disappointing for evolutionists, because of the thoroughness of the excavation of the fossil record, with *still* not a single transitional form found, bodes worse and worse for Darwinism every year. More about this in Chapter 7: "The Fossil Record."

Darwin also stated:

> If it could be demonstrated that any complex organ existed, which could not possibly have been formed by numerous, successive, slight modifications, **my theory would absolutely break down.**[109]

This has indeed been demonstrated numerous times and in numerous ways, since 1859, as you will read in the following chapters.

And, Darwin also said:

> **To suppose that the eye,** with all its inimitable contrivances for adjusting the focus to different distances, for admitting different amounts of light, and for the correction of spherical and chromatic aberration, **could have been formed by natural selection, seems, I freely confess, absurd in the highest possible degree.**[110]

The more we learn about the eye, or any other organ or organism, the more absurd it becomes to believe that such intricacies would fall together by accident, especially when things that are many orders of magnitude more trivial—like laptops and smartphones—don't fall together by accident.

So in every case, further study has shown the theory of evolution to be even more lacking than previously thought. Evolution is not being more and more confirmed as time goes on; it is being more and more refuted. Any appearance to the contrary is because of a pre-existing assumption that evolution is unquestionable, so therefore all discoveries are interpreted in such a way as

---

[109] Charles Darwin, *On the Origin of Species by Means of Natural Selection, or the Preservation of Favoured Races in the Struggle for Life*, 1859, Chapter VI: "Difficulties on Theory," p. 80.

[110] Charles Darwin, *On the Origin of Species by Means of Natural Selection, or the Preservation of Favoured Races in the Struggle for Life*, 1859, Chapter VI: "Difficulties on Theory", p. 80.

to support the theory—they are not permitted to do otherwise. More about this in several of the later chapters.

The Scientific Method

You may remember from school the definition of the "scientific method"—it is the iterative process that underlies scientific discovery and advancement. Simply stated, it is a procedure for discovering knowledge. Once it is recognized that there is a problem to be solved or a question to be answered, the scientific method can be put to use.

Here are the steps of the scientific method:

1. Develop a hypothesis: propose a process to solve the problem, or an explanation to answer the question.
2. Formulate a way to test the process, or make some predictions based on the explanation.
3. Using repeatable experiments and/or repeatable real-world observations, collect data that pertains to the predictions.
4. Compare the results: if the predictions closely match the actual results/observations, the hypothesis is a good one; if not, revise the hypothesis and go to Step 2.

Note especially Step 3 in the list above: it requires *repeatable* experiments or observations. Since we can't repeatedly create the whole universe to see how it's done, or repeatedly create life where there was none, it is not possible to test evolution by experiment. Thus, it remains a hypothesis—a theory.

Of course, we are also unable to repeatedly cause the creation of the universe by a supernatural God, so creation too is a theory, in that sense. Since we can't experiment on the full-blown processes themselves—creation and evolution, in this case—the best we can do, in either case, is one or both of these two approaches:

- Do repeatable experiments on small subsets of the overall hypotheses and see if the results of the experiments support or undermine the overall hypotheses.
- Make predictions based on the hypotheses, and see how well the predictions match repeatable observations.

We will see many examples of these actions in the following chapters, and the results are very eye-opening.

It is very common in mainstream scientific literature to see evolutionists dismiss creationism as "pseudoscience," and unworthy of even refuting. But it's interesting to note that of the *Skeptic's Dictionary*'s ten characteristic indicators of pseudoscience, the theory of evolution exhibits nine of them.[111] I wonder why we never hear about this in mainstream media. . .

The Second *Law* of Thermodynamics

In contrast to the *theory* of evolution, there is another principle that plays an enormous role in the topic we're investigating. It is a law of physics called the "Second Law of Thermodynamics." Note that this is a *law*, not a *theory* like evolution is.

But what makes it a law? Is it a law simply because some person declared it to be one? No. It is a law because it is repeatedly observable and verifiable. Because the laws of physics are not actual entities, they can't actually *cause* anything to happen. In reality, scientific laws are simply descriptions of what we observe happening consistently, just as the continental outline on a map is simply a description of the shape of a coastline. Treating scientific laws as prescriptive, as if they are the *cause* of the observed regular behavior, is like claiming that the continental outline drawn on the map *caused* the actual coastline to be that shape.

So even though the Second Law of Thermodynamics law does not *cause* things to happen, the manifestations predicted by this law are things that we all have to deal with on a daily basis. It is one of the most undeniable, obviously valid, and repeatedly observable laws in nature, because we all—literally, *all* of us—have to deal with its effects every—literally, *every*—single day. Moreover, there is nothing in this physical universe above the atomic scale that has been observed to violate this law.

So what is this law? What does the Second Law of Thermodynamics actually say? Technically, as its name implies, the field of thermodynamics deals with how heat flows, and it pertains to how work can be done by energy flowing from one place to another.

Specifically, the Second Law of Thermodynamics is:

> "Heat cannot spontaneously flow from a colder location to a hotter location."

---

[111] Mark Johansen, "Is evolution pseudoscience?", *Creation* 29(4):25–27, September 2007. [Online at Creation.com/is-evolution-pseudoscience; accessed November 25, 2023.]

CHAPTER 2: THE THEORY OF EVOLUTION

Why is this important to the subject matter at hand? Essentially, the upshot of the Second Law of Thermodynamics is that in systems left to themselves, **entropy** (i.e., disorder or chaos) is always increasing; in other words, *things are always running down.*

Think about it:

- If you put lukewarm water into a teapot and an ice-cube tray sitting side by side on the kitchen counter, does the water in the teapot spontaneously heat to the boiling point, while the water in the ice-cube tray spontaneously freezes? No.
- Does the paint job on your house or car spontaneously grow more thick, more waterproof, and more lustrous and attractive as the years go by? No.
- Does your computer or mobile device spontaneously gain new features and capabilities? Does its battery continuously get better at holding a charge? Does it continuously increase in energy efficiency, memory capacity, and processor speed, all by itself? No.
- Does your garden, left unattended, become more orderly, more productive, more attractive, and less weedy? Do the rows become straighter, and the plants organize themselves by kind? Does it spontaneously outgrow the need for cultivation and watering? No.
- Do the roads on which you drive spontaneously become more solid, more smooth, their cracks and potholes healing themselves even better than new? Do new intersections, overpasses, tunnels, and bridges form all by themselves, making your travel shorter, faster, easier, and more enjoyable every year? No.
- Do your light bulbs spontaneously grow ever brighter, more energy-efficient, and less likely to burn out with each passing year? No.
- Does forgotten food in your refrigerator spontaneously get more fresh, more nutritious, and more appealing with every passing month? No.
- Does your body spontaneously—with no time or effort spent—become ever more supple, toned, limber, strong, attractive, and healthy as the years go by? No.

Clearly, the list of such questions could grow very long, and in every case, the answer would be a resounding "No!" Things *don't* spontaneously get better, stronger, healthier, fresher, more organized and orderly. We need to put much time and effort into keeping them that way, and even so, it is a losing battle unless God intervenes. Why? Because of the Second Law of Thermodynamics.

The Bible acknowledges the existence of what we now call the Second Law of Thermodynamics. The current state of affairs, where everything ages, decays, wears out, and breaks, was not God's original plan; it came into effect when Adam and Eve sinned in the Garden of Eden. By taking their delegated authority over creation and turning it over to Satan, who steals, kills, and destroys, the physical universe was put into the current state where (unless God intervenes) the Second Law of Thermodynamics reigns supreme.

Here's how the Bible describes it:

> Romans 8:20–21 (TPT): For against its will **the universe itself has had to endure the empty futility** resulting from the consequences of human sin. But now, with eager expectation, [21]**all creation longs for freedom from its slavery to decay** and to experience with us the wonderful freedom coming to God's children.

So yes, the entire universe has suffered because of man's sin, but that is not the end of the story. God's redemptive plan has been in full swing for quite a while now, many of its manifestations have already been revealed, and the Church is waking up to its part in bringing the Kingdom of Heaven to earth (Matthew 6:10).

The Psalms comment on the Second Law of Thermodynamics, as well as the fact that God originally created everything:

> Psalm 102:25–26 (NIV): In the beginning you laid the foundations of the earth, and the heavens are the work of your hands. [26]**They will perish,** but you remain; **they will all wear out like a garment.** Like clothing you will change them and they will be discarded.

Solomon noticed the effects of the Second Law of Thermodynamics, too, and commented on it:

> Ecclesiastes 10:18 (NKJV): Because of laziness **the building decays,** and through idleness of hands **the house leaks.**
>
> NLT: When men are lazy, **the roof begins to fall in.** When they will do no work, **the rain comes into the house.**
>
> Proverbs 24:30–31 (ESV): I passed by the field of a sluggard, by the vineyard of a man lacking sense, [31]and behold, it was **all overgrown with thorns; the ground was covered with nettles, and its stone wall was broken down.**

CHAPTER 2: THE THEORY OF EVOLUTION

Isaiah the prophet perceived it too:

Isaiah 51:6 (NIV): Lift up your eyes to the heavens, look at the earth beneath; the heavens will vanish like smoke, **the earth will wear out like a garment and its inhabitants die like flies.** But my salvation will last forever, my righteousness will never fail.

People have been noticing for millennia that it takes work to resist the effects of the Second Law of Thermodynamics, and even then, we are not overcoming it; we are only delaying the inevitable.

The Second Law of Thermodynamics is of paramount importance to the discussion at hand, because the theory of evolution completely ignores it: the theory of evolution pretends that the Second Law of Thermodynamics—which applies to *everything*—doesn't apply to evolution! Even while everything in the observable universe is running down, wearing out, and decaying, somehow evolution magically sidesteps the laws of physics!

Here's an example of how evolution is allegedly exempted from following the laws of physics. This first quote describes the Second Law of Thermodynamics, which every one of us encounters every day:

> There is a **general natural tendency of all observed systems to go from order to disorder,** reflecting dissipation of energy available for future transformation—the law of increasing entropy.[112]

Then here is evolution, according to evolutionary biologist and eugenicist Julian Huxley:

> Evolution in the extended sense can be defined as **a directional and essentially irreversible process** occurring in time, which in its course **gives rise to an increase of variety and an increasingly high level of organization in its products.** Our present knowledge indeed forces us to the view that the whole of reality is evolution—a single process of self-transformation.[113]

---

[112] R. B. Lindsay: "Physics—To What Extent Is It Deterministic?" *American Scientist*, Vol. 56, Summer 1968, p. 100. Emphasis added.
[113] Julian Huxley: "Evolution and Genetics" in *What is Man?* (Ed. by J. R. Newman, New York, Simon and Schuster, 1955), p. 278. Emphasis added.

OH, EVOLVE! (GOOD LUCK WITH THAT...)

There are two very significant points here. First, on the topic of entropy: evolution somehow completely ignores the Second Law of Thermodynamics. This random and undirected process called evolution causes an *increase* in organization. It's nice to have a theory where you can disregard the laws of physics when they become inconvenient.

The second point is that Huxley says that evolution is a *directional* process! One of the most sacred and fundamental creeds of evolution is that it's undirected. If it really is undirected, why would there be a tendency toward greater complexity? There wouldn't be, even without the Second Law of Thermodynamics. And *with* the Second Law of Thermodynamics, there would be a definite tendency toward *dis*order.

So how, you may ask, can it simultaneously be directional and undirected? That is an excellent question, which we are apparently not supposed to ask. But if we ask it anyway, we realize that evolution *cannot* be simultaneously directional and undirected. If it were directional, it wouldn't be evolution (because it wouldn't be random), and if it were undirected, even ignoring the Second Law of Thermodynamics, things would not increase in order and complexity—the process would stall out, bouncing aimlessly around its starting point: with nothing, no life. And if we *don't* ignore the Second Law of Thermodynamics, there would be a distinct tendency for things to run down, decay, and fall apart. Not only that, but the rate of decay would far outstrip the rate of any accidental assembly of components that might eventually turn out to be useful once thousands of other compatible and required components are also accidentally formed. Again: no life.

So how does evolution get away with disregarding the Second Law of Thermodynamics? Aye, there's the rub. Freeman Dyson was an English-American theoretical physicist and mathematician known for his work in quantum electrodynamics, solid-state physics, astronomy and nuclear engineering. He was professor emeritus in the Institute for Advanced Study in Princeton, a member of the Board of Visitors of Ralston College and a member of the Board of Sponsors of the Bulletin of the Atomic Scientists.[114] In other words, he was a very smart guy.

So again, how does evolution sidestep the Second Law of Thermodynamics? Here's what Freeman Dyson said about it:

> **We are still at the very beginning** of the quest for understanding of the origin of life. **We do not yet have even a**

---

[114] en.wikipedia.org/wiki/Freeman_Dyson; accessed May 27, 2020.

rough picture of the nature of the obstacles that prebiotic evolution has had to overcome. **We do not have a well-defined set of criteria by which to judge** whether any given theory of the origin of life is adequate.[115]

Basically, "We haven't a clue about how evolution could ever have happened." Then Angrist and Hepler, in their book *Order and Chaos,* acknowledge the infinitesimal likelihood of it all happening:

> Life, the temporary reversal of a **universal trend toward maximum disorder,** was brought about by the production of **information mechanisms.** In order for such mechanisms to first arise it was necessary to have matter capable of forming itself into a self-reproducing structure that could extract energy from the environment for its first self-assembly. **Directions** for the reproduction of plans, for the extraction of energy and chemicals from the environment, for the growth of sequence and the mechanism for translating **instructions** into growth all had to be simultaneously present at that moment. This combination of events has seemed an **incredibly unlikely happenstance** and often divine intervention is prescribed as the only way it could have come about.[116]

Interesting: there is a "universal trend toward maximum disorder" (the Second Law of Thermodynamics), but evolution "temporarily reverses" it. Apparently, the alleged 3.8 billion years of continuous evolution is considered "temporary." And notice that Angrist and Hepler acknowledge that *information* and *directions* are needed. This will be covered in more detail later on, but at this point, suffice it to say that DNA actually has *information*—instructions on creating proteins of a very precise shape and function, and these proteins are indispensable for the processes of life.

The obvious question that arises is, "Where did the information come from?" In essence, DNA is a computer program, vastly more complex than anything mankind has ever written, but evolution claims nobody wrote it.

Note also the reference above to life's "first self-assembly." Do you see the problem with this idea? Life can't do anything before it exists, so in order to

---

[115] Freeman Dyson, "Honoring Dirac," *Science,* Vol. 185, Sept. 27, 1974, p. 1161. Emphasis added.

[116] Stanley W. Angrist and Loren G. Hepler, *Order and Chaos* (New York: Basic Books, Inc., 1967), pp. 203–204. Emphasis added.

assemble itself the first time—which is what "first self-assembly" means—life would have to exist (or else it couldn't do anything) and simultaneously *not* exist (because it hadn't been assembled yet). Do people really not notice the self-contradiction?

As we proceed through the rest of this book, you will see many examples of "rescue devices." These rescue devices are supplemental theories that are attached to already-existing theories that are in trouble because experimentation and observations don't support them, and in fact, often directly refute them. A most telling feature about these rescue devices is that the objects about which they theorize have never been observed, and usually there is absolutely no reason for them to exist, other than to bolster the underlying presupposition of deep time—the "billions and billions of years" that evolutionary theory absolutely depends on to offer even a caricature of credibility.

If you come across any statements in this book that you find hard to believe, or that are just surprising, please don't take my word for it. There are many citations in each section, and following those references to their source material will allow you to pursue further detail on the topic of that section.

The "God of the Gaps"

Sometimes evolutionists accuse creationists of invoking "the God of the gaps." That idea says that, when creationists don't know what would be required to go from one stage to another in the development of the universe, we invoke "the God of the gaps." Supposedly, creationists do this in order to fill in the blanks where we don't see how the laws of physics alone could have resulted in the universe we see today. That is to say, we "plug the gaps" in our theory by inserting God wherever we don't understand something.

But the actuality is the reverse of that: creation scientists *do* know numerous repeatable, observable, testable laws of physics that would necessarily *prevent* what evolution requires or assumes. Therefore, the only logical conclusion, since the laws of physics by themselves are insufficient, is that there is an intelligent Creator.

But while we're on the subject, let's also also include how naturalists regularly invoke "evolution of the gaps." That is, "We don't know how the laws of physics could have resulted in what we see today, but we know that evolution is unquestionable, so evolution must have done it. Somehow. We'll simply adjust the theory to explain whatever we find."

Examples of such blind devotion to evolution, in the face of conflicting evidence, include the following quotes from evolutionists who remain staunch evolutionists in spite of experiments and repeatable observations to the contrary. More detailed discussion of *all* of these quotes, and many more, you'll find in the following pages of this book:

- On the Big Bang: ". . .the dominance of the big bang within the field has become self-sustaining, **irrespective of the scientific validity of the theory.**"
- On hot blue stars: "**It is not clear** how blue stragglers form."
- On spiral galaxies: "Galaxy rotation and how it got started is **one of the great mysteries of astrophysics.** In a Big Bang universe, linear emotions are easy to explain: They result from the bang. But what started the rotary motions?"
- On the solar system's formation: "About 5 billion years ago, and **for reasons that are not yet fully understood,** this huge cloud of minute rocky fragments and gases began to contract under its own gravitational influence."
- On planetary formation: "The discovery of thousands of star systems wildly different from our own **has demolished ideas about how planets form.** Astronomers are searching for a whole new theory."
- On planetary formation: "That finding *[of retrograde orbits]* is **inconsistent with the view that planets are formed by the condensation of dust** from a disk surrounding a newly formed star."
- On Mercury's volatiles: "The presence of potentially recent surface modification implies that **Mercury's nonimpact geological evolution may still be ongoing.**"
- On Venus' surface features: "The geologic rule of **uniformitarianism**—'the present is the key to the past'—**does not apply to Venus. . .**"
- On Jupiter's moon Io: "**Complete elucidation of the heat source remains a significant outstanding problem** resulting from the discovery of active volcanism on Io."
- On Saturn's rings: "Part of the reluctance for everyone to leap off this bridge into the unknown is **we haven't had any kind of feasible explanation** [for how the rings could form recently]."
- On Saturn's moon Enceladus' geysers: "What causes and controls the jets **is a mystery.**"

- On Saturn's moon Enceladus' liquid core: "There is **no possible combination of parameters** that allow for a thermally stable ocean." Yet there it is.
- On Saturn's moon Titan's equatorial methane lakes: "Lakes at the poles are easy to explain, but **lakes in the tropics are not.**"
- On Pluto's geological activity: "scientists **don't quite know how that heat has lasted** over 4 billion years."
- On the solar system's precision: ". . .move any piece of the solar system today, or try to add anything more, and the whole construction would be thrown fatally out of kilter. So **how exactly did this delicate architecture come to be?**"
- On the Moon's precision: "**By some absolutely incomprehensible quirk of nature,** the Moon also manages to precisely imitate the perceived annual movements of the Sun each month."
- On the Moon's formation: "The whole subject of the origin of the moon **must be regarded as highly speculative**" and ". . .the time scale of the earth-moon system **still presents a major problem.**"
- On the Moon's orbit: "The Moon's orbit **is fiendishly difficult to explain. . .**"
- . . .and *numerous* others, on the topics of geology, biology, genetics, and more.

In all of these cases, the evolutions quoted were absolutely unwilling to consider that evolution itself might not be real. In each case, a new feature or capability of evolution was considered to be the only option. This redefinition of evolution whenever needed, so it can explain everything—even why experimentation and observation disagrees with it—ultimately shows that evolution proves nothing. It's like using an elastic tape measure: the "measurement" you get can be whatever you want it to be; it is "evolution of the gaps" at its best.

Division in the Camp

In decades past, creationists often felt like they were so few, that their voice was being lost in the cacophony of scientists, media, educators, and the general public pledging allegiance to evolution. But as the years go by, that is becoming less and less the case.

Sometimes it is because of evolutionists honestly examining the questions and challenges of creationists, and realizing that they make some very good points, and sometimes it is because the evolutionists themselves notice that evolutionary predictions have a rather dismal success rate. In any case, Darwinian evolution is coming under fire from people who are not Christians, nor are they creationists. But they are intellectually honest, and they are starting to challenge the entrenched evolutionary doctrines.

Some critiques of evolution are more blunt and frank than others:

> The Theory of Evolution is no longer with us, because neo-Darwinism is now acknowledged as being **unable to explain anything more than trivial change,** and in default of some other theory we have none. . . Despite the **hostility of the witness provided by the fossil record,** despite the **innumerable difficulties,** and despite the **lack of even a credible theory,** evolution survives. Can there be any other area of science, for instance, in which a concept as intellectually barren as embryonic recapitulation could be used as evidence for the theory?[117]

As you remember, "embryonic recapitulation" is the discredited theory, discussed earlier in this chapter, that claimed the growth stages of a developing embryo illustrates successive stages of evolution.[118]

More and more evolutionists are beginning to doubt the ability of neo-Darwinism to answer the question of origins. So heated is the contention between the "old guard" and these new upstarts, that it is even getting media attention. In a *Newsweek* article entitled "Science Contra Darwin," describing such contention, writer Sharon Begley states:

> The great body of work derived from Charles Darwin's revolutionary 1859 book, *On the Origin of Species,* is under increasing attack—and not just from creationists. . . So heated is the debate that one Darwinian says there are times when he thinks about going into **a field with more intellectual honesty: the used-car business.**[119]

Michael Denton holds an M.D. and Ph.D. from British universities. He is neither a Christian nor a professing creationist, so no one would categorize him as a religious nut. However, he has written a scathing critique of modern

---

[117] R. Danson, *New Scientist* 49:35 (1971). Emphasis added.
[118] en.wikipedia.org/wiki/Recapitulation_theory; accessed June 20, 2020.
[119] Sharon Begley, *Newsweek,* April 8, 1985, p. 80. Emphasis added.

evolutionary theory. Consider this excerpt from the inside flap of the book cover:

> **The theory of evolution,** as propounded by Darwin and elaborated into accepted "fact" by biologists, **is in serious trouble.** This sober, authoritative, and responsible book by a practicing scientist presents an accurate account of the **rapidly accumulating evidence which threatens to destroy almost every cherished tenet of Darwinian evolution...** At a fundamental level of molecular structure, each member of a class seems equally representative of that class and **no species appears to be in any real sense "intermediate" between two classes.** Nature, in short, appears to be profoundly discontinuous. Furthermore, advances in biochemistry are making the existence of a "prebiotic soup"—the supposed primordial broth in which life began on earth—look **highly unlikely if not completely absurd.**[120]

Søren Løvtrup is a well-known Swedish biologist, and a committed evolutionist. Nevertheless he utterly rejects Darwinian evolution, saying that mutations and natural selection have had little, if anything, to do with evolution. In his book, *The Refutation of a Myth,* he states:

> I suppose that nobody will deny that it is a great misfortune if **an entire branch of science becomes addicted to a false theory. But this is what has happened in biology:** for a long time now people discuss evolutionary problems in a peculiar "Darwinian" vocabulary—"adaptation," "selection pressure," "natural selection," etc.—thereby believing that they contribute to the *explanation* of natural events. **They do not,** and the sooner this is discovered, the sooner we should be able to make real progress in our understanding of evolution. **I believe that one day the Darwinian myth will be ranked as the greatest deceit in the history of science.**[121]

Fascinating: Darwinism, the "greatest deceit in the history of science." So why is Løvtrup still an evolutionist? That is a very good question.

---

[120] Michael Denton, *Evolution: Theory in Crisis* (London: Burnett Books, 1985) Emphasis added.

[121] Søren Løvtrup, *Darwinism: The Refutation of a Myth* (New York: Croom Helm, 1987). Bold emphasis added; italic emphasis in original.

Ernst Mayr, of Harvard University, is one of the most respected evolutionists in recent times, admits:

> One must grant Darwin's opponents the validity of two of their objections. First, Darwin produced **embarrassingly little concrete evidence** to back up some of his most important claims...[122]

The famous British astronomer Sir Fred Hoyle researched the probability of life "arising" spontaneously in an evolutionary way, and commented that the probability of an evolutionary, naturalistic origin of life anywhere in the universe in only the 20 billion years evolutionists claim for the universe is comparable to the probability of a tornado crossing a junkyard and accidentally assembling a functional Boeing 747. Sir Fred had been an atheist, but after studying it, announced that life *had* to be created, and therefore God must exist. The presence of design and purpose we see everywhere we look, and in even the smallest detail of the structure and function of living organisms speak eloquently of the existence of the Designer.[123]

Sherwood Washburn, from the University of California, Berkeley says it well:

> It may be more useful to regard the study of evolution as **a game rather than as a science.**[124]

If Darwinism is described—even by non-creationists—as "unable to explain anything more than trivial change," and has less intellectual honesty than "the used-car business," and as being "in serious trouble" because of the "rapidly accumulating evidence which threatens to destroy almost every cherished tenet of Darwinian evolution," and "a false theory" to which biology has become "addicted," doesn't it seem reasonable that we should stop believing in it? When non-creationists state that "the Darwinian myth will be ranked as the greatest deceit in the history of science" and that it has "embarrassingly little concrete evidence to back up some of his most important claims," and that evolution should be considered "a game rather than as a science," is it not astonishing that it still has such a devoted following?

---

[122] *Nature* 248, 22 March 1974, p. 285.

[123] ICR.org/article/science-education-subject-origins; accessed July 4, 2020.

[124] Sherwood L. Washburn, *Abstracts of the 71st Annual Meeting, American Anthropological Association,* Washington, DC, 1972, p. 121. Emphasis added.

If only Charles Darwin had taken his own advice to heart when, in *The Descent of Man,* he said:

> **False facts** are highly injurious to the progress of science, **for they often endure long.** . .[125]

Agreement in the Other Camp

At the same time the theory of evolution is coming under more and more suspicion from evolutionists, there is a ever-accelerating upsurge in creationism, as well as in the number of scientists that are willing to acknowledge that Creation requires a Creator.

For example, in 1984, New York's American Museum of Natural History brought in the "Ancestors" exhibit. This was a showcase of more than 40 original fossils, supposedly of early man, being displayed to the public. Security was extreme: fossils were put on display behind one-inch-thick laminate acrylic panels that were bulletproof and crushproof, as well as electronically monitored. Several countries didn't participate, because of the risk of breakage if original fossils were transported.

But if the risk was so great, why do it in the first place? Because of the "rising threat of creationism." The American Museum scientists who were largely responsible for the Ancestors exhibit—Eric Delson, John Van Couvering, and Ian Tattersall—acknowledged that the creationists' "assault" on evolutionary biology was a matter of "great and growing concern" at the museum. They also stated that the primary purpose of the exhibit was to show people, lay and professional, the evidence for evolution. They refrained from making any kind of political statement regarding human evolution lest they "dignify" the challenge of creation science.[126]

More and more people are seeing the light, so to speak, and realizing that evolutionary doctrine is smoke and mirrors. It answers so few questions about how we got here, and where things came from, that people are seeing more and more design in everything, and they are realizing that there has to be a Designer behind the design.

---

[125] Geoffrey Simmons, M.D., *What Darwin Didn't Know,* 2004 (Harvest House Publishers, Eugene, Oregon), p. 26.

[126] Eric Delson, ed., *Ancestors: The Hard Evidence* (New York: Alan R. Liss, Inc., 1985), pp. 1–2.

For example:

> Scientists who utterly reject evolution may be one of our **fastest-growing controversial minorities.** . . . Many of the scientists supporting this position hold impressive credentials in science.[127]

The seven-part PBS television series *Evolution* made the statement that "all known scientific evidence supports [Darwinian] evolution" as does "virtually every reputable scientist in the world." This created quite an uprising because there are numerous world-class scientists who strongly disagree with those statements. Scientists like Nobel nominee Henry F. Shaffer, the third-most-cited chemist in the world, James Tour of Rice University's Center for Nanoscale Science and Technology, Fred Figworth, professor of Cellular and Molecular Physiology at Yale Graduate School, and many others.

Together, they wrote a detailed, 151-page critique of the PBS series, in which they claimed that it "failed to present accurately and fairly the scientific problems with the evidence for Darwinian evolution." In addition, the series ignored "disagreements among evolutionary biologists themselves."[128]

New York Times reporter John Horgan conceded that scientists have no idea how the universe was created or. . .

> . . .how inanimate matter on our little planet coalesced into living creatures. . . Science, you might say, has discovered that **our existence is infinitely improbable, and hence a miracle.**[129]

Francis Crick was one of the co-discoverers of the molecular structure of DNA, and shared a Nobel prize for the discovery. He also acknowledges:

> An honest man, armed with all the knowledge available to us now, could only state that in some sense, **the origin of life appears at the moment to be almost a miracle, so many are the conditions which would have had to have been satisfied to get it going.**[130]

---

[127] Larry Hatfield, "Educators Against Darwin," *Science Digest* (Winter 1975). Emphasis added.

[128] *Getting the Facts Straight* (Seattle: Discovery Institute Press, 2001), pp. 9–11.

[129] John Horgan, "A Holiday Made for Believing," New York Times (December 25, 2002). Emphasis added.

[130] Francis Crick, *Life Itself* (New York: Simon and Schuster, 1981), p. 88. Emphasis added.

Note that Crick says an honest man would state that the existence of life "appears at the moment to be almost" a miracle. I would propose that, based on evidence already presented in this book, as well as more to be presented in the following pages, that an honest man would be driven to conclude that the existence of life is *most definitely and undeniably* a miracle. As you keep reading, please evaluate the plausibility of that statement for yourself.

More and more scientists are examining the scientific evidence, and coming to the eminently logical conclusion that physical laws, unaided, could *not* have resulted in the universe, or life, as we observe it. They are cautiously entertaining the idea that the universe may have actually been created. By a Creator, yet!

Chapter 2 Summary

Here are the most important points in this chapter:

- Darwinian evolution requires **undirected, accidental, random factors** to create the order, complexity, life, and interdependency we observe today.
- The word "evolution" is **used very inconsistently** in common language, which distracts people from its intrinsic randomness.
- Evolution is a **theory** at best; it is most definitely *not* a proven fact.
- Evolution does not satisfy the requirements of the **scientific method**, which intrinsically includes observation and repeatable experimentation.
- Evolution defies the easily verifiable **Second Law of Thermodynamics**, which states that **entropy** (disorder and chaos) is always increasing (which is why physical things wear out, crumble, die, and decay).
- Evolution is not at all universally accepted, and it is **forcefully refuted even by many who are not creationists,** nor are they Christians.
- **Information** is needed for life to exist and continue, but evolution says that the information didn't come from anywhere.

Oh, Evolve! (Good Luck With That. . .)

Chapter 3:

A Young Universe

In the beginning was the Big Bang. At least, that how adherents of evolution explain it. In a nutshell, the Big Bang is the hypothetical event that began the universe: an infinitesimally small, infinitely dense singularity exploded, and the entire universe came out. After the explosion, the energy became subatomic particles, and then atoms, which coalesced and organized themselves into the galaxies, stars, life, and everything else we see today.

The Big Bang

Wikipedia confidently states that ". . .a wide range of empirical evidence has strongly favored the Big Bang, which is now universally accepted."[131]

Well, maybe acceptance of the Big Bang is not *quite* universal. A web search turned up these articles (and many more), all of which cast doubt on the validity of the Big Bang as the Source Of All Things™:

- "Down With The Big Bang," by John Maddox: "Apart from being philosophically unacceptable, **the Big Bang** is an oversimple view of how the Universe began, and it **is unlikely to survive the decade ahead.**"[132]

---

[131] en.wikipedia.org/wiki/Big_Bang; accessed May 27, 2020.
[132] Maddox, John; "Down with the Big Bang," *Nature,* 340: 425, 1989. Emphasis added.

- "Big Bang, Deflated? Universe May Have Had No Beginning," by Tia Ghose: "If a new theory turns out to be true, **the universe may not have started with a bang.**"[133]
- "Science News: Stephen Hawking proved WRONG! Big Bang was NOT the origin of the universe," by Paul Baldwin: "Stephen Hawking's view that the universe started with a huge Big Bang **has been disproved** by a team of mathematicians."[134]
- "Everything we thought we knew about the Big Bang theory may be WRONG according to this amazing new theory," by Maryse Godden: "Brazilian physicist Juliano Cesar Silva Neves claims **the Big Bang never happened** some 13.8 billion years ago."[135]
- "The Big Bang never happened?" by Steve Deace: "Here's something you're not seeing reported in the news: **the Big Bang may not have ever happened.**"[136]
- "No Big Bang? Quantum equation predicts universe has no beginning," by Lisa Zyga: "**The universe may have existed forever,** according to a new model that applies quantum correction terms to complement Einstein's theory of general relativity."[137]

This list could have been much longer, but the above is enough to make the point: even among those who want nothing to do with the Creation model, there is serious disagreement and doubt as to whether the Big Bang is even a credible theory.

Evolutionists don't believe that the universe was created by God, because they don't believe that God exists. But the universe is clearly here. And of course, God couldn't have created the universe without first existing, because then you'd have an effect without a cause, an action with no agency, something done without a doer. That is clearly a self-contradiction, so evolution disallows it.

---

[133] LiveScience.com/49958-theory-no-big-bang.html; accessed May 28, 2020. Emphasis added.

[134] Express.co.uk/news/science/928537/science-news-stephen-hawking-big-bang-proved-wrong-physics-general-relativity; accessed May 28, 2020. Emphasis added.

[135] TheSun.co.uk/tech/5017045/big-bang-theory-wrong-new-universe-space-big-crunch; accessed May 28, 2020. Emphasis added.

[136] GlennBeck.com/2015/02/10/watch-the-big-bang-never-happened; accessed May 28, 2020. Emphasis added.

[137] Phys.org/news/2015-02-big-quantum-equation-universe.html; accessed May 28, 2020. Emphasis added.

Unless, of course, evolution needs it, like for the Big Bang. Where did the primordial singularity come from? What caused it to explode? Apparently, for those who believe in the Big Bang, it's okay for it to be a causeless effect.

Creation *Ex Nihilo*

The Latin phrase *ex nihilo* translates into the English as "out of nothing." So the phrase "creation *ex nihilo*" means "creation out of nothing." But there are two ways of understanding this phrase, and they need to be understood before going further into detail.

Some people understand *creation ex nihilo* to mean "creation coming from absolutely nothing, not even God." Others take the phrase to mean "God creating our universe out of nothing." I think you'd agree, especially if you know God, that the *presence* of God is pretty dramatically different than the *absence* of God.

The Big Bang could be considered in both ways as well. Did God create the universe in a creative act we call the Big Bang? Or do we disallow the God concept and imagine that the universe created itself, for no reason, with no causation, in an event we call the Big Bang?

Because this book is discussing the plausibility of evolution, and because evolution disallows God, let's pick the Godless version and examine that.

For example, think about this quote from the evolutionist George Wald:

> Time is in fact the hero of the plot. . . What we regard as impossible on the basis of human experience is meaningless here. Given so much time, the "impossible" becomes possible, the possible probable, and the probable virtually certain. One has only to wait: **time itself performs the miracles.**[138]

Notice that Wald is attributing miracle-working power to time itself. The problem is that time is not an entity; it is not an agent, a doer, or even a *thing*. It is merely the continuum—the expanse—in which events occur, conditions exist, and processes continue. But it is not an entity itself, which means it cannot *cause* anything to happen. This means that Wald is claiming that everything was caused by *nothing at all:* creation ex nihilo, without even God existing.

---

[138] George Wald, "The Origin of Life," p. 48. Emphasis added.

However, this notion violates another fundamental principle of reality. It is *"ex nihilo, nihil fit,"* which is Latin for "out of nothing, nothing comes." In other words, if you actually could have *nothing*—not even God—nothing else would ever exist, because if nothing—not even God—exists, there could be no causation (because there's no agency, or First Cause), and therefore there could never be an effect of any kind. Like, for example, anything existing.

But let's assume for a moment that the universe did create itself. We'll just gloss over the intrinsic self-contradiction in that approach: The universe couldn't "do" anything before it existed, so it had to exist before it could perform the act of creation that brought it into existence. Thus, it had to exist and *not* exist at the same time. But we'll let that one pass.

If the entire universe exploded into existence from the aforementioned hypothetical singularity—this "primeval atom" or "cosmic egg," as it is sometimes called—that immediately raises questions, even without getting into the scientific details.

For example:

- If the original singularity really existed, how did it get there? If someone answers "It always was," how is that any less a statement of faith than positing a God who always was?
- If the original singularity exploded, what made it explode? Every effect must have a cause, so what was the cause that effected the explosion?
- If space itself was created by the Big Bang, where did the singularity exist before it exploded, if not in space? If it existed at all, it must have been somewhere.
- If the singularity didn't exist in any location, what exploded?

The answers to these question are more in the realm of ontology (the study of "being"), which is a very valid and fascinating field of inquiry, but let's discuss some of the aspects of physics that pertain to the Big Bang.

Some Problems with the Big Bang Theory

Keep in mind that the following discussion, concerning the doubtfulness of the Big Bang, refers to a naturalistic, undirected process. The universe did indeed have a beginning; there is no question of that. The question is: could it have happened without a transcendent First Cause outside of our laws of physics? The answer to that is no, and the following information shows that physical laws alone would not result in what we observe today. Therefore, the necessity of a supernatural beginning.

For example, here are some of the problems with the Big Bang idea:

- **Magnetic Monopoles.** The kind of magnets people are familiar with in everyday life have two poles: a north pole and a south pole. Like poles repel, and opposite poles attract. In the high temperatures theorized for the Big Bang, magnetic monopoles, having a north pole or a south pole, *but not both,* should have been created. They are stable, so they should still be around, but they are nowhere to be found. This indicates that the universe was never as hot as the Big Bang requires.

- **The Flatness Problem.** The expansion rate of the universe is *extremely* well balanced with the force of the gravity of the matter in the universe: the expansion rate is "flat." If the universe is a random accident caused by the Big Bang, such flatness is an astonishingly rare coincidence: less than one part in 10^{62} away from the critical density.[139] It is much more likely that the mass of the universe would be too large for the expansion rate, causing the universe to collapse again, perhaps within seconds of the Big Bang (a closed universe), or if the mass is too small for the expansion, the universe would fly apart so quickly that stars and galaxies could not have formed (an open universe).

- **Inflation.** Endeavoring to explain the flatness problem and the magnetic monopole problem, cosmologists have conjectured a process called "inflation" that says the universe went through a time of greatly accelerated expansion before slowing down to its current rate. According to this theory, all the particles in the universe, for some unknown reason, flew apart from each other at speeds billions or trillions of times faster than the speed of light, and then a moment later, for some other unknown reason, inflation just stopped.[140] It's just a rescue device—there's no evidence for it, so in that sense, it's kind of like the Big Bang itself. How and why would inflation start? How and why would it stop? The whole inflation idea is just storytelling with no supporting evidence—like all rescue devices, it is just speculation designed to preserve a pet theory in spite of contradictory observations.

- **Antimatter.** When energy condenses into matter, 50% of it is regular matter, and the other 50% is antimatter (matter with similar properties except opposite charge). There are no known exceptions. If the Big Bang were a real thing, half of our universe should be composed of antimatter, yet only tiny traces of it are found—this very much defies the

---

[139] en.wikipedia.org/wiki/Flatness_problem; accessed October 30, 2020.
[140] Walt Brown, *In the Beginning: Compelling Evidence for Creation and the Flood,*, Eighth edition, 2008 (Center for Scientific Creation, Phoenix, Arizona, CreationScience.com), p. 31.

odds. If half the universe were composed of antimatter, life would not be possible.

- **The Event Horizon.** When large-enough stars die, they sometimes turn into black holes, which are objects that have such intense gravity that even light can't escape. The event horizon is that distance from the center of the black hole, at which the escape velocity is exactly c—the speed of light. If any particle falls below the event horizon, it would have to travel faster than light to escape, and we have no evidence that that is possible, so such a particle could never escape the gravitational field. The singularity that allegedly exploded in the Big Bang would have had all the mass of the universe in it, and its stupendous gravity would have had an enormous event horizon, so how could it even explode?

- **Acceleration in the Wrong Direction.** In the universe today, we can see significant redshift, which implies that stellar objects are receding from each other. In the Big Bang model, since the mass of the universe exerts gravitational force on all its constituent pieces, the rate of recession should be slowing down, much like a rock, thrown straight up, slows before it stops and falls back down. Cosmologists have been trying for years to measure the deceleration, and when they finally did, they were astonished to find—and this has been confirmed multiple times—that the recession is *accelerating,* not decelerating![141] Something is overpowering gravity and pushing galaxies away from each other faster and faster.[142]

- **Population III Stars.**[143] If the Big Bang happened, the matter condensing out of the energy would be only hydrogen, helium, and trace

---

[141] Walt Brown, *In the Beginning: Compelling Evidence for Creation and the Flood,*, Eighth edition, 2008 (Center for Scientific Creation, Phoenix, Arizona, CreationScience.com), p. 31.

[142] Ron Cowen, "A Dark Force in the Universe," *Science News,* Vol. 159, April 7, 2001, p. 218.

Adam Reiss, as quoted by James Glaz, "Astronomers See a Cosmic Antigravity at Work," *Science,* Vol. 279, February 27, 1998, p. 1298.

National Sciences Foundation Advertisement, "Astronomy: Fifty years of Astronomical Excellence," *Discover,* September 2000, p. 7.

Gordon Kane, "The Dawn of Physics Beyond the Standard Model," *Scientific American,* Vol. 288, June 2003, p. 73.

[143] As early as 1926, astronomer Jan Oort noticed that bluer (hotter) stars predominate in the spiral arms of our galaxy—the Milky Way—and the yellower (cooler) stars predominate in the galactic center. In 1944, astronomer Walter Baade defined the terms **Population I** for the blue stars because, based on the presumption of deep-time chronology, he assumed they were younger. The yellow stars he named **Population II** because, based on the same deep-time presumption, he assumed they were older. In 1978, a third category, **Population III,** was defined as those stars theorized to have very low metal content, because they were the

amounts of lithium; no heavier elements at all. So all the first stars would have only these elements. The heavier elements would form in stellar cores, and then supernovas would blow them out into space, "seeding" second- and third-generation stars with these elements. But some of the first-generation stars (named "Population III") have life spans that would allow them to still be around today, so we should be able to find some stars with only the hydrogen, helium, and lithium. But in our galaxy of 100 billion stars, *not one* Population III star has been found. This too seriously defies the odds.

And the critics of the Big Bang theory are not just writing in non-technical fluff pieces; here are some quotes from non-creationist scientists in various disciplines, and as usual, many more could have been included if space allowed:

> Astronomy, rather cosmology, is in trouble. It is, for the most part, beside itself. It has departed from the scientific method and its principles, and drifted into the bizarre; it has **raised imaginative invention to an art form;** and has shown **a ready willingness to surrender or ignore fundamental laws, such as the second law of thermodynamics and the maximum speed of light,** all for the apparent rationale of saving the status quo. Perhaps no "science" is receiving more self-criticism, chest-beating, and self-doubt; none other seems so lost and misdirected; **trapped in debilitating dogma.**[144]

> Despite the widespread acceptance of the big bang theory as a working model for interpreting new findings, **not a single important prediction of the theory has yet been confirmed, and substantial evidence has accumulated against it.**[145]

> The big bang theory can boast of **no quantitative predictions that have subsequently been validated by observation.**[146]

---

first stars supposed to have "formed" after the Big Bang. Population III stars might really have existed if deep time were real and the Big Bang had actually happened.

[144] Roy C. Martin Jr., *Astronomy on Trial: A Devastating and Complete Repudiation of the Big Bang Fiasco* (New York: University Press of America, 1999), p. xv.

[145] Tom Van Flandern, "Did the Universe Have a Beginning?" *Meta Research Bulletin,* Volume 3, Number 3, September 15, 1994, p. 33.

[146] Eric J. Lerner, et al., "Bucking the Big Bang," *New Scientist,* Volume 182, May 22, 2004, p. 20. (This article was originally signed by 33 scientists from ten countries. Later, 374 other scientists, engineers, and researchers also endorsed the article. See CosmologyState-

And those are just some of the physics-related difficulties. But there are also other troubles with the Big Bang and related theories. For example, are we supposed to assume that Nothing, when it exploded, using its profound absence of mind, morals, conscience, and intention, actually created logic, reasoning, awareness, perception, imagination, understanding, laws and moral codes, the "motivational imperative" (the awareness of right and wrong), music, art, theatre, literature, dance, and belief systems that include God? Are we really supposed to buy that? It's not a bad résumé for Nothing, eh?

Image by Dave Arns

And even more problems will be discussed below, in a bit more detail. But the point, as expressed in an open letter from dozens of scientists repudiating the Big Bang,[147] is that ever-increasing numbers of rescue devices are needed, fabricated, and being employed to prop up the sagging Big Bang theory: inflation, dark matter, dark energy, etc. No field of science—except evolution-based theories, including the Big Bang—would tolerate such a plethora of supplemental hypothetical theories in such a desperate attempt to avoid the obvious conclusion that *evolution didn't happen.*

The open letter referenced above is copied in its entirety on rense.com, and is introduced with the following paragraph:

> Our ideas about the history of the universe are dominated by big bang theory. But **its dominance rests more on funding decisions than on the scientific method,** according to Eric J Lerner, mathematician Michael Ibison of Earthtech.org, and dozens of other scientists from around the world.[148]

---

ment.org for details.)

[147] For an official statement on the growing doubt as to the plausibility of the Big Bang, see "Open Letter on Cosmology," published by E. Lerner in *New Scientist,* May 22, 2004. [Online at Cosmology.info/media/open-letter-on-cosmology.html; accessed July 16, 2020.]

[148] "Big Bang Theory Busted By 33 Top Scientists" [Online at rense.com/general53/bbng.htm;

The open letter itself shows how this happens:

> Today, virtually all financial and experimental resources in cosmology are devoted to big bang studies. Funding comes from only a few sources, and all the peer-review committees that control them are dominated by supporters of the big bang. As a result, the dominance of the big bang within the field has become self-sustaining, **irrespective of the scientific validity of the theory.**

An unfortunate fact: money speaks louder than truth. . .

Before we go on, a note on terminology is in order. In the following sections and chapters are several mentions of items with names like "M51" or "NGC 6384." These are entry names in two of the most common astronomical catalogues. The "M" stands for "Messier," after French astronomer Charles Messier (1730–1817) who founded the first catalog of astronomical objects,[149] and the "NGC" stands for the much more voluminous "New General Catalog," which was developed in 1888 by John Louis Emil Dreyer.[150] Both are still in use today.

The Cosmological Principle

The Cosmological Principle is the idea that, because of the directional uniformity of the Big Bang explosion, the resulting matter would have been flung out in all directions of space equally. An experiment to support this theory would be to look in multiple areas of the sky, and see virtually identical densities of galaxies in every square degree of sky, regardless of which direction you looked.

This overall "sameness" of the universe from one place to another is called **homogeneity,** and the "evenness" in all directions is called **isotropy.** Both of these were predictions of Big Bang theory.

But that's not what we observe. Recent observations with ever-more precise telescopes have revealed that there are massive superclusters (clusters of clusters) of galaxies, separated by vast areas of comparatively empty space. Far from being homogeneous, the universe is "clumpy" and, according to the Big Bang theory, it ought not be that way.

---

accessed May 4, 2022.] Emphasis added.

[149] For more information see en.wikipedia.org/wiki/Charles_Messier.

[150] For more information, see en.wikipedia.org/wiki/New_General_Catalogue.

(By the way, the following discussion includes mention of units called **light-years**. Contrary to what it might seem, this is not a unit of time, but a unit of *distance:* A light-year is the distance that light travels in one year, or about 6 trillion miles. This practice of using a time measurement to indicate distance is common in other areas as well: with a given or implied speed, time is implicitly converted to distance, as in "I live about an hour north of Denver.")

For example, R. Brent Tully, an astronomer at the University of Hawaii, discovered in 1986 that there were long "ribbons" of superclusters 300 million light-years long and 100 million light-years thick. The enormous ribbons were separated from each other by voids about 300 million light-years across.[151] Because of the Cosmological Principle of directional uniformity, such structures are far too large for the Big Bang to produce in the "mere" 13.8-billion-year age of the universe. For such enormous structures to be formed under the influence of only the laws of physics would take 80 billion years—more than *five times* the postulated age of the universe!

Then in 1989, Margaret Geller and John Huchra, of the Harvard-Smithsonian Center for Astrophysics, came across even larger structures. Their research discovered what they called "the Great Wall," which was an enormous sheet of galaxies 200 million light-years across and 700 million light-years long.[152] And even larger structures than this have been reported which, based on their location and velocity, would have taken *150 billion years* to assemble, assuming the Big Bang cosmology.[153]

Then, a team of nine astronomers led by Will Saunders finished and published their full-sky redshift survey of galaxies that were found by the Infrared Astronomical Satellite (we'll elaborate on what "redshift" is and how it's used, later in this chapter). This too, revealed many more superclusters of galaxies than the Big Bang could account for.[154] The universe is seriously *non*-homogeneous.

So the observations are not working out well for the Big Bang cosmology; there's just not enough gravity to attract the galaxies into clusters like they are, at least within the time constraints that the Big Bang offers.

---

[151] R. B. Tully, *Astrophysics Journal* 303:25–38 (1986).
[152] M. J. Geller and J. P. Huchra, *Science* 246:897–903 (1990).
[153] E. G. Lerner, *Aerospace America,* March 1990, pp. 38–43.
[154] Will Saunders, et al, *Nature* 349:32–38 (1991).

Something else was needed, in order to keep the Big Bang theory alive. (Remember, the Big Bang theory is the leading contender to counter the Creation model.) So, the advocates of evolution came up with a rescue device: "Maybe the other 90%–99% of the mass of the universe is undetectable!" They called this imaginary substance Cold Dark Matter, or CDM (popularly known simply as "dark matter"). This dark matter *had* to be cold, or else it would emit electromagnetic radiation, at least in the infrared range, and it would therefore no longer be "dark." But the hope was that all this extra mass of the CDM might supply enough gravity to coax the galaxies into clusters and superclusters. (Although, it seems like gravitational effects would make it detectable, doesn't it? After all, Neptune was detected by its gravitational effect on Uranus before it was visually discovered to be a planet.[155])

As it turns out, the galactic structures that astronomers are seeing are *so* big that even if CDM *did* exist, it wouldn't be enough to cause their formation. The calculations of Saunders and his team resulted in a 97% confidence level in the conclusion that the CDM hypothesis is invalid.[156] That same month, in the same publication, was an article by David Lindley entitled "Cold Dark Matter Makes an Exit." And cosmologist S. George Djorgovski from Caltech predicts that the abandonment of CDM is "inevitable," simply because the observations don't match the theory.[157]

E. G. Lerner, writing in *Aerospace America,* notes:

> No energetic processes, even unknown ones, could have occurred that were vigorous enough to either create the large-scale structures astronomers have observed or stop their headlong motion once created. **There is simply no way to form these structures in the 20 billion years since the Big Bang.**[158]

Cosmic Microwave Background

Another predicted by-product of the Big Bang theory is that there should be a leftover "warmth" everywhere as the universe cools down.

---

[155] en.wikipedia.org/wiki/Neptune#Discovery; accessed September 18, 2020.
[156] Will Saunders, et al, *Nature* 349:32–38 (1991).
[157] T. H. Maugh, II, *Los Angeles Times,* San Diego Edition, January 5, 1991, p. A29.
[158] E. G. Lerner, *Aerospace America,* March 1990, p. 42. Emphasis added.

When astronomers using optical telescopes look at the spaces between stars or between galaxies, all they see is darkness. However, when radio telescopes (which see wavelengths besides those visible to the human eye) are used to look into those spaces, they pick up a background noise that is not associated with any star or galaxy. NASA's Cosmic Background Explorer (COBE) satellite, observing from above the atmosphere, was able to tell us much more than ground-based radio telescopes, and sure enough, COBE detected the telltale glow. This "glow" is strongest in the microwave section of the electromagnetic spectrum, so it was dubbed Cosmic Microwave Background, or CMB.[159]

The discovery of CMB delighted the Big Bang enthusiasts, because it was something they predicted. (Although history shows that some of the cosmologists' "predictions" were adjusted after the fact, to agree with observations.[160,161]) They had originally expected it to be homogeneous, like they expected the galactic distribution to be, and they discovered that indeed, the CMB was the same everywhere. You might expect the Big Bang cosmologists to be overjoyed, but remember, they expected the CMB distribution and the galactic distribution to *both* be homogeneous. Since they had already discovered, to their dismay, that galactic clusters, superclusters, and even larger structures existed, they found the matter density to be very non-homogeneous. Therefore, the CMB ought also to be non-homogeneous.

But alas, it wasn't. The CMB was almost perfectly homogeneous,[162] in spite of the fact that the mass distribution wasn't that way at all! They were supposed to be *both* homogeneous! Or at least both *non*-homogeneous. But how could one be homogeneous and the other not, if they came from the same singularity explosion? It was looking like, far from the Big Bang theory hitting a home run, it was turning out to be an embarrassing whiff.

So, in spite of Wikipedia's cheery pronouncement that the Big Bang theory is "universally accepted," it appears to be on its last legs. Already, other theories are popping up to replace it, including some dealing with plasma processes and revamped versions of the steady-state theory.[163,164] It seems like

---

[159] en.wikipedia.org/wiki/Cosmic_microwave_background; accessed May 28, 2020.

[160] William C. Mitchell, "Big Bang Theory Under Fire," *Physics Essays*, Vol. 10, No. 2, June 1997, pp. 370–379.

[161] Dominic Statham, "Is the Big Bang Really Scientific?" *Creation* 41(1):48–50, January 2019. [Online at Creation.com/big-bang-scientific-theory; accessed May 1, 2022.]

[162] E. G. Lerner, *Aerospace America,* March 1990, p. 41.

[163] A. L. Peratt, *The Sciences,* January/February 1990, p. 24.

[164] H. C. Arp, G. Burbidge, F. Hoyle, J. V. Narlikar, and N. C. Wickramasinghe, *Nature*

the prevailing attitude among evolutionary cosmologists is *"Anything* is better than the Creation model!"

The Expanding Universe

One of the observed characteristics of the universe is that it seems to be expanding in all directions. And indeed, it may be; there is nothing in the Creation model that says such a thing couldn't or shouldn't happen.

Indeed, read this rather tantalizing passage:

Isaiah 40:22 (AMP): It is God Who sits above the circle (the horizon) of the earth, and its inhabitants are like grasshoppers; **it is He Who stretches out the heavens like [gauze] curtains and spreads them out like a tent to dwell in.** . .

Note that this theory of the expanding universe says that *the universe itself* is expanding. It's not that empty space is there, and stars and galaxies are flying into it; space itself is expanding and the contents of the universe are being "pulled along" with it as it expands. Perhaps the main observation supporting this apparent expansion is the redshift that is observable for all but the closest astronomical objects.

Redshift and Blueshift

But what in the world is redshift? **Redshift**, and its counterpart **blueshift**, are manifestations of the wave nature of light, and are both examples of the Doppler effect.[165] You are more likely to have experienced the Doppler effect in the acoustic arena, because sound is also a wave motion.

346:807–812 (1990).

[165] For more information, see en.wikipedia.org/wiki/Doppler_effect.

Oh, Evolve! (Good Luck With That...)

Imagine for a moment a motorcycle driving fast toward one man and away from another man:

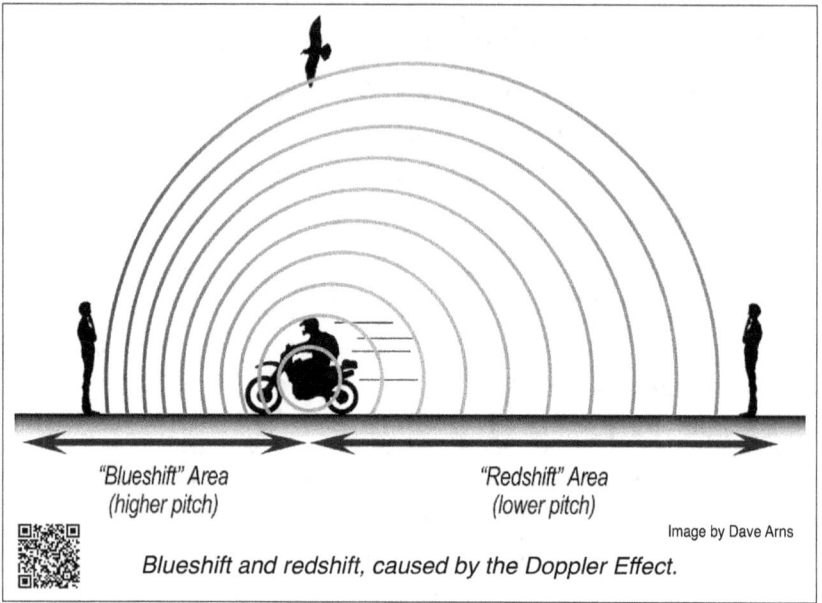

Blueshift and redshift, caused by the Doppler Effect.

The motorcycle engine is putting out a sound predominantly in one frequency, related to the RPMs the engine is going. If the motorcycle is driving along at a constant speed and not changing gears, the sound coming out of it stays at the same pitch. However, because the motorcycle is moving, every new sound wave is generated from a different point. The sound waves "pile up" in front of the motorcycle, so the man in front hears them closer together than they were originally generated. Thus, he hears the motorcycle's engine at a higher pitch than the motorcyclist does.

The opposite happens behind the motorcycle: those sound waves are "stretched out," because each successive one is generated a bit farther away than the previous one. This is heard as a lower-than-actual pitch. The Doppler effect is also why the pitch of a vehicle seems to go down as it passes you: you're in front of it, so the pitch seems higher, it goes past, and then you're behind it, so the perceived pitch becomes lower.

The bird flying above the motorcycle, if it keeps flying in the same direction and speed as the motorcycle, will hear the actual pitch of the motorcycle's engine. This is because its movement is perpendicular to the direction of the sound source.

128

A comparable thing happens with light, but its effect is so subtle it is not detectable with the naked eye. There is a device called a **spectrometer**[166] that breaks light down into its color components. The resulting display would look something like the illustration to the right.

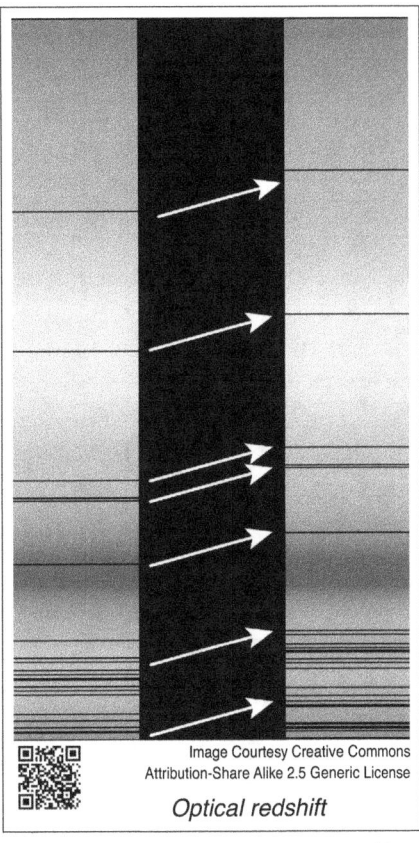

Image Courtesy Creative Commons
Attribution-Share Alike 2.5 Generic License
Optical redshift

Pure light shows up as a continuous, unbroken spectrum, like the image to the right, only without the dark lines. But pure light is pretty uncommon in nature. When pure light passes through a substance—like the gases in the corona of a star—each gas absorbs and scatters light in a very specific and well-defined way. The **absorption lines** appear as dark lines in a spectrogram (see the illustration). Each gas has its own particular "signature," and those lines show up in very specific places, at very specific intervals, in the electromagnetic spectrum.

When the source of the light that is passing through those gases is receding from you, that pattern of dark lines moves towards the red end of the electromagnetic spectrum: it is **redshifted**, as shown in the illustration. Notice that absorption lines stay at the same intervals relative to each other, but the whole pattern is offset, as indicated by the white arrows.

The amount that the absorption lines move toward the red end—that is, the amount of redshift—depends on the speed at which the light source is moving away from you. Thus, by noticing how large the redshift is, one can deduce the speed at which the light source is receding. In support of the expanding universe idea (which does *not* automatically presume support for the Big Bang theory), the objects that appear to be the farthest away exhibit the greatest redshift, implying that they are moving away from us more rapidly than the closer objects are.

---

[166] For more information, see en.wikipedia.org/wiki/Optical_spectrometer.

Blue Stars

Stars come in lots of different temperatures: there are cool red ones, medium-hot yellow ones (like our sun), and very hot blue ones. This is very understandable, since the same thing can be seen in everyday life with little difficulty.

For example, the flame of a burning log in the fireplace is yellowish-orange and relatively cool, but the flame of an oxyacetylene torch in a welder's shop is blue and much hotter.

Stars are the same way: the blue ones are much hotter than yellow or red ones, and in order to maintain that temperature, blue stars must consume their nuclear fuel at a much faster rate. And of course, the faster a star uses its fuel, the sooner its fuel runs out. And this is where deep time runs into yet another snag.

Image courtesy ESA/Hubble and Creative Commons CC BY 4.0

A Hubble Space telescope image of hot, short-lived blue stars in NGC 6397, whose lifespans are short. How can they still be there?

According to astronomical calculations, such stars as shown in the accompanying photo would exhaust their nuclear fuel in only a couple million years or slightly longer. So if the universe is 13.8 *billion* years old, how are these blue stars still here? They should have burned out eons ago! We shouldn't still be seeing any of these blue stars, but we do. The observations are not in agreement with the theories—they haven't burned out yet, even though they "should" have, billions of years ago. These blue stars are really straggling behind the evolutionary end-of-life curve for some unknown reason; hence their nickname: "blue stragglers."

To address this huge monkey wrench in the works, deep-time cosmologists came up with another rescue device: they invented the "stellar nursery" idea. These are gas clouds in space, where these cosmologists imagine new stars are being born. There's no observational evidence that any stars are actually being born there, but if stars are *not* being born in such nurseries, blue stars are another nail in the coffin of long-age cosmology.

In these gas clouds in which long-age cosmologists sure hope stars are born, the star-formation models come up short. True, a slight gravitational collapse would draw some of the gas together, but as anyone who has pumped up a bicycle tire knows, compressed air gets hot, and so does the bottom of the air pump. And when gases get hot, they expand and the atoms push away from each other, halting any further collapse, and preventing a potential star from igniting.[167]

Then the long-age cosmologists say, "Well, a shock wave from an exploding star might push the gases together with sufficient force to overcome the gas pressure and ignite a new star." But there's a problem here too: how did the exploding star initially form? By having its gases compressed by another star exploding even earlier? The obvious problem with this scenario is that the first-ever star could never have formed. Therefore, none of the rest of them could have either.[168]

Not only that, but if stars could form only when gases were hit by shockwaves of star matter, the visible universe would be cluttered with the debris from innumerable stellar explosions. But it's not.[169] Again, those pesky observations are conflicting the deep-time beliefs.

Another rescue-device hypothesis that has been put forward is that maybe these blue stars are really "older stars that acquire a new lease on life when they collide and merge with other stars."[170] Or perhaps they siphon matter off of nearby stars through a "mass transfer."[171] During a study of the star cluster NGC 188 (in the constellation Cephus), 21 blue stragglers were found. Sixteen of these were binary stars that interacted with their companion stars. Even though the mass distribution of the blue stragglers, as compared with their companion stars, matched predictions well, it still didn't solve the problem. The reason was because if this process had been happening during the billions of years that uniformitarian Big Bang cosmology assumes, these "blue straggler stars should already have evolved into giant stars and stellar remnants."[172] Which they haven't.

---

[167] Faulkner, D. 2010. "Blue Stars—Unexpected Brilliance" [online at AnswersInGenesis.org/articles/am/v6/n1/blue-stars]. *Answers.* 6 (1): 50–53.

[168] Lisle, J. 2012. "Blue Stars Confirm Recent Creation." *Acts & Facts.* 41 (9): 16 [online at ICR.org/article/6943].

[169] Thomas, B. "Rare Supernova Recalls Missing Remnants Mystery." *Creation Science Update.* Posted on ICR.org September 6, 2011. [Online at ICR.org/article/rare-supernova-recalls-missing-remnants; accessed August 11, 2020.]

[170] "Omega Centauri: Colorful Stars Galore Inside Globular Star Cluster Omega Centauri." NASA press release, September 9, 2009. [Online at NASA.gov/mission_pages/hubble/multimedia/ero/ero_omega_centauri.html.]

[171] Geller, A. M. and R. D. Mathieu. 2011. "A mass transfer origin for blue stragglers in NGC 188 as revealed by half-solar-mass companions." *Nature.* 478 (7369): 356–359.

And there's another problem as well: not all the blue stragglers are binary stars, which means for these, there is no companion star from which to siphon fuel. How are *they* still around?

It seems that no matter what rescue devices evolutionary cosmologists come up with, repeatable observations end up militating against the billions and billions of years required by the deep-time cosmologies. As mentioned, blue stragglers should burn out in only one or two million years.[173] According to University of South Carolina astronomer Danny Faulkner, "In fact, the hottest blue stars could last only a few million years at best. Both creationists and evolutionists acknowledge this fact."

So evolutionists hypothesized another rescue device: maybe new blue stars are constantly being created; that would account for their present existence *and* their short lifespan. But of course, that would mean blue stars would be forming all the time, which means we should be able to see it happening. "Despite their diligent search, however, [astronomers] have never observed one of these blue stars forming—or any other star, for that matter. . ."[174] But these blue stragglers in NGC 188 are supposed to be 7 billion years old, in spite of their maximum life span of a couple million years.

Then in another seemingly desperate attempt to prop up Big Bang cosmology, evolutionary astronomers are suggesting yet another rescue device: that maybe these stars burned fuel at a more "normal" rate for billions of years, and then, almost as if on cue, all 12 of them recently began siphoning stellar material from their companion stars and became blue, so therefore they only *look* young, but are really old stars. That just seems a bit too convenient, doesn't it? That almost sounds like Someone was orchestrating them, doesn't it? But no, we can't have that; orchestration implies *intention* in the universe, and intention requires an entity taking deliberate steps to accomplish a goal. No, we *certainly* can't have that.

All of the above rescue devices have at least two things in common. First, there is absolutely no actual evidence for them; what they propose has never actually been observed. And second, their *raison d'être*—their sole reason for existence—is to preserve the deep-time presupposition. After all, we *must* pre-

---

[172] Geller, A. M. and R. D. Mathieu. 2011. "A mass transfer origin for blue stragglers in NGC 188 as revealed by half-solar-mass companions." *Nature.* 478 (7369): 356–359.

[173] Thomas, B. "Young Blue Stars Found in Milky Way" *ICR News.* June 9, 2011. [Online at ICR.org/articles/view/6194/245; accessed August 11, 2020.]

[174] Faulkner, D. 2010. "Blue Stars—Unexpected Brilliance" [online at AnswersInGenesis.org/articles/am/v6/n1/blue-stars]. *Answers.* 6 (1): 50-53.

serve deep time, because if that goes away, all we'll have left is the Creation model, and that, according to evolutionist Sir Arthur Keith, "is unthinkable."

Professor D.M.S. Watson, Jodrell Professor of Zoology and Comparative Anatomy at University College, London from 1921 to 1951,[175] joins Sir Arthur Keith in his horror of the thought that a Creator may have been involved in the Creation. C. S. Lewis quotes Professor Watson's defense of evolution, and then comments on it:[176]

> Evolution itself is accepted by zoologists not because it has been observed to occur or. . . can be proved by logically coherent evidence to be true, but **because the only alternative, special creation, is clearly incredible.**

After quoting Watson's statement of faith in how nothingness created everything, Lewis comments:

> Has it come to that? Does the whole vast structure of modern naturalism depend not on positive evidence but simply on an *a priori* metaphysical prejudice? **Was it devised not to get in facts but to keep out God?**

And actually, as you'll see in numerous places in the following pages, evolution's main purpose is indeed to keep God out—out of science, out of education, out of the media, out of government, out of public discourse, and out of people's lives. And don't be fooled by Watson's use of the word "incredible." He does not mean that special creation is "very good and wonderful," but rather that it is "impossible to believe." It's cute, and more than a little bit tragic, to see people who don't know God trying to authoritatively expound upon what things are or are not "impossible," as if they are the standard against which everything is measured.

All these rescue devices remind me of when, years ago, I heard a radio talk-show host interviewing a creationist. The creationist was describing a discussion he had had with a skeptic who was offering alternatives to Creation, and the skeptic went on for quite some time in this vein: "You could say that *(some fanciful thing)*, and then you could say that *(some other fanciful thing)*, and then you could say that *(some other fanciful thing). . .*" The creationist finally replied, "Of course you 'could say' that. You could say *anything*. But

---

[175] wikipedia.org/wiki/D._M._S._Watson; accessed December 30, 2023.
[176] C.S. Lewis, *The Weight of Glory*, chapter "Is Theology Poetry?" (HarperCollins e-books, 1949, 1976).

why should I believe your theory is better than mine? What evidence do you have that your theory is plausible?" The proponents of any theory have the burden of proof, or at least the burden of plausible *evidence*, using the operation of known laws of physics, to present a feasible mechanism whereby their proposed theory could have actually happened.

It's also revealing, and a bit saddening, that in 1998, when University of South Carolina astronomer Danny Faulkner cited the nine studies that confirmed the data on blue stragglers, only one of them acknowledged that observations directly contradicted the deep-time theory required by evolution.[177] It seems that the remaining eight studies couldn't bring themselves to admit that the billions of years postulated by evolutionary cosmology could be mistaken.

The study cited above was looking at just the open star cluster NGC 188, but blue stragglers have been found in other places in the Milky Way as well. As of 2011, the Hubble Space Telescope had found more than 40 of them just in our own galaxy.[178]

Well, maybe they got their new lease on life by colliding with other stars, and that influx of material rejuvenated them. When this rescue device was put forward, it was quickly shot down. Even though stars are quite large in comparison to Earth, they are quite tiny in comparison to the vast distances between them. Which means that statistically, it's *very* unlikely that stars collide with each other. Christopher Tout, a University of Cambridge astronomer, summarized Geller and Mathieu's paper and stated, "Thus, in this cluster *[NGC 188]*, a collisional origin for blue stragglers is much rarer than expected, and the authors' study casts doubt on whether it occurs at all."[179] Again, observations don't match the speculations put forward to salvage deep-time cosmology.

In short, they haven't a clue. As NASA phrases it, "It is not clear how blue stragglers form."[180] Notice the choice of words: it's not clear how they

---

[177] Faulkner, D. 1998. "The Current State of Creation Astronomy." *Proceedings of the Fourth International Conference on Creationism*. Pittsburgh, PA: Creation Science Fellowship [online at ICR.org/i/pdf/technical/The-Current-State-of-Creation-Astronomy.pdf].

[178] Thomas, B. "Young Blue Stars Found in Milky Way" *ICR News*. June 9, 2011. [Online at ICR.org/articles/view/6194/245; accessed May 30, 2020.]

[179] Tout, C. 2011. "Astrophysics: Stars acquire youth through duplicity." *Nature*. 478 (7369): 331–332.

[180] Gundy, C. "NASA's Hubble Finds Rare 'Blue Straggler' Stars in Milky Way's Hub." NASA news release, May 25, 2011. [Online at NASA.gov/mission_pages/hubble/science/blue-straggler.html.]

"form," as if it was accomplished by unaided natural processes. Like driving past a beautiful home and thinking, "I wonder how that home formed," not even considering that perhaps a homebuilder was involved.[181]

The end result is that deep-time cosmologists have resorted to non-explanations that claim blue stars "somehow" formed in the "distant past" by the action of "unknown, unobserved" processes. Which is basically an appeal to magic. (Because evolution disallows the existence of God, this could not have been a miracle, even though it sure sounds like one. Therefore it must have been magic.) And *creationists* are accused of faith-based explanations?

The simplest explanation, which is too often rejected outright, is this: maybe the blue stars look young because they *are* young.[182] Remember Occam's Razor?

Supernova Remnants

A supernova is an exploding star; it suddenly increases greatly in its brightness, which may cause it to be noticed for the very first time, and during the explosion, most of its mass is thrown out into space.

Photo courtesy NASA, ESA, J. Hester and A. Loll (Arizona State University)

The Crab Nebula, resulting from a supernova explosion.

One of the most famous supernovae is the one that created the Crab Nebula, shown at right. The light from the supernova that caused the Crab Nebula was seen by Chinese astronomers in 1054AD, who recorded its precise location relative to surrounding stars. The resulting nebula, composed of the star's mass thrown into space when it exploded, was first observed by the British astronomer John Bevis in 1731.[183] So what we see now as the Crab Nebula has been expanding for almost a thousand years. Based on the quantity and velocity of the stellar mass composing the nebula, it is estimated that the supernova shone, for a short time, 400 million times brighter than our sun![184] For 23 days, observers

---

[181] Unsurprisingly, the Bible confirms this self-evident fact: "For every house is built by someone, but God is the builder of everything" (Hebrews 3:4, NIV).

[182] Amos 5:8a (NLT) It is the Lord who created the stars, the Pleiades and Orion...

[183] en.wikipedia.org/wiki/Crab_Nebula; accessed May 29, 2020.

[184] JPL.Nasa.gov/news/press_kits/spitzer/gallery/multi-observatory-images; accessed May 29,

on earth could see the supernova even during the day, and at its brightest, it would have been brighter (although much smaller in apparent size) than the full moon![185]

How often do supernovae happen? According to astrophysicist D. Russel Humphreys:

> According to astronomical observations, galaxies like our own experience about one supernova (a violently exploding star) **every 25 years.** The gas and dust remnants from such explosions (like the Crab Nebula) expand outward rapidly and **should remain visible for over a million years.** Yet the nearby parts of our galaxy in which we could observe such gas and dust shells contain only about 200 supernova remnants. That number is consistent with only **about 7,000 years' worth of supernovas.**[186,187]

Well, isn't that interesting? Observations reveal another supernova every 25 years or so, and its remnants wouldn't dissipate for about a million years, which means that we should be able to see about 40,000 supernova shells in our galaxy. But we see only about 200, which (assuming the uniformitarianism that evolutionary cosmology demands) indicates supernovas have been happening only for the last 7000 years or so. Notice that this correlates well with the Creation model's stated age of the universe of 6000 years.

2020.

[185] Walt Brown, *In the Beginning: Compelling Evidence for Creation and the Flood,*, Eighth edition, 2008 (Center for Scientific Creation, Phoenix, Arizona, CreationScience.com), p. 41.

[186] Humphreys, D. R. 2005. "Evidence for a Young World." Acts & Facts. 34 (6) [online at ICR.org/article/evidence-for-young-world; emphasis added.

[187] See also Sarfati, J. 1997. "Exploding stars point to a young universe." *Creation.* 19 (3): 46-48 [online at Creation.com/exploding-stars-point-to-a-young-universe#calculations]; and Clark, D. H. and J. L. Caswell. 1976. A Study of Galactic Supernova Remnants, Based on Molonglo-Parkes Observational Data. *Royal Astronomical Society,* Monthly Notices. 174: 267-305 [online at adsbit.harvard.edu/cgi-bin/nph-iarticle_query?bibcode=1976MNRAS. 174..267C&db_key=AST&page_ind=0&plate_select=NO&data_type=GIF&type= SCREEN_GIF].

Spiral Galaxies

Spiral galaxies are disk-shaped collections of stars with recognizable "arms" going out from the center. These arms "wrap around" the center of the galaxy for some angular distance; the total angular wraparound from the innermost end to the outermost end varies from one spiral galaxy to another. Our own Milky Way is such a spiral galaxy.

Spiral galaxy NGC 6384 in the constellation Ophiuchus.

Spiral galaxies are aesthetically beautiful, as shown by these Hubble Space Telescope photos of NGC 6384 and NGC 5194 (also known as M51a). The graceful arms extending and curling around the center are sources of delight for many. But what causes the arms to curl around like that?

The wrapping around of a spiral galaxy's arms are caused by two basic, well-known, and solidly confirmed laws of physics:

Spiral galaxy NGC 5194 or M51a—the "Whirlpool Galaxy"—in the constellation Canes Venatici.

- The law of gravity (okay, "the curvature of spacetime," for you technical folks), which, unless prevented by something else, causes objects to move toward each other.

- The law of conservation of angular momentum, which causes an ice skater, slowly spinning with his arms extended, to spin faster when he brings his arms in.

Even before we talk about how the spiral arms still exist, we first need to ask: how did such galaxies start spinning in the first place? Since evolutionists disallow God as a Creator, we're stuck with only natural laws to create everything. Here's how that is working out:

> Galaxy rotation and how it got started is one of the great mysteries of astrophysics. In a Big Bang universe, linear emo-

tions are easy to explain: They result from the bang. But what started the rotary motions?[188]

So we'll assume, like evolutionists do, that "somehow" the galaxies were set into rotational motion billions of years ago. According to evolutionary cosmologists, the arms that wrap around the cores of spiral galaxies are 10 billion years old.[189] Because of the conservation of angular momentum, the stars closer to the galactic core would orbit faster than those farther out. You can see this same effect in soapsuds in the whirlpool of a draining bathtub: the closer they are to the center, the faster they go around.

In a spiral galaxy, because the constituent stars at the inside end of each arm are orbiting faster, they would get angularly ahead of those at the outside end, eventually lapping them once, twice, three times, and more. But here's a question: how many laps ahead of the outside ends would the inside ends have to be before the individual arms are no longer distinguishable? It wouldn't take too many laps before the arms would all blend together into a homogeneous disk.

Calculations show that it would only take about 100 million years for the spiral arms to completely blend together, creating the disk mentioned above. So how are the arms still present after the posited 10 *billion* years? Not only that, but what laws of physics could have formed the (straight?) arms in the first place?

To address the first question, the traditional explanation for how the arms could last 100 times longer than they "should" have (assuming deep time) invoked the existence of a rescue device called "density waves," which shepherded the stars in such a way that the appearance of arms was preserved. However, this rescue device didn't really answer the question; it just pushed it one step farther away. Specifically, what created the density waves?

Postgraduate student Robert Grand of the Mullard Space Science Laboratory, University College London, developed a computer model to simulate the behavior of a galaxy full of stars acting on each other, taking into account gravitation, the conservation of angular momentum, and other factors. Using known laws of physics, Grand's digital model attempted to simulate the origin of the arms in spiral galaxies. Presenting his findings to the Royal Astronom-

---

[188] William R. Corliss, *Stars, Galaxies, Cosmos: A Catalog of Astronomical Anomalies* (Glen Arm, Maryland: The Sourcebook Project, 1987), p. 177.

[189] Humphreys, D. R. 2005. "Evidence for a Young World." *Acts & Facts*. 34 (6). [Online at ICR.org/article/evidence-for-young-world.]

ical Society's National Astronomy Meeting in Wales, he stated "We have found it impossible to reproduce the traditional theory, but stars move with the spiral pattern in our simulations at the same speed."[190]

Grand developed another model, this time based on the density-wave idea. But again the arms broke up, due to shear forces, after 100 million years. This result was very consistent with the analysis done by Dr. D. Russell Humphreys, who said:

> The stars of our own galaxy, the Milky Way, rotate about the galactic center with different speeds, the inner ones rotating faster than the outer ones. The observed rotation speeds are so fast that if our galaxy were more than **a few hundred million years** old, it would be a featureless disc of stars instead of its present spiral shape.[191]

Grand was apparently unwilling to give up the idea of long-age cosmology, so he came up with the third attempt:

> We find that the spiral arms are **recurrent** material features, and the pattern speed generally decreases with radius, in such a way that the pattern speed almost equals the rotation speed of stars at all radii.[192]

In Grand's simulation, the spiral arms were "recurrent," but if this were a real phenomenon, we would see spiral galaxies in various stages of arm formation, arm collapse, and arm re-formation. Observation of actual galaxies has not revealed this to be true. Not only that, but Grand's third model did not produce the clearly defined arms that we see in the sky.[193]

---

[190] "NAM 21: New theory of evolution for spiral galaxy arms." Royal Astronomical Society press release, April 20, 2011. [Online at RAS.org.uk/news-and-press/217-news2011/1967-new-theory-of-evolution-for-spiral-galaxy-arms.]

[191] Humphreys, D. R. 2005. "Evidence for a Young World." *Acts & Facts.* 34 (6). [Online at ICR.org/article/evidence-for-young-world.] Emphasis in original.

[192] Grand, R. "Analysing stellar motions and spiral arm formation in spiral galaxies." Presented at the Royal Astronomical Society National Astronomy Meeting 2011 in Llandudno, North Wales, April 17-21. [Online at RAS.org.uk/component/db/?task=viewrecord&report_id=1097&recid=165.] Emphasis added.

[193] Brian Thomas, "New Galaxy Model Leaves Old Questions Unanswered" May 5, 2011. [Online at ICR.org/article/new-galaxy-model-leaves-old-questions; accessed May 30, 2020.]

Chapter 3 Summary

Here are the most important points in this chapter:

- The **Big Bang**, as a naturalistic process, is not a foregone conclusion; even many evolutionists dispute its validity.
- The Big Bang would produce a universe exhibiting **homogeneity**, but observations of stellar mass distribution—galactic superclusters—show the opposite. (Rescue device: Dark matter.)
- The **Cosmic Microwave Background** should be very non-homogeneous, because the galaxies are, but the CMB *is* homogeneous. According to Big Bang, the homogeneity of stellar mass distribution and CMB should be the same: both homogeneous, or both not.
- The **Doppler Effect** indicates movement toward or away from the observer.
- **Blue stars** shouldn't be here anymore; they should have burned out eons ago. (Rescue devices: stellar nurseries, shock waves from previous supernovas, mass transfer from nearby stars, stellar collisions, blueness a new change to old stars.)
- **Supernova remnants** are too rare: given the observed supernova rate (one every 25 years), we should see 40,000 supernova shells, assuming deep time. We see 200. This works out to an age of about 7,000 years.
- **Spiral galaxies** shouldn't exist anymore—their arms would have wound up into featureless disks after deep time. (Rescue devices: density waves, arm recurrence.)

Chapter 4:

A Young Solar System

In this chapter on the solar system, Earth will not be included. There is *so* much information that pertains specifically to the Earth, that the Earth (and Earth's moon) will be covered in the next few chapters after this one. So this chapter will include planets and their moons, asteroids, comets, and so forth.

The solar system consists primarily of the sun and eight planets, as shown below. (Pluto was demoted in 2006 to the status of "dwarf planet.") Other objects in the solar system include moons around most of the planets, rings around the gas giants, asteroids primarily between Mars and Jupiter, and comets on highly elliptical orbits.

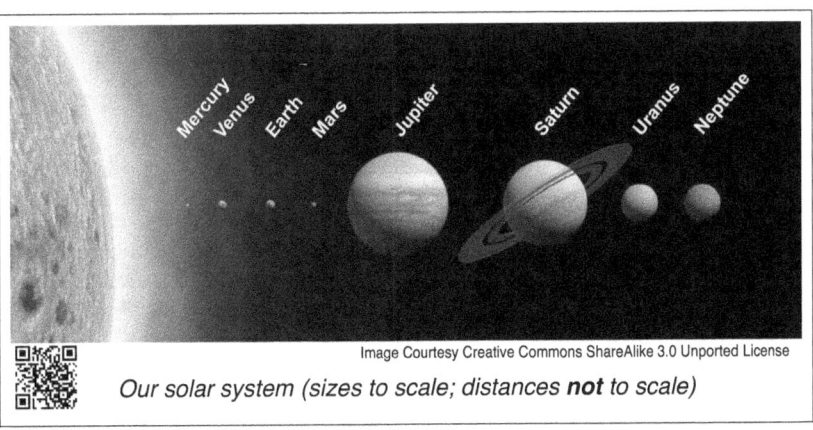

Image Courtesy Creative Commons ShareAlike 3.0 Unported License

*Our solar system (sizes to scale; distances **not** to scale)*

Oh, Evolve! (Good Luck With That. . .)

Where Did the Solar System Come From?

This is a very fair question, and much research has been put into answering it. The usual answer is that the solar system—the sun and everything in orbit around it—started out as an enormous cloud of dust and gas in interstellar space.

The Nebular Hypothesis, as this theory is called, was developed by Immanuel Kant and promoted, starting in 1796, by Pierre-Simon Laplace. They postulated that a huge cloud of "spinning gas supposedly threw off rings that eventually condensed to become the planets."[194] And even though astronomers have been finding evidence to the contrary ever since the late 1700s, this is still being taught!

For example, a college textbook from 1993 says this:

Image courtesy NASA/Jet Propulsion Laboratory
Artist's rendition of dust disk that hypothetically can spontaneously turn into a star and its planets.

> About 5 billion years ago, and **for reasons that are not yet fully understood,** this huge cloud of minute rocky fragments and gases began to contract under its own gravitational influence.[195]

Note that it is stated as an established fact. (It isn't.) Note also the emphasized phrase in the above quote. In other words, "We have no idea what could make this gas cloud contract, because it defies the laws of physics." Think about it: If the Nebular Hypothesis followed the laws of physics, the processes of sun-formation and planet-formation *would* be understood. The above quote is a tacit admission that the Nebular Hypothesis is just make-believe, because it doesn't adhere to the laws of physics.

Here's how the story of the Nebular Hypothesis goes: Random bits of this dust/gas cloud stuck together, resulted in increased gravity in some areas,

---

[194] Pasachoff, J. M. 1998. *Astronomy: From the Earth to the Universe.* Fort Worth, TX: Saunders College Publishing, p. 124.

[195] Tarbuck, E. J. and F. K. Lutgens. 1993. *The Earth: An Introduction to Physical Geology,* 4th ed. New York: Macmillan, 10, 12. Emphasis added.

which attracted more matter into the clump, and the self-reinforcing process eventually caused so much pressure in the center of the clump that nuclear fusion began, and the sun was born. Smaller clumps of matter in the growing disk of orbiting material (the **accretion disk**) condensed and formed the planets.

At least, that's how the story goes. Even today, Wikipedia says it this way: "The Solar System formed 4.6 billion years ago from the gravitational collapse of a giant interstellar molecular cloud."[196] Very matter-of-fact, as if everyone knows that.

But there are many hard-core scientists, both creationists *and* evolutionists, that are not all that confident in the Nebular Hypothesis. Three problems with the theory are 1) death spirals (matter falling into the star instead of coalescing into planets), 2) accretion (getting dust and gas to actually stick together), and 3) turbulence (collisions or gravitational perturbations scattering anything that was "trying" to stick together).[197]

Let's look more closely at these problems with the Nebular Hypothesis:

- **Death Spirals:** As clumps of dust and gas supposedly formed from the large cloud, the gases condensed. But planets could only form if the temperature was cool enough for the gases to condense into liquid, thus accumulating enough mass to gravitationally pull more gases in, and enough "stickiness" for incoming particles to not simply bounce off. But this couldn't have happened close to the sun, because it's too hot for gases to condense, so the gas giants far away from the sun must have somehow been pulled in closer to become the terrestrial planets (how they were pulled in is not explained). But if they were pulled closer to the sun, they would have encountered much uncondensed gas, which would cause friction, which would have slowed down their orbital velocity, and their orbits would have gotten progressively smaller (hence the spiral), and they would have crashed into the sun. And even for the outer planets, their increasing size would have caused friction with the gases through which they traveled, causing the same result. Strike one.
- **Accretion:** The Nebular Hypothesis surmises that dust particles and gas atoms would actually stick together spontaneously, although this is

---

[196] en.wikipedia.org/wiki/Solar_System; accessed June 2, 2020.
[197] Asphaug, E. 2009. "Growth and Evolution of Asteroids." *Annual Review of Earth and Planetary Sciences.* 37: pp. 413–448.

extremely unlikely (as in, impossible). If you have two grains of dust, or two atoms of gas, coming close together, their force of gravity would be negligible, so the chances of the first two particles sticking together is so small, it wouldn't have happened, even in the paltry 4.5 billion years deep time assumes. And if their relative velocities weren't so precisely matched to each other's, they would simply bounce off each other. Strike two. (Furthermore, the chances of *three* particles sticking together is even smaller, because of the next problem.)

- **Turbulence:** Suppose enough particles stuck together to actually form a mass big enough to pull more mass in. There are two problems with this scenario, both having to do with preventing other clumps from forming too near the original one. First, if two clumps collided, even at a low speed, their gravitational fields would be no match for even trivial momentum, so the colliding masses would scatter each other's particles and the process would have to start all over. And second, even if they didn't collide, but just passed closely enough to each other, the same gravity that pulled particles in from the gas cloud would pull on the other clump's particles, and again, both clumps would disintegrate and the particles would disperse. Strike three. You're out.

Planetary scientists are acknowledging the troubles: ". . .many aspects of the formation of the giant planets remain unresolved." [198]

One evolutionary astronomer conceded:

> The discovery of thousands of star systems wildly different from our own has demolished ideas about how planets form. Astronomers are searching for a whole new theory.[199]

Yet another problem is that these disks of dust and gas, which evolutionary cosmologists hope will accrete, apparently don't have enough mass to build a solar system. Astronomers have detected **exoplanets** (planets around other stars), and they've seen dust disks around stars, but the dust disks aren't enough to build solar systems.[200] We've seen disks, and we've seen stars, but we've never seen evidence of a disk *turning into* a star. (Come to think of it, that is

---

[198] Chaisson, E. and S. McMillan. 2014. *Astronomy Today.* Boston: Pearson Publishers, p. 154.
[199] Finkbeiner, A. 2014. "Astronomy: Planets in chaos." *Nature.* 511 (7507): pp. 22–24.
[200] Mann, A. "Cosmic conundrum: The disks of gas and dust that supposedly form planets don't seem to have the goods." [Online at ScienceMag.org/news/2018/09/cosmic-conundrum-disks-gas-and-dust-supposedly-form-planets-don-t-seem-have-goods; accessed August 11, 2020.]

very similar to a concept we'll examine in a later chapter: we've seen apes and we've seen people, but we've never seen evidence of apes *turning into* people.)

How would a gas cloud turn into a star? Supposedly, by gravitational collapse, heating the core of the growing ball of matter until nuclear fusion starts. But as we saw earlier, gases, when compressed, heat up, and this causes *expansion*, not contraction.[201] Even if some outside force compressed the gases enough to initiate fusion, magnetic fields and angular momentum would oppose any further collapse, and these would prevent the formation of a functioning star.

At Chile's Atacama Large Millimeter Array (ALMA) radio telescope,[202] a team of astronomers surveyed hundreds of "young" star systems, which they estimated to be between one and three million years old, and which were surrounded by these disks of dust particles. They compared the masses of the dust disks to the total planetary masses of other stars of comparable size, and there just wasn't enough material to build the planets. In some cases, up to 99% of the required mass was missing!

---

[201] Jason Lisle, Ph.D., "The Solar System: The Sun." [Online at ICR.org/article/solar-system-sun; accessed June 2, 2020.]

[202] See en.wikipedia.org/wiki/Atacama_Large_Millimeter_Array for details.

The Roche Limit

These dust disks, if they are not stable features around the stars that they orbit, are actually more likely to be the result of a *destructive* process, rather than a constructive one, because of, if nothing else, the Second Law of Thermodynamics. But actually, there *is* something else: the **Roche Limit**.

In orbital mechanics, the Roche Limit is the distance within which an orbiting object, held together only by its own gravity, will disintegrate because of the **tidal forces** exerted by the gravity of the body being orbited. Just like the moon's gravity causing a "stretching out" of the oceans toward it, which we perceive as tides flowing in and out, a star exerts tidal forces on objects orbiting it. A collection of objects outside the Roche Limit would be able to stay collected in a clump because of their mutual gravity. But *inside* the Roche Limit, such a clump would be torn apart by the tidal forces exerted by the star. In other words, the parts closest to the star are more affected by the star's gravity than they are by the gravity of the other members in the clump, so the clump disintegrates, as shown in the illustrations at right.

Which means that this spinning disk of dust and gas. . . but that brings up another point. Evolutionary cosmologists always assume this disk is spinning, even though they haven't been able to figure out how it started spinning. Where did the energy come from to start it spinning? Anyway, let's just assume (like evolutionists do) that "some-

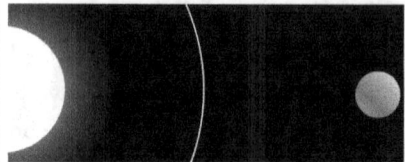
Far outside the Roche Limit, a clump could hang together by its own gravity.

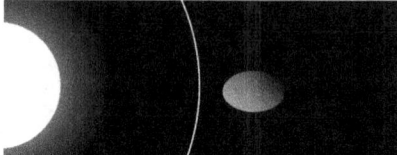
Close to the Roche Limit, a clump is distorted, but not pulled apart, by the star's gravity.

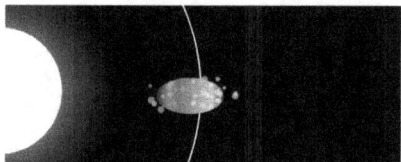
At the Roche Limit, a clump would be pulled apart because the nearest parts of the clump are attracted more by the star's gravity than the clump's own gravity.

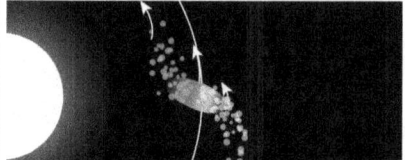
Because the clump was orbiting the star already, the pieces continue to do so, but the closer ones orbit faster, getting ahead of the farther pieces.

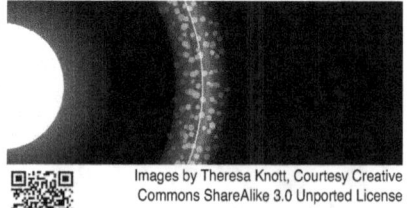
Images by Theresa Knott, Courtesy Creative Commons ShareAlike 3.0 Unported License
Eventually, the pieces settle into a ring around the star.

how" it got started spinning: before the clumps form, any localized gravitational field would be negligible, so why would planets *ever* form around its central star? Again, assuming the star *did* form. But why would the star itself form? The Nebular Hypothesis can't explain this.

Orbital Considerations

According to the Nebular Hypothesis, because the star and all its planets and moons came from the same whirlpool of dust, the star in the middle, all its planets, and all their moons should be orbiting the star in the same direction (this is called a **prograde** orbit), which would be the same direction as the original cloud of dust and gas.

Not only that, but they should orbit the star within the plane of the star's equator. But many of the exoplanets don't do as predicted: some orbit backwards (this is called a **retrograde** orbit), and others have orbits that are highly inclined, relative to their star's equator.[203,204] The Nebular Hypothesis can't explain this. As Amaury Tiraud, co-author of the original study, says: "This is a real bomb we are dropping into the field of exoplanets." [205]

In the same article, Andrew Cameron, an astronomer at the University of St. Andrews, described the recent retrograde discoveries at a meeting of the Royal Astronomical Society in Glasgow:

> That finding is inconsistent with the view that planets are formed by the condensation of dust from a disk surrounding a newly formed star.[206]

In other words, by observing exoplanets, we're realizing that the Nebular Hypothesis doesn't reflect reality. And this isn't just a feature of exoplanets, either—astronomers are realizing more and more, from observing exoplanets,

---

[203] Brian Thomas, Ph.D, 2010. "Planet Formation Theory Collides with Backward-Orbiting Planets." *Creation Science Update.* [Online at ICR.org/article/planet-formation-theory-collides-with; accessed June 1, 2020.]

[204] Brian Thomas, Ph.D, 2011. "Exoplanet Discoveries Demolish Planet Formation Theories." *Creation Science Update.* [Online at ICR.org/article/exoplanet-discoveries-demolish-planet; accessed June 1, 2020.]

[205] Maugh II, T. H. "Distant planets' orbits rattle theories." *Los Angeles Times.* Posted on LATimes.com April 14, 2010 [Online at LATimes.com/news/science/la-sci-planets14-2010apr14,0,964698.story.]

[206] Maugh II, T. H. "Distant planets' orbits rattle theories." *Los Angeles Times.* April 14, 2010. [Online at LATimes.com/archives/la-xpm-2010-apr-14-la-sci-planets14-2010apr14-story.html; accessed July 26, 2020.]

that the Nebular Hypothesis doesn't work in our own solar system either. Mike Brown, an astronomer from Caltech says it this way:

> Before we ever discovered any *[planets outside the solar system]* we thought we understood the formation of planetary systems pretty deeply. It was a really beautiful theory. And, clearly, thoroughly wrong.[207]

And this isn't true only for exoplanets either; planets in our own solar system defy the predictions of the Nebular Hypothesis. Neptune's moon Triton exhibits a retrograde orbit—it orbits the "wrong" way as compared to Neptune—as do about half of the comets we've discovered so far, and Venus has a retrograde axial rotation—it spins the "wrong" way on its own axis. So on Venus, the sun (if you could see it through all the cloud cover) would rise in the west and set in the east.[208] The Nebular Hypothesis can't explain this either.

The Nebular Hypothesis, if it were true, would result in virtually identical compositions for all the planets in our solar system, since they all came from the same dust cloud. But each planet has a unique composition.[209] Now how could that happen, if all the planets coalesced from the same dust cloud?

Much more could be said about the problems with the Nebular Hypothesis, but the above is enough to get the point across. As usual, if you want to get into more detail, peruse the publications listed in the footnotes referenced above.

Far from an accidental, undirected, random clumping of dust and gas, the solar system exhibits incredible precision of construction. Such precision is no accident, as creationists have been saying all along. But evolutionists, unwilling to acknowledge the existence of God, still marvel at how "lucky" we are for everything to have fallen together so exquisitely:

> You might also think that these disparate bodies are scattered across the solar system without rhyme or reason. But move

---

[207] Krulwich, R. "Our Very Normal Solar System Isn't Normal Anymore." *National Public Radio.* Posted on NPR.org May 7, 2013. [Online at NPR.org/sections/krulwich/2013/05/06/181613582/our-very-normal-solar-system-isn-t-normal-anymore; accessed June 28, 2020.] Emphasis added.

[208] Brian Thomas, Ph.D, "Planet Formation Theory Collides with Backward-Orbiting Planets." Online at ICR.org/article/planet-formation-theory-collides-with; accessed June 2, 2020.

[209] Brian Thomas, Ph.D, "Planet Formation Theory Collides with Backward-Orbiting Planets." Online at ICR.org/article/planet-formation-theory-collides-with; accessed June 2, 2020.

any piece of the solar system today, or try to add anything more, and the whole construction would be thrown fatally out of kilter. So **how exactly did this delicate architecture come to be?**[210]

More on this in "Chapter 5: A Young Earth." Stay tuned.

The Sun

Our sun, around which everything in our solar system orbits, is basically a hydrogen bomb that is constantly exploding. Every second, it gives off a thousand times more energy than a million large cities would consume in an entire year.

Image courtesy NASA/Jet Propulsion Laboratory

Our sun, photographed in ultraviolet light, taken by NASA's Solar Dynamics Observatory satellite.

The sun is by far the largest object in our solar system, comprising 99.86% of the mass of the system. It is 109 times the diameter of the Earth, and more than 1¼ million Earths could fit inside the sun. The sun is composed mostly of hydrogen and helium, which were identified by their characteristic absorption lines as shown in a spectrometer (discussed in the "Redshift and Blueshift" section above).

The equator of the sun rotates faster than the latitudes farther north and south, so the surface is constantly being stretched and wrapped around. This, interacting with magnetic field lines, is thought to be the reason for the 11-year sunspot cycle. The temperature at the sun's core is estimated to be 27,000,000°F—that's 27 *million* degrees! At the surface, it's "only" about 11,000°F.

The "Young Faint Sun" Paradox

In any nuclear fusion process, as is happening in the middle of our sun, there has to be a source of fuel. This fuel is hydrogen, and as it is fused into

---

[210] Webb, R. 2009. "Unknown solar system 1: How was the solar system built?" *New Scientist*. 2693: p. 31.

helium, energy is released, and the sun shines. However, as the process continues, the hydrogen gradually gets used, and at the same time, the amount of helium increases. And as this ratio changes, the star gets brighter.

This seems counterintuitive—for the sun to get *brighter* as it uses its fuel—but that actually is the case. Here's how it works:[211]

1. In the process of nuclear fusion, like what our sun uses to generate heat and light, four hydrogen nuclei are fused into one helium nucleus, generating energy.[212]
2. One helium nucleus is smaller—it occupies less space—than the four hydrogen nuclei it was made from, so helium nuclei can be packed together more tightly.
3. The tighter packing results in more mass per unit of volume—a greater density.
4. The greater density causes a stronger gravitational field. This allows the radius of the star to shrink more rapidly than its mass, resulting in greater compression at its center.
5. The greater compression causes higher temperatures in the star's interior.
6. Higher temperatures make it easier for further fusion to happen more readily, generating more heat.

This process continues, with the star growing ever brighter until the nuclear fuel runs out. After that, what happens depends on the mass of the star.

In the timeframe of creation, the amount of change would be negligible, unnoticeable. But in the deep time of evolutionism, the sun would have been 30% dimmer 4.3 billion years ago, when life is supposed to have begun. All the water on Earth would have been frozen solid, so how could life have evolved without liquid water?[213] But evolutionists claim that the Earth of 4.3 billion years ago must of necessity have been comparable to today's temperature.

---

[211] Jonathan Sarfati, "Our Steady Sun: A Problem for Billions of Years", *Creation* 26(3):52–53, June 2004. [Online at Creation.com/our-steady-sun-a-problem-for-billions-of-years; accessed March 27, 2021.]

[212] See en.wikipedia.org/wiki/Nuclear_fusion#Nuclear_fusion_in_stars for details; accessed March 27, 2021.

[213] Faulkner, D. 1998. "The Young Faint Sun Paradox and the Age of the Solar System." *Acts & Facts.* 27 (6); online at ICR.org/article/young-faint-sun-paradox-age-solar-system/. See also Goldblatt, C. and K. J. Zahnle. 2011. "Faint Young Sun Paradox Remains." *Nature.* 474 (7349): 744-747.

What is needed here is another rescue device: Let's just say that there was 1000 times more carbon dioxide in the air back then than there is now; that will create enough of a greenhouse effect for life to have developed. Is there any evidence for that much atmospheric carbon dioxide in the past? Well, no, but it *must* have been there; evolution depends on it!

Is it possible that the sun is *not* 4.5 billion years old? Could be.

Mercury

Going out from the sun, Mercury is the first planet in our solar system. It is also the smallest of the eight planets: only 38% of the Earth's diameter, which means it is about 5% of the Earth's volume. It takes 87.97 Earth-days to orbit the Sun, and its distance from the Sun is 36 million miles.

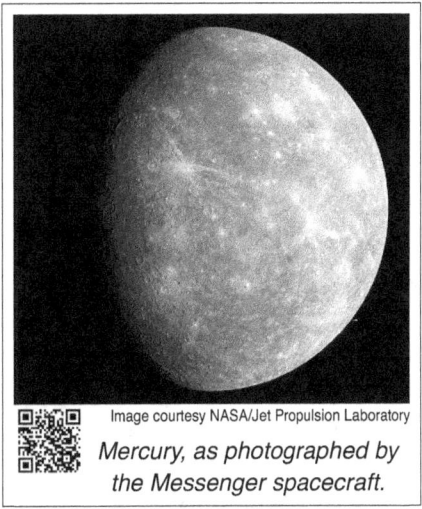

Image courtesy NASA/Jet Propulsion Laboratory

Mercury, as photographed by the Messenger spacecraft.

Many distances within the solar system use a unit of measure called a **Astronomical Unit**, or AU, and it is defined as the Earth's distance from the sun, which is about 93 million miles. So if Mercury's distance from the sun is 36 million miles, its distance from the sun could also be expressed as $^{36}/_{93}$ AU, or 0.39 AU. Another example: Mars is 1.52 AU from the sun, which means it's 1.52 times as far from the sun as Earth is.

Because Mercury is so close to the sun, and does not have any atmosphere to distribute heat, the daylight side of Mercury gets up to 800°F. And on the night side, it can get down to −280°F! But for many years, photographs were poor: astronomers' knowledge of Mercury was limited to Earth-based telescopes, whose results were less than spectacular. But then NASA launched

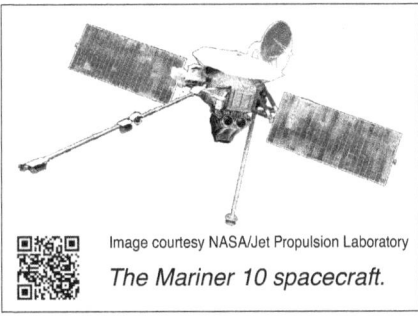

Image courtesy NASA/Jet Propulsion Laboratory

The Mariner 10 spacecraft.

the Mariner 10[214] spacecraft for three flybys of Mercury in 1974, and then the Messenger probe[215] (launched in 2008), which orbited Mercury for *four years*, studying it up close. What was discovered greatly surprised many astronomers.

The Unexpected Magnetic Field

One thing that was discovered by Mariner 10 and confirmed by Messenger is that Mercury has a magnetic field. This was disturbing to evolutionists because if the solar system is really 4.5 billion years old, Mercury's magnetic field should have dissipated eons ago, especially since Mercury is so small.

Ten years later, D. Russell Humphreys, a physicist who takes the Genesis account of creation seriously, developed a model to predict planetary field strengths. Two of the factors he used in his model were that the planet started out as water (as in Genesis 1:2), and that the age of the solar system is roughly 6,000 years (based on adding contiguous durations from Genesis). Humphreys' predictions lined up very well with Mariner 10's magnetic field strength readings.[216]

Image courtesy NASA/Jet Propulsion Laboratory
Artist's conception of the Messenger spacecraft orbiting Mercury.

Then Humphreys also predicted, based on the same model, how much Mercury's field strength should decay over time, and then when Messenger

---

[214] See en.wikipedia.org/wiki/Mariner_10 for details.

[215] Technically, the name of the probe is "MESSENGER" (with all caps), not "Messenger," because MESSENGER is an acronym for "MErcury Surface, Space ENvironment, GEochemistry, and Ranging." See en.wikipedia.org/wiki/MESSENGER for details.

[216] Humphreys, D. R. 1984. "The Creation of Planetary Magnetic Fields." *Creation Research Society Quarterly.* 21 (3): 140–149. [Online at CreationResearch.org/crsq/articles/21/21_3/21_3.html.]

got there in 2011, his predictions were again borne out.[217] Here's how he described the basic principle of his model:

> Electrical resistance in a planet's core will decrease the electrical current causing the magnetic field, just as friction slows down a flywheel.[218]

Even if evolutionists could explain how Mercury's magnetic field still existed after billions and billions of years, they would have even a harder time explaining a detectable decrease within a single person's lifetime—that would be even stronger evidence for a recent creation. The uniformitarianist expectations of deep time would not predict any noticeable decrease in the mere 33 years between the two spacecrafts' observations.

Humphreys predicted a 4% decrease in Mercury's magnetic field strength in the 33 years between the Mariner 10 and the Messenger crafts, but he was also aware of the possibility of even a greater drop than 4%. He wrote:

> My predicted 4% decrease in only 33 years would be very hard for evolutionary theories of planetary magnetic fields to explain, but a greater decrease would be even harder on the theories.[219]

So when Messenger's measurements of Mercury's magnetic field came, evolutionists were dismayed that the drop was even larger than Humphreys predicted. The writers of the *Science* magazine acknowledged that Mercury's field strength was:

> . . .≈27% lower in magnitude than the centered-dipole estimate implied by the polar Mariner 10 flyby.[220]

Naturally, this is no small source of consternation for the secularists, to whom deep time is extremely important.

---

[217] Thomas, B., "Mercury's Fading Magnetic Field Fits Creation Model." *Creation Science Update.* [Online at ICR.org/article/6424.]

[218] Humphreys, D. R. 2008. "Mercury's magnetic field is young!" *Journal of Creation.* 22 (3): 8–9. [Online at Creation.com/images/pdfs/tj/j22_3/j22_3_8-9.pdf.]

[219] Ibid.

[220] Anderson, B. J. et al. 2011. "The Global Magnetic Field of Mercury from MESSENGER Orbital Observations." *Science.* 333 (6051): 1859–1862.

And, by the way, Humphreys' field-strength model, based on a 6,000-year planetary age, also correctly predicted the field strength of Uranus and Neptune.[221]

Outgassing of Volatile Compounds

Mercury's magnetic field was certainly not the only unexpected thing Messenger observed. The data returned from the spacecraft showed that Mercury is geologically active, which is very unexpected, assuming deep time. It "should" have gone geologically dead long ago.

But it's not. On the sunward side of Mercury, it's hot enough to melt lead. And like so many other hard-surfaced objects in the solar system, many craters pockmark its surface. Messenger sent back photographs of unprecedented detail, and many of these craters were surrounded by very bright, highly reflective material, called "high reflectance haloes."

Astronomers say these craters:

> . . .appear fresh and lack superposed impact craters, implying that they are relatively young.[222]

Because fresh-looking craters imply young age, which is unacceptable, evolutionists developed a rescue device to bolster the long-age worldview. The most likely explanation for the fresh-looking craters, they say, is outgassing of volatile materials from beneath the surface of the planet. But like so many such rescue devices, this does not solve the problem; it simply attempts to avoid it by pushing it further away.

In this case, let's suppose that it is indeed the outgassing of volatile material that keeps the craters looking fresh. But if that is true, where did Mercury get the volatile materials? That close to the sun, any volatile material would have burned away during the formation of the planet (assuming the Nebular Hypothesis). But even if the volatile material somehow remained inside the planet during its formation, why is there any left to still be outgassing nowadays? These materials are volatile, after all. Mercury would have exhausted its supply of such material billions of years ago, assuming deep time. And how does the planet still have the internal energy to be expelling gases after all this time?

---

[221] Humphreys, D. R. 1990. "Beyond Neptune: Voyager II Supports Creation." *Acts & Facts*. 19 (5). [Online at ICR.org/article/beyond-neptune-voyager-ii-supports-creation.]

[222] Blewett, D. T. et al. 2011. "Hollows on Mercury: MESSENGER Evidence for Geologically Recent Volatile-Related Activity." *Science*. 333 (6051): 1856–1859.

The photo at right shows indentations—"hollows"—in Mercury's surface. Astronomers are at a loss to explain them, because they insist on the solar system being billions of years old. Mercury has no atmosphere, so there is no wind and no rain to create these depressions. As close to the sun as Mercury is, volatile substances like potassium and sulfur should be gone eons ago, assuming the Nebular Hypothesis, which is how secular astronomers assume the solar system "formed."

Image courtesy NASA/Johns Hopkins University Applied Physics Laboratory/Carnegie Institution of Washington

Hollows, containing volatile compounds, inside Mercury's Raditlati impact basin.

Here is example of the puzzlement resulting from presuming deep time:

> MESSENGER has indeed proven Mercury unexpectedly rich in sulfur. That in itself is a surprise that's forcing scientists to rethink how Mercury was formed. The prevailing models suggest that either (1) very early in Solar System history, during the final sweep-up of the large planetesimals that formed the planets, a colossal impact tore off much of Mercury's rocky outer layering; or (2) a hot phase of the early Sun heated up the surface enough to scorch off the outer layers. **In either case, the elements with a low boiling point—** volatiles like sulfur and potassium—**would have been driven off.**
>
> **But they're still there.**
>
> **The old models just don't fit with the new data, so we'll have to look at other hypotheses.**[223]

Mercury's observed geological activity contradicts deep-time predictions, much to the puzzlement of evolutionary researchers. They write:

> Mercury is a small rocky-metal world whose internal geological activity was generally thought to have ended long ago. The presence of potentially recent surface modification im-

---

[223] "Strange Hollows Discovered on Mercury" Oct 24, 2011. [Online at https://science.nasa.gov/science-news/science-at-nasa/2011/24oct_sleepyhollows;; accessed May 29, 2022.] Emphasis added.

plies that Mercury's nonimpact **geological evolution may still be ongoing.**[224]

And even still, they desperately hang onto the evolutionary view of the solar system. Amazing.

Venus

Venus is about the same size as Earth (95% of Earth's radius), and is sometimes called Earth's "sister planet." Its mass is about 81½% that of Earth, and it is about 0.72 AU from the Sun. Venus' composition is similar to that of Earth, and of the orbits of all the planets, Venus' orbit is the physically closest to Earth's.

But despite the similarities, there are far more differences between Venus and Earth. For many years, we couldn't see the planet Venus at all; we could only see the top of the cloud cover that shrouds the entire planet. The high reflectivity of the cloud cover is what makes Venus so bright in the night sky. For all their beauty, however, those Venusian clouds are lethal. Composed mostly of carbon dioxide, there are also significant amounts of sulfuric acid and sulfur dioxide—both highly toxic.

Image courtesy NASA/Jet Propulsion Laboratory

False-color radar image of the surface of Venus.

Venus' atmosphere is also the densest of four **terrestrial planets** (Mercury, Venus, Earth, and Mars): at the surface of Venus, the atmospheric pressure is 92 times more than that of Earth at sea level! The carbon dioxide and other gases in this poisonous atmosphere cause an extreme greenhouse effect, and the surface of Venus gets up to 900°F—even hotter than Mercury! Combine that with the supersonic winds,[225] and you have the makings of a very uninviting vacation spot.

---

[224] Blewett, D. T. et al. 2011. "Hollows on Mercury: MESSENGER Evidence for Geologically Recent Volatile-Related Activity." *Science*. 333 (6051): 1856–1859. Emphasis added.

[225] David Coppedge "Venus vs. Uniformitarianism" July 1, 2007 [Online at ICR.org/article/venus-vs-uniformitarianism; accessed June 4, 2020.]

CHAPTER 4: A YOUNG SOLAR SYSTEM

Many exploratory spacecraft were sent to Venus by the United States and other countries in the past few decades,[226] but as of this writing, the most recent US mission was the Magellan. This was NASA's first interplanetary mission to be launched from the Space Shuttle and it orbited Venus for four years so, needless to say, enormous quantities of observations were gathered.

Image courtesy NASA/Jet Propulsion Laboratory

The deployment of the Magellan spacecraft from the Space Shuttle Atlantis.

Retrograde Axial Rotation

Perhaps the most startling finding was its retrograde axial rotation, as mentioned in the "Orbital Considerations" section above. The fact that Venus rotates on its axis the "wrong" way is a deathblow—another one—to the Nebular Hypothesis of solar system formation. How in the world could Venus have gotten spinning the wrong way? In the uniformitarian, evolutionary, Nebular Hypothesizing view of the solar system, it couldn't.

Surface Features

According to the readings from a variety of instruments aboard Magellan, the crustal rocks all appear to be about the same age. Not only that, they appear to be less than 20% of Venus' hypothetical age (based on deep time, of course).

Not only that, but the craters, mountains, and volcanoes on the surface of Venus show approximately the same amount of erosion, indicating that they've been there about the same amount of time, as if all of them were formed almost simultaneously.[227]

When confronted with these observations, evolutionary cosmologists (with their deep-time assumptions) have no choice but to admit that 90% of

---

[226] en.wikipedia.org/wiki/List_of_missions_to_Venus; accessed June 4, 2020.
[227] Coppedge, D. 2007. "Venus vs. Uniformitarianism." Acts & Facts. 36 (7). [Online at ICR.org/article/venus-vs-uniformitarianism.]

157

Venus' history is absent. In the 4th edition to *The New Solar System*, R. Stephen Saunders states:

> The geologic rule of uniformitarianism—'the present is the key to the past'—does not apply to Venus...

And as we'll see in a future chapter, it doesn't apply to Earth either.

Earth

As mentioned earlier, there is *so* much Earth-specific evidence that supports creationism substantially better than it does evolutionism, that there is way too much for it to fit into one chapter. Therefore, various aspects of Earth will be covered in separate chapters after the Solar System chapter you're reading now. But for the moment, enjoy the photo at the right.

Image courtesy NASA/Jet Propulsion Laboratory
The "Blue Marble" photo of earth, taken from the command module during the Apollo 17 lunar mission.

Mars

Mars is the next planet out from Earth, and it is the last of the terrestrial planets—those with rocky, solid ground. Its radius is 53% that of Earth, its mass less than 11% of Earth, its distance from the Sun is about 1.52 AU, and it orbits once every 687 Earth-days.

Mars appears in the night sky as a distinctly red dot of light. Its nickname—the "Red Planet"—is very apropos; the reddish color comes from large quantities of iron oxide. Basically, rust. Although significantly thinner than Earth's, Mars has an atmosphere, so there is weather: seasons, clouds, wind, dust storms, and sometimes even fog and frost. But the atmosphere is 96% carbon dioxide;[228] you wouldn't want to go outside without a breather unit.

Image courtesy ESA—European Space Agency & Max-Planck Institute for Solar System Research for OSIRIS Team (Creative Commons ShareAlike 3.0 Unported License)

Mars, the "Red Planet:" rusty red with dark blotches, and icecaps visible at both poles.

Mars rotates on its axis at a rotational speed very close to that of Earth: a day on Mars is just a tad less than 24 hours and 40 minutes. A human's sleep-wake cycle would be hardly disrupted at all from the Earth schedule.

Liquid Water on Mars

Numerous spacecraft have been sent to Mars,[229] and much has been discovered. For example, even though there is very little liquid water there now, Mars once had rivers; we know this because we've seen dry riverbeds and deltas. This is quite surprising, considering the thinness of the Martian atmosphere.

So what happened there? Was Mars' atmosphere denser or thicker in the past? It would need to have been for liquid water to stay around very long and not immediately evaporate. Or another possibility is a catastrophic release

---

[228] Thomas, Brian. "Mars Atmosphere Could Be Young." [Online at ICR.org/article/mars-atmosphere-could-be-young; accessed June 4, 2020.]

[229] en.wikipedia.org/wiki/List_of_missions_to_Mars; accessed June 4, 2020.

of water that washed over the entire surface of the planet briefly before evaporating into space.

It's intriguing to note that there are evolutionists who are willing to entertain the possibility of a planet-wide flood on Mars, which is incapable of supporting liquid water, while simultaneously rejecting as impossible the idea of the worldwide flood on Earth as described in Genesis. And Earth's surface is 71% water! Curious, isn't it?

The "Unbalanced" Atmosphere

As mentioned above, the atmosphere of Mars is predominantly carbon dioxide. Chemically speaking, carbon dioxide is CO_2, which indicates that in each molecule of carbon dioxide, there is one carbon atom and two oxygen atoms.

We'll need to take a diversion here so we can understand why Mars' atmosphere indicates a young age, and not the long ages evolutionists are so fond of. Carbon, like most elements, has different **isotopes**. An isotope is one of two or more forms of an atom, in which the number of *protons* in the nucleus is the same for all of them (they have the same **atomic number**), but the number of *neutrons* in the nucleus is different (they have a different **atomic weight**). Chemically, they operate identically, but of course, if there is an extra neutron or two in the nucleus of an atom, it's a tiny bit heavier.

Carbon's most common form is Carbon-12, whose chemical symbol is ^{12}C; its nucleus is composed of 6 protons and 6 neutrons. Then there is Carbon-13, or ^{13}C, with 6 protons and 7 neutrons. Both of these are stable, in the nuclear sense; they are not radioactive, so they do not decay into other substances.

But since they operate identically in the chemical sense, both ^{12}C and ^{13}C can bond with two oxygens to make carbon dioxide, or CO_2. Since ^{13}C is a tiny bit heavier than ^{12}C, CO_2 made with ^{13}C is a tiny bit heavier than CO_2 made with ^{12}C.

And here's where the above technicalities become significant. Normally $^{12}CO_2$ constitutes about 98.9% of all CO_2, with the other 1.1% being $^{13}CO_2$. But since $^{12}CO_2$ is little bit lighter than $^{13}CO_2$, the $^{12}CO_2$ would tend to escape Mars' gravity more easily than its heavier version. Which means that the ratio of the two would change over time, with $^{13}CO_2$ becoming more predominant all the time, since, being heavier, it escapes the atmosphere more slowly.

If Mars was indeed billions and billions of years old, there would have been ample time for much of the $^{12}CO_2$ to escape into space, and the 13-to-12 ratio would be much bigger than the quantity actually measured.[230]

Well, we can't have that. Observations are clearly contradicting the cherished deep-time assumptions, which must be preserved at all costs. So a rescue device was invented, and put into the same press release:

> The results imply Mars has replenished its atmospheric carbon dioxide relatively recently, and the carbon dioxide has reacted with liquid water present on the surface.

Now keep in mind that there is no evidence for this "replenishment," no observations of it, and no experimental result to indicate it. It was entirely a fabrication whose sole purpose was to rescue the teetering deep-time assumption. In the words of evolutionist Sir Arthur Keith, evolutionists believe in evolution because special creation is "unthinkable."

Also note that even if it were true that Mars' atmospheric CO_2 was constantly being replenished via "active and ongoing" volcanic "mantle degassing,"[231] it still doesn't solve the deep-time problem; it merely pushes it one more step farther away, as so many other rescue devices do.

In this case, the rescue device merely pushes the problem from the atmosphere into the planet itself. Specifically, if Mars really is billions and billions of years old, and the mantle is constantly outgassing CO_2 to the atmosphere, why hasn't the mantle exhausted its supply? That should have happened *eons* ago. So the mantle outgassing is not a solution; it's just another layer of obfuscation that helps disguise the shakiness of the long-age timeframe.

Jupiter

Jupiter is the largest planet in the solar system: its mass is 318 times that of Earth, its diameter about 11 times that of Earth (so 1,321 Earths could fit inside it). It orbits the Sun at 5.20 AU, and takes 11.86 years per orbit.

---

[230] "NASA Data Shed New Light About Water and Volcanoes on Mars." NASA press release, September 9, 2010. [Online at NASA.gov/mission_pages/phoenix/news/phx20100909.html.]

[231] Niles, P. B. et al. 2010. "Stable Isotope Measurements of Martian Atmospheric CO_2 at the Phoenix Landing Site." *Science.* 329 (5997): 1334–1337.

After the four terrestrial planets are two **gas giants:** Jupiter and Saturn; the first of the gas giants is Jupiter.[232] A gas giant is a planet that has no rocky surface at all, as far as we can tell. Instead, it is composed primarily of an enormously deep atmosphere, made of primarily hydrogen and helium. The outermost, visible cloud layers are mostly water and ammonia, and the core is likely to be composed of metallic hydrogen—hydrogen under such tremendous pressure that it becomes an electrical conductor.

The largest planet in the solar system, it has 2½ times the mass of all the other planets combined, which is about one-thousandth that of the Sun. For all that size and mass, it rotates remarkably fast: its day is less than 10 Earth hours. So even though Jupiter's diameter is more than 11 times that of Earth, its equator travels 45 times faster than Earth's! Its great mass, combined with its gaseous composition and its fast rotation rate make Jupiter noticeably oblate: its diameter at the equator is more than 2300 miles greater than its diameter pole to pole.

Image courtesy NASA, ESA, and A. Simon (Goddard Space Flight Center)

Jupiter, showing the turbulent cloud cover. In the lower right, the Great Red Spot—a hurricane-like storm that has lasted for hundreds of years, and in the lower left, the shadow of one of the four Galilean moons.

Jupiter is also a planet with many moons: as of this writing, we know of 79,[233] and we'll probably discover more with each new probe sent to investigate. Jupiter also has rings, but they are faint and dusty, as compared with Saturn's large, bright, icy rings.

Jupiter is indeed a puzzle for evolutionists. Space probes have observed argon, krypton, and xenon in Jupiter's atmosphere—*three times* as much as evolutionists predicted.[234] But at Jupiter's distance from the sun, those gases could have could not have condensed at such high concentrations. Well, maybe Jupiter "formed" (all by itself, of course) farther away from the sun,

---

[232] See en.wikipedia.org/wiki/Gas_giant for details.

[233] "Planets: Jupiter. Solar System Exploration." NASA. Posted on SolarSystem.nasa.gov; accessed June 5, 2020.

[234] Russell Grigg, "The Planets are Young, Part 3: Jupiter" [Online at Creation.com/planets-jupiter; accessed May 29, 2022.]

where those gases could have condensed, and later on, the planet moved to where it is now. But there isn't enough mass that far away from the sun to build Jupiter. Plus, what could have moved such a huge mass closer to the sun and put it into such a precise orbit as we see today? Plus, again, even if those gases *did* collect in Jupiter as it "formed" farther out, they would have evaporated by now at Jupiter's current distance, if deep time was a thing. Oh, dear—evolution keeps running into reality, and having strong disagreements. . .

Retrograde Moons

Jupiter's moons are fascinating in their arrangement. Based on their orbits, they naturally fall into eight groups. The Galilean moons—those discovered in 1610 by Galileo Galilei—are the largest, which is why those were the first discovered; they were the only ones that could be seen with Galileo's crude telescope. These Galilean moons, plus four smaller moons that orbit inside them, all have prograde orbits—they orbit the same way Jupiter spins (counterclockwise, as viewed from above the north pole). And, their orbits are in the plane of Jupiter's equator.

The next several beyond Callisto (the outermost of the Galilean moons) are also prograde, but their orbits are substantially tilted, relative to Jupiter's equator. The Nebular Hypothesis fails in explaining why the orbits of this set of moons could become so tilted, if they all coalesced out of the same cloud of dust and gas.

Of the remaining moons, most are retrograde; they revolve *backwards* around Jupiter. Again, that shouldn't happen if the Nebular Hypothesis was a real thing.

Jupiter's Moon Io

The innermost Galilean moon is Io, and it is full of surprises. Every spacecraft that went by showed us more, and raised more and more prickly problems for the uniformitarian approach of evolutionism.

Specifically, Io has more than 400 volcanoes,[235] and some of them are real doozies, sending clouds of sulfur and sulfur dioxide 300 *miles* above the surface.

Of course, every time a volcano erupts, it radiates heat into space, and the heat remaining in the moon is that much less. With 400 active volcanoes, it is losing heat pretty fast, so how long could this volcanic activity last? And Io is pretty small; its diameter is 28.6% that of Earth, so its volume is less than 2.5% that of Earth.

Jupiter's volcanically active moon Io.

An astronomy video summarizes the problem:

> If Io is young, it could still be cooling off from its initial formation. But if it's really billions of years old, that energy would have dissipated long ago.[236]

Data received from spacecraft telemetry confirms that Io has a huge ocean of magma under the surface.[237] Measurements of Io's magnetic field were puzzling, given the initial assumption that Io was a solid ball; the magma ocean with its fluidic flow explains the magnetic fields. But the magma ocean also raises another question: where is all the heat coming from?

---

[235] en.wikipedia.org/wiki/Io_(moon); accessed June 5, 2020.

[236] *What You Aren't Being Told about Astronomy,* Volume 1: Our Created Solar System. 2009. DVD. Directed by Spike Psarris. Creation Astronomy Media.

[237] "Galileo Data Reveal Magma Ocean Under Jupiter Moon." Jet Propulsion Laboratory News & Features. [Online at JPL.NASA.gov/news/news.cfm?release=2011-141.]

As mentioned above, hundreds of active volcanoes radiate away a lot of heat. And again, we come up against the age-old problem (pun intended): if Io were billions and billions of years old, the heat would long ago have dissipated, and it would no longer be volcanically active.

And not only would the *heat* have dissipated long ago if deep time were real, but Io doesn't have enough *mass* to keep erupting at that rate. One astronomer calculated that if Io had been erupting—at only *one tenth* its current rate—throughout the alleged 4.5-billion-year age of the solar system, it would have erupted its *entire mass 40 times over* by now!²³⁸

Image courtesy NASA/Jet Propulsion Laboratory

Image of active volcano on Io. This one is throwing volcanic debris "only" 86 miles above the surface.

Authors Pearl and Sinton published a compilation of research on Jupiter's moons, entitled *Satellites of Jupiter*. In it, they wrote:

> Complete elucidation of the heat source **remains a significant outstanding problem** resulting from the discovery of active volcanism on Io.²³⁹

Note that the presumption deep time is absolutely non-negotiable. All these problems would simply evaporate if the blind allegiance to deep time would be discarded. But that is not considered an option, because it would make it even more obvious that a Creator was required in creation.

Io's lava is hotter than Earth's typical magma—3100°F compared to 1700°F, so it is less viscous, or more runny, than Earth-based magma. Not only that, but Io's lava is **ultramafic**; that is, rich in dark-colored minerals

---

²³⁸ McEwan, A.S., "Active Volcanism on Io," *Science* 297 (5590):2220–21, September 27, 2002.

²³⁹ Pearl, J. C., and W. M. Sinton. 1982. "Hot Spots of Io." *Satellites of Jupiter*. D. Morrison, ed. Tucson, AZ: Arizona University Press, 724–755. Cited in Spencer, W. 2003. Tidal Dissipation and the Age of Io. In *Proceedings of the Fifth International Conference on Creationism*. R. L. Ivey, ed. Pittsburgh, PA: Creation Science Fellowship, Inc., 585–595.

based on iron and magnesium.[240] That is strange in itself because, assuming the Nebular Hypothesis, such heavy elements would have sunk to the core during the moon's formation, leaving a hard crust thick enough that the heavy elements and minerals could not get through and be present in volcanic eruptions. But we've seen that ultramafic minerals *do* get through, so its seems like the crust must either be thin, or the interior soft.

But if the interior of Io were soft, or the crust were thin, Io wouldn't be able to support the mountains that we've seen—mountains that rival Earth's Himalayas. Also, it is clear that Io's volcanoes are not a new phenomenon, because Io *has no impact craters,* like so many other rocky planets and moons do—every square foot of Io has been resurfaced by the material coming out of its volcanoes. There are just all sorts of problems and contradictions that arise if we presuppose an evolutionary long-age timeframe.

Io continuously emits 10^{14} watts of power, equal to about 33 billion electric stoves with all the burners on full blast. Considering how much heat it is constantly emitting, the energy has to come from somewhere; the magma ocean will only last so long.

What is usually suggested as the source of Io's long-lasting heat output is **tidal flexing**. Tidal flexing is caused by mechanical deformation, and the deformation creates heat from internal friction. It is a real effect, and you can test it yourself. Take an old wire hanger and stretch it out as straight as you can, and then repeatedly bend it back and forth at the same spot in the wire. The spot experiencing the bending will get hot. Could the same thing be happening with a whole moon?

In the two illustrations to the right (not to scale), note that in both cases, Io is being attracted by multiple entities: on the one hand, it is being attracted by other moons (here represented by Europa), and on the other hand, by Jupiter.

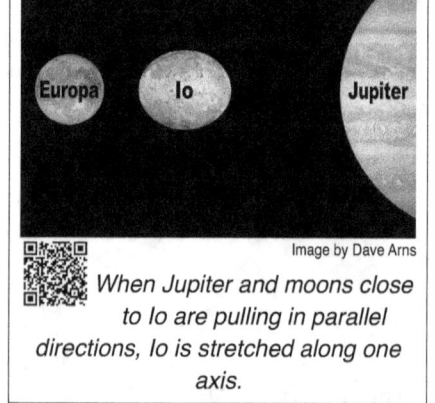

Image by Dave Arns

When Jupiter and moons close to Io are pulling in parallel directions, Io is stretched along one axis.

When the attractions are in parallel directions (as in the first panel), Io is stretched horizontally, but when the attractions are in perpendicular

---

[240] David Coppedge, "Spewing Hot Rocks on Old Ideas" October 1, 2007. [ICR.org/article/spewing-hot-rocks-old-ideas; accessed June 6, 2020.]

directions (as in the second panel), Io is stretched vertically. Repeatedly being stretched this way and that, as Io and the other moons orbit Jupiter, indeed generates heat in the crust of Io.

It's a good idea, and it's a real effect, but it's not enough. The German planetary scientist Tilman Spohn describes it this way. He:

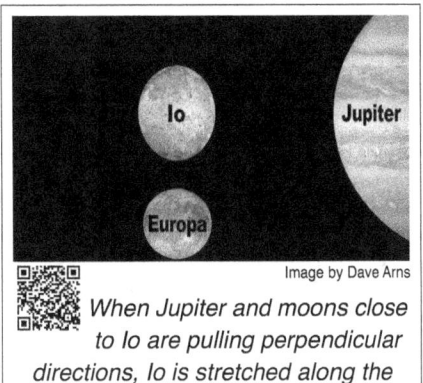

When Jupiter and moons close to Io are pulling perpendicular directions, Io is stretched along the other axis.

> . . .acknowledges that there is a gap of about one order of magnitude between the observed heat flow from infrared measurements and the heat flow theoretically determined from tidal [friction] dissipation models.[241]

One "order of magnitude" means one "power of ten." In other words, the amount of heat emitted by Io is *ten times more* than the heat generated by tidal flexing.

And there is another problem, too: if tidal flexing were the actual source of the heat (ignoring the fact that its contribution is 90% too small), the volcanoes would be in certain regions, as determined by the where the tidal stresses would occur within Io. In a nutshell, the volcanoes are in the wrong places to use tidal flexing as the source of heat. A 2012 press release from the Jet Propulsion Laboratory states that the locations of the volcanoes "disposes of the generally accepted model of internal heating."[242] A NASA press release in 2013 stated that the volcanoes are "significantly displaced" from where tidal-flexing heat would be generated, where "significantly" means they are *30°–60° away* from where they should be, if tidal flexing were actually the source of the heat Io is dissipating so prodigiously.[243]

---

[241] Spencer, W. 2003. "Tidal Dissipation and the Age of Io." In *Proceedings of the Fifth International Conference on Creationism*. R. L. Ivey, ed. Pittsburgh, PA: Creation Science Fellowship, Inc., 585–595.

[242] Russell Grigg, "The Planets are Young, Part 3: Jupiter" [Online at Creation.com/planets-jupiter; accessed May 29, 2022.]

[243] Neal-Jones, N., and Steigerwald, B., "Scientists to Io, Your Volcanoes are in the Wrong Place," NASA.gov, 4 April, 2013. [Online at NASA.gov/topics/solarsystem/features/io-volcanoes-displaced.html; accessed May 29, 2022.]

So deep time is still in trouble: if Io is as old as evolutionists claim, it would have run out of heat ages ago, in spite of the contribution made by tidal flexing. *But it hasn't.* Why not? Is it possible it's not as old as evolutionists claim?

Jupiter's Moon Europa

Europa is the next Galilean moon out from Io. Europa is the smoothest of any solid body in the solar system, and it has very few craters, giving it a "youthful" appearance. Why is it called "youthful?" Because of its smoothness.

With the Nebular Hypothesis and its supposed, and requisite, "billions and billions of years," the expectation is that impact craters would continually build up on any solid object, resulting in a heavily cratered surface. Then, by counting the craters, we can estimate the age of the object. But this approach makes several fundamental assumptions:

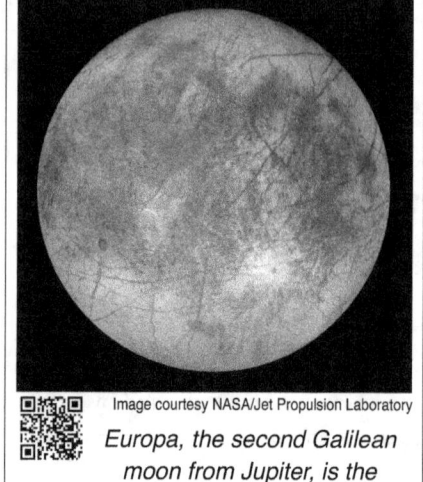

Image courtesy NASA/Jet Propulsion Laboratory

Europa, the second Galilean moon from Jupiter, is the smoothest objects in the solar system.

- That the rate of cratering has remained constant since the formation of the moon, and
- That impact craters are formed only by impacts from free-floating objects in space (impactors create one crater per hit), and
- That the age of the solar system actually *is* 4.5 billion years, or thereabouts.

If all three of the above speculations were valid, then crater-counting might be a valid analytical tool, if other relevant factors (erosion, geologic activity, gravity, atmosphere, etc.) are taken into account. But since no human beings were around during the formation of Europa or any other planetary body, we must consider what we *can* observe, then perform repeatable, verifiable experiments using relevant laws of physics, and see which model—creation or evolution—predicts and explains most accurately what we actually observe.

Crater-count dating, like other dating methods, is dependent upon its lowest-level assumption. If that foundational assumption, upon which everything else rests, is shown to be flawed, everything built upon it comes crashing down. **Secondary cratering** is one of those things that can cause crater-count dating to collapse. If a big enough object impacts on the surface of a planet or moon, a lot of material is thrown upward. When that material comes back down, many of those pieces are big enough to cause craters of their own. In *Science,* one writer estimates that a single large impact on Mars could cause *ten million* secondary craters, and adds that secondary cratering could be responsible for 95% of the craters on Europa, few though it has.[244]

Planetologists have known about secondary cratering for some time, but new data has revealed that it is more prevalent than previously thought. This whole dating method was developed from counting the craters on Earth's moon, and its primary assumption was that we knew how old the moon was: 4.5 billion years. We've already seen in numerous ways that such a conjecture is very wobbly indeed.

The May 26, 2006 issue of *Science* reported that, at a conference a few months earlier, 125 planetary scientists were "deadlocked" in their attempt to see how, and even *if,* crater-counting is at all related to age. They stated that planetary ages that were estimated through this method might be "off by orders of magnitude."[245]

So all this points to the implication that, with up to 95% of its already-few craters being secondaries, Europa might be a lot younger than previously thought.

---

[244] Coppedge, David "Crisis in Crater Count Dating." [Online at ICR.org/article/crisis-crater-count-dating; accessed June 6, 2020.]

[245] As you remember from the discussion of Io above, an "order of magnitude" is a "power of ten," so 1 order of magnitude is 10^1 or 10, 2 orders of magnitude is 10^2 or 100, 3 orders of magnitude is 10^3 or 1000, and so forth. It builds up fast.

Jupiter's Moon Ganymede

The third of the Galilean moons is Ganymede; it is the biggest moon of any planet in the solar system. It is so large—it's bigger than the planet Mercury—that it could be considered a planet itself, if it were orbiting the Sun instead of Jupiter.

Ganymede is 8% larger in diameter than Mercury, but only 45% as massive,[246] so it is made up of much lighter materials. About half of Ganymede is made up of silicate rock and trace amounts of other substances, and the other half is water ice, which explains its lightness of mass.

Image courtesy NASA/Jet Propulsion Laboratory

Ganymede, Jupiter's (and the solar system's) largest moon.

The water ice is the reason that its craters, seen especially well in the lower half of this image, are white: an object impacted on Ganymede, and threw up lots of ice, which would have almost instantly melted because of the heat generated by the impact. As it fell, it re-froze, forming snow-like particles.

One of the things that planetary scientists learned when the various spacecraft were studying it was that Ganymede has a magnetic field. Like Mercury, it "shouldn't," if deep time is valid and the solar system is indeed 4.5 billion years old. In all that time, any magnetic field would have died out.[247]

---

[246] en.wikipedia.org/wiki/Ganymede_(moon); accessed June 6, 2020.

[247] Williams, M. "Jupiter's moon Ganymede." *PhysOrg*. Posted on phys.org October 16, 2015; accessed June 6, 2020.

Saturn

Saturn is perhaps the most spectacular-looking planet in the solar system, primarily because of its ring system. But in addition to its rings, Saturn has a variety of fascinating moons as well. Saturn orbits the Sun at about 9.58 AU, taking 29.46 years per orbit.

Saturn is the second-largest planet in our solar system; its average radius is about 9 times larger than Earth's. The "average radius" is used because, like Jupiter, it rotates on its own axis so fast that it distorts into an oblate spheroid—it bulges at the equator. Its mass is about 95 times that of Earth, but its density is only about 13% that of Earth.[248] In fact, it is so light, density-wise, that the whole planet would float on water (assuming you could find a big enough bathtub).

Image courtesy NASA/Jet Propulsion Laboratory

Saturn with its complex, intricate ring system, composed of trillions of small moonlets.

As of this writing, Saturn has 82 moons, 53 of which are officially named. But the most obvious difference between Saturn and the rest of the planets is its elaborate ring system.

Saturn's Rings

Saturn's rings are composed of mostly ice crystals, with some rocky material and dust. Although the rings extend about 70,000 miles from the inside edge to the outside edge, their thickness is only about 70 feet! A DVD, to the same scale as the rings, would have to be 1,000 times thinner than DVDs actually are.[249] Saturn has been known since antiquity, but it wasn't until Galileo looked at it with his telescope in 1610 that the rings were first discovered.

---

[248] en.wikipedia.org/wiki/Saturn; accessed June 6, 2020.

[249] ICR.org/article/solar-system-saturn; accessed June 7, 2020.

Image courtesy NASA/Jet Propulsion Laboratory

Diagram of Saturn's rings, along with the names of the rings, as well as some of the gaps and moons.

As you can see, the rings are named in what appears to be a random order. Rings A, B, C, and D, going from outermost inward, make sense. But then the E ring is *outside* the A ring, the F ring is between A and E, and then the G ring is between F and E. The explanation is that the rings were named in order of discovery, not in their distance from Saturn, as one might expect.

Since the rings' names are A, B, C, D, E, F, and G, does that imply that there are exactly seven rings? Not exactly. As mentioned, the rings were named in order of discovery, and discovery accelerated as people invented better and better telescopes, and as we sent more spacecraft to look close up. And what we discovered is there are *thousands* of rings, but they are close enough together that we couldn't see the individual rings until we sent interplanetary probes to investigate.

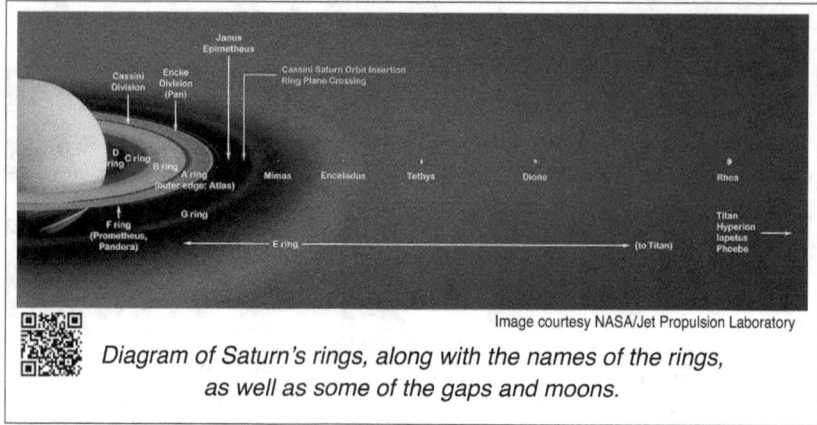

Image courtesy NASA/Jet Propulsion Laboratory

A closer view of Saturn's rings, showing more of the great number of thin rings that, from far away, look like fewer, larger rings.

How Old Are the Rings?

According to the Nebular Hypothesis, the rings should be about the age of the solar system: about 4.5 billion years. But we've already seen (repeatedly) that the Nebular Hypothesis, with its suppositions of uniformitarianism and deep time, is insufficient to explain what we observe.

CHAPTER 4: A YOUNG SOLAR SYSTEM

Data returned by the Cassini space probe have convinced planetary scientists that the rings can be no more than a few hundred million years old. For some strange reason, they look "young," even to evolutionary scientists.[250]

For example, the rings look too clean. If Saturn's rings were as old as the solar system is claimed to be, micrometeorite bombardment over all those eons would have darkened the rings to a smudgy, sooty appearance. But they're not; they are still bright and shiny pieces of ice. One of the instruments aboard the Cassini spacecraft was the Cosmic Dust Analyzer, which allowed scientists to measure the rate of micrometeorite bombardment.

After analyzing the data, one of the researchers stated:

> The flux [of micrometeorite dust], about 10 times higher than estimates from before the Cassini mission, suggest a ring age of between 150 million and 300 million years, or even younger. "Our measurement is the most direct way you can measure it," [space physicist Sascha] Kempf adds. "There's not much you can do about it. It has to be young."[251]

Evolutionary scientists, now forced to acknowledge the young age of the rings, now have to explain it, given their assumption of deep time. NASA Ames Research Center's Jeff Cuzzi, an expert on Saturn's rings, states it this way:

> Part of the reluctance for everyone to leap off this bridge into the unknown is we haven't had any kind of feasible explanation [for how the rings could form recently].[252]

Note that Cuzzi, because of his faith-based commitment to the philosophical presumption that God does not exist, arrives at the only conclusion possible. And it's true: if you've already decided that God cannot exist, there *is* no feasible explanation for the existence of the universe as we encounter it.

---

[250] Hebert, J. "Youthful Solar System Bodies Puzzle Evolutionary Scientists." Creation Science Update. [Online at ICR.org/article/youthful-solar-system-bodies-puzzle; accessed June 7, 2020.]

[251] Voosen, P. "Saturn's rings are solar system newcomers." *Science.* 358 (6370): 1513–1514. [Online at Science.ScienceMag.org/content/358/6370/1513.full

[252] Ibid.

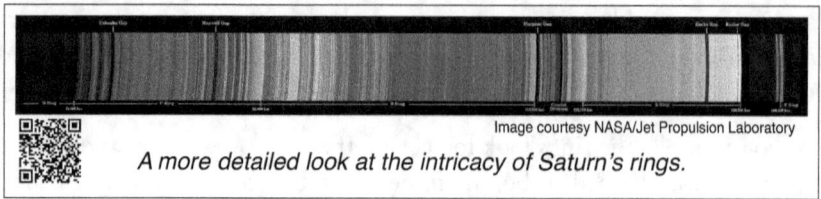

A more detailed look at the intricacy of Saturn's rings.
Image courtesy NASA/Jet Propulsion Laboratory

In 2002, Cuzzi stated, "After all this time we're still not sure about the origin of Saturn's rings. . . There's a growing awareness that Saturn's rings can't be so old." He continued, "There are two reasons to believe the rings are young: First, they are bright and shiny like something new. It's no joke." As mentioned above, after the long ages evolutionists assume, the rings should have collected so much micrometeorite dust that they would look like charcoal by now. Not only that, but after only a few million years, the tiny moons embedded within the rings should have "flung away. This is a young dynamical system."[253]

In an attempt to avoid the Biblical explanation at all costs, various non-explanations are suggested, such as:

> It was then that **some sort of catastrophe** befell the gas giant [Saturn]. **Perhaps** a stray comet or asteroid struck an icy moon, tossing its remnants into orbit. Or **maybe** the orbits of Saturn's moons **somehow** shifted, and the resulting gravitational tug-of-war pulled a moon apart.[254]

David Coppedge was the systems administrator and team leader on the Cassini project, and he is a creationist. Referring to conversations he had had with scientists from the JPL (Jet Propulsion Laboratory), he comments:

> It was clear to me that nothing would dislodge their belief in billions of years, but there was a subtext that it would be very troubling to them if the rings turned out to be young. These quotes [shown above] show that to be the case: they are flummoxed and dumbfounded by the evidence. They have no explanation, and they admit it.[255]

---

[253] "The Real Lord of the Rings." *Science @ NASA*. [Online at Science.NASA.gov/science-news/science-at-nasa/2002/12feb_rings; accessed June 7, 2020.]

[254] Voosen, P. "Saturn's rings are solar system newcomers." *Science*. 358 (6370): 1513–1514. [Online at Science.ScienceMag.org/content/358/6370/1513.full.] Emphasis added.

[255] Coppedge, David F. "It's Official: Saturn's Rings Are Young." *Creation Evolution Headlines*. [Online at CrEv.info/2018/01/official-saturns-rings-young/; accessed June 7, 2020.]

By the way, Coppedge was demoted and eventually fired from JPL in a blatant example of "viewpoint discrimination." Being a creationist, he had unacceptable opinions.[256, 257]

In 2006, another report came out of NASA, pertaining to the "Age Problem:" the rings are spreading out, and the observed rate of spread doesn't fit the long ages supposed by evolutionist planetary scientists. Not only that, but some of the moons are being chipped away so fast by collisions with ring particles that they shouldn't even exist anymore.[258] It just seems like *nothing* is working out for the beleaguered deep-time notion. . .

Image courtesy NASA/Jet Propulsion Laboratory

Cassini photo of Saturn from behind, looking toward the sun. The upper part of the rings is closer to the camera, and the lower part closer to the sun. Note that the upper hemisphere of the spaceward side of Saturn is lit by sunlight reflecting from the rings, and that some rings, more opaque, block that reflected illumination from the camera's view. Also note the bluish, diaphanous E ring (contrast enhanced for better visibility).

But wait: there's more! Saturn's gravity is constantly pulling its own rings into the planet; that is, the rings are being dismantled as we speak. NASA's Goddard Space Flight center reports that *every hour,* the rings lose enough ma-

---

[256] Klinghoffer, D. "NASA Versus David Coppedge: Most Reprehensible Case of Anti-Intelligent Design Persecution Yet?" *Evolution News and Science Today.* [Online at EvolutionNews.org/2016/12/david_coppedge/; accessed June 7, 2020.]

[257] Coppedge, David F. "Wheel of Fortune: The Liberal Culture of Shunning." Footprints of David Coppedge. [Online at DavidCoppedge.com/jpl-trial-blog.html.]

[258] Colwell, J. "Unraveling the Twists and Turns of Saturn's Rings." NASA/JPL multimedia report. Cassini-Huygens Analysis and Results from the Mission (CHARM) Presentation, April 25, 2006, Slide 33. [Online at SolarSystem.NASA.gov/missions/cassini/overview/; accessed June 7, 2020.]

terial to fill six Olympic-sized swimming pools. James O'Donoghue stated that with the rings losing volume at this pace, "the rings have less than 100 million years to live. This is relatively short, compared to Saturn's age of over 4 billion years."[259] Again, notice the absolute unwillingness to entertain the idea that deep time might be wrong. The rate of loss of ring material comes from repeatable observations, but Saturn's age from dubious speculations of a theory that already has hundreds of strikes against it.

The removal of the icy ring material is through the attraction of charged water molecules, dubbed **ring rain**, and it happens at a rate of about 1,000 to 6,000 pounds of water *per second*.[260] Planetary scientists who are unwilling to consider the possibility of a young universe feel compelled to invent more rescue devices to preserve deep time. One of them is: Maybe Saturn captured a comet. The likelihood of a comet settling into a circular, wafer-thin orbit perfectly aligned with Saturn's equator is laughably infinitesimal.

Not only that, but the comet would have to be larger than Earth's Moon to create even the B ring, let alone all the other ones.[261] And comets are pretty tiny in comparison to our Earth's Moon: their sizes have been observed as "up to 30 kilometers (19 miles)" in radius.[262] In other words, the comet Saturn supposedly captured and tore apart was more than *194,000 times larger* than the largest comet ever observed! (And remember, this is just for the B ring.) Such a comet-capture "explanation" just leads to more problems requiring further assumptions and rescue devices to prop up already-existing assumptions and rescue devices.[263]

---

[259] "NASA Research Reveals Saturn is Losing Its Rings at 'Worst-Case-Scenario' Rate." [Online at NASA.gov/press-release/goddard/2018/ring-rain.]

[260] O'Donoghue, J., et al. "Observations of the chemical and thermal response of 'ring rain' on Saturn's ionosphere." *Icarus*.

[261] "Evidence for a primordial origin of Saturn's rings." Laboratory for Atmospheric and Space Physics. University of Colorado Boulder Press Release. [Online at LASP.Colorado.edu/home/2007/10/15/origin-of-saturns-rings; accessed June 7, 2020.

[262] en.wikipedia.org/wiki/Comet#Nucleus; accessed August 8, 2020.

[263] Brian Thomas, "Saturn's Ring Rain Rates Run Fast" January 15, 2019. [Online at ICR.org/article/saturns-ring-rain-rates-run-fast; accessed June 7, 2020.]

Saturn's Moon Enceladus

Enceladus is Saturn's sixth-largest moon, about 10% the diameter (and thus only one thousandth the volume) of Saturn's largest moon, Titan. It is the most highly reflecting body in the solar system—it looks like a giant snowball.

Enceladus looks like a giant snowball because, for the most part, it *is* a giant snowball. The surface is covered by water ice, and because it is so brilliantly white, a large percentage of the incoming solar radiation is reflected back into space instead of being absorbed into the surface as heat. This means it gets *really* cold there: its coldest temperature is –400°F, and its warmest temperature is a balmy –200°F!

Image courtesy NASA/Jet Propulsion Laboratory

This photograph of Enceladus may look like a black-and-white photo, but even in full color, Enceladus really is that white.

Enceladus orbits Saturn within the E ring, but because the E ring is so rarified, there is little to no problem with collisions of Enceladus and the E ring's material. Questions that had puzzled scientists for some time were "Why is Enceladus so smooth?" and "Why is Enceladus so white?" The Cassini spacecraft answered both of these questions in 2005. One of the reasons it is so smooth is that there are more than 100 active geysers constantly shooting material into space.[264]

Image courtesy NASA/Jet Propulsion Laboratory

A photo, taken by the Cassini spacecraft, of several geysers simultaneously erupting close to the south pole of Enceladus.

The makeup of the geysers' plumes is mostly water ice, liquid water coming from a pressurized underground ocean (which powers the geysers), molecular hydrogen, methane, tiny bits of silica, and crystals of sodium chloride

---

[264] en.wikipedia.org/wiki/Enceladus; accessed June 7, 2020.

(table salt).²⁶⁵ If the speed of any particle coming from the geysers is less than Enceladus' escape velocity, it falls back to the surface as ice and snow, with a dash of salt. If the particulates are thrown faster than the escape velocity, they escape the moon's gravity altogether, and become part of Saturn's E ring—it is likely that these geysers are the source of the entire E ring.

As you might expect, from previous discussions, one very problematic question concerning the geysers of Enceladus is, "Why are they still going?" Indeed, after the imagined 4.5 billion years the solar system has supposedly been around, all the small bodies in the solar system should be long since dead: no geological activity, no magnetic field; just inert lumps. Certainly, there should no longer be any liquid water. So why is it still there?²⁶⁶

The usual go-to answer is that tidal flexing provides the heat necessary to maintain liquid water with enough pressure to power more than 100 geysers for billions of years. But, as in the case of Io's volcanic activity, tidal flexing provides barely 10% of the energy output needed to maintain Enceladus' geysers. A writer for the *Astronomical Journal* says "If regional tidal heating is occurring today, it may be responsible for some of the erupting water and heat."²⁶⁷

Image courtesy NASA/Jet Propulsion Laboratory

Some of Enceladus' geysers are enormous, as these photos show. The image on the left is a black-and-white photo, and the image on the right is a false-color version to more easily see the magnitude of the eruption.

So, tidal flexing "may" cause enough heat for "some" of Enceladus' geyser activity. Yep, about 10% at best.

About the size of Arizona, Enceladus simply does not have enough material to keep these geysers going for the presumed age of the solar system—4.5

---

²⁶⁵ Brian Thomas, "Saturn's Enceladus Looks Younger than Ever" [Online at ICR.org/article/saturns-enceladus-looks-younger-than; accessed June 7, 2020.]

²⁶⁶ Thomas, B. "Heat of Saturn Moon Far Surpasses Long-age Expectations." *Creation Science Update*. [Online at ICR.org/article/heat-saturn-moon-far-surpasses-long/; accessed June 7, 2020.]

²⁶⁷ Porco, C., D. DiNono, and F. Nimmo. 2014. "How the Geysers, Tidal Stresses, and Thermal Emission across the South Polar Terrain of Enceladus Are Related." *The Astronomical Journal*. 148 (3): 45.

billion years. It's just not big enough.²⁶⁸ An article written by NASA in 2008 stated: "What causes and controls the jets is a mystery."²⁶⁹

A JPL study calculated the amount of energy that tidal flexing could offer Enceladus, and they estimated that tidal flexing could yield about 1.1 gigawatts of heat, and there may also be a tiny bit of heat provided by radioactivity beneath the surface. But when the Cassini spacecraft measured the amount of heat Enceladus was emitting, a JPL press release, a photo caption stated "the south polar terrain of Saturn's moon Enceladus emits much more power than scientists had originally predicted."²⁷⁰ and it is in the neighborhood of 15.8 gigawatts—over *14 times* what tidal flexing could contribute!²⁷¹

Even with the heat generated by tidal flexing, Enceladus would have frozen solid after only about 30 million years—less than 1% of its alleged evolutionary age. But it has a stable, liquid ocean in its core, even after "billions of years!" How can that be?

The presence of the geysers, as well as observed tidal flexing, shows that Enceladus is still a flexible moon with a liquid core, in spite of the fact that it would have frozen solid long ago if deep time were an actual thing. As James Roberts of the University of California, Santa Cruz, puts it:

> **There is no possible combination of parameters** that allow for a thermally stable ocean.²⁷²

So "no possible combination of parameters" would allow for Enceladus to still have a liquid core, nor to have enough energy to have these geysers, after billions of years. Yet there they are.

It's very scary for an evolutionist to admit that something that looks young, and couldn't possibly be old, actually *is* young. Because, just think of what he would be acknowledging: that creation has a Creator. So the *Astro-*

---

²⁶⁸ Thomas, B. "Solar System Geysers—Each a Fountain of Youth." Creation Science Update. [Online at ICR.org/article/solar-system-geyserseach-fountain-youth; accessed June 7, 2020.]

²⁶⁹ "Enceladus Jets—Are They Wet or Just Wild?" *Cassini Equinox Mission News.* [Online at SolarSystem.NASA.gov/news/12901/enceladus-jets-surprises-in-starlight; accessed June 7, 2020.]

²⁷⁰ "Cassini Finds Enceladus is a Powerhouse." Jet Propulsion Laboratory press release, March 7, 2011. [Online at JPL.NASA.gov/news/news.php?feature=2926.]

²⁷¹ Howett, C. J. A. et al. 2011. "High heat flow from Enceladus' south polar region measured using 10–600 cm-1 Cassini/CIRS data." *Journal of Geophysical Research.* 116: E03003.

²⁷² Schirber, M., "Frigid Future for Ocean in Saturn's Moon," *Astrobiology Magazine,* June 19, 2008. [Online at space.com/scienceastronomy/080619-am-enceladus-ocean.html; accessed May 29, 2022.]

nomical Journal deferred: "Future Cassini observations may settle the question."

Saturn's Moon Titan

Titan lives up to its name: it is Saturn's largest moon, and is the second-largest in the solar system (after Jupiter's moon Ganymede). It is about 50% larger in diameter than Earth's moon, and 80% more massive.

Titan is large enough to have an atmosphere—mostly nitrogen—and therefore it has weather. But because of its cold temperatures, it's not pleasant weather, from Earth's standpoint. Its weather includes seasons, wind, and rain, and therefore Titan exhibits surface features like dunes, rivers, deltas, lakes, and seas. However, the rain is composed of liquid methane

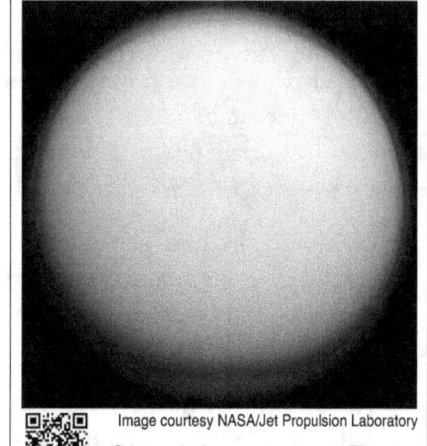

Image courtesy NASA/Jet Propulsion Laboratory

Saturn's largest moon, Titan. Its orange color, fuzzy edges, and apparent featurelessness are caused by its atmosphere, rich in nitrogen, methane, and ethane.

at about –290°F! Titan's atmosphere (mostly nitrogen, and about 5% methane) is thick and opaque, and there is a layer of haze that extends 60–120 miles above the surface.[273]

Titan does have water, but it is apparently frozen (frozenness of water around Saturn is not a foregone conclusion, as we saw in the discussion of Enceladus, above).

Titan's Atmosphere

One of the puzzling things for evolutionists when we sent probes to study Titan was that its atmosphere is unstable. In contrast to the atmosphere of Earth, where the nitrogen, carbon, and oxygen are recycled, when Titan's methane goes away, it doesn't come back.

---

[273] en.wikipedia.org/wiki/Titan_(moon); accessed June 7, 2020.

CHAPTER 4: A YOUNG SOLAR SYSTEM

On the other hand, ethane is very stable, and scientists were expecting it to have been building up for the last 4.5 billion years. So when the Huygens probe landed on Titan in 2005, scientists were astonished that it measured only trace amounts of ethane. Also, it found that liquid methane not only saturated the surface, but vast lakes also existed—one liquid-methane lake is 40 miles long and 25 miles wide. River channels revealed signs of erosion of icy mountains caused by cloudbursts of methane rain. So the prediction, based on deep-time models, was much ethane and little methane, but the observations revealed exactly the opposite.

Image courtesy NASA/Jet Propulsion Laboratory

Radar image of Titan's hydrocarbon lake Ligeia Mare. With an average depth of 100 feet, and a maximum depth estimated to be more than 650 feet, this lake contains enough liquid methane to fill three Lake Michigans. The four magnified portions show the liquid level changing over time.

The problem is that lakes "shouldn't" exist along the equator: solar energy would have evaporated them eons ago. Under the influence of solar wind and cosmic rays, methane in the upper atmosphere gets changed into other hydrocarbons such as ethane, propane, acetylene, and benzene, so when rain falls, the methane is not replenished. University of Arizona planetary scientist Caitlin Griffith is the lead author of the study results published in *Nature*.[274] She told *Nature News*, "Lakes at the poles are easy to explain, but lakes in the tropics are not."[275] The problem is clear: Why is there any methane left at the equator if it has been evaporating for billions of years? Why is there so much?

Time for another rescue device to prop up the deep-time billions-and-billions-of-years assumption evolutionists cherish so. Caitlin says further: "Because tropical lakes on Titan should evaporate over a period of just a few thousand years, the researchers argue that these ponds and lakes are being re-

---

[274] Griffith, C. A. et al. 2012. "Possible tropical lakes on Titan from observations of dark terrain." *Nature*. 486 (7402): 237–239.

[275] McKee, M. "Tropical lakes on Saturn moon could expand options for life." *Nature News*. [Online at Nature.com/news/tropical-lakes-on-saturn-moon-could-expand-options-for-life-1.10824; accessed June 7, 2020.]

plenished by subsurface oases of liquid methane." Like so many other rescue devices, this hypothesized subsurface reservoir has not been observed, and there is no actual evidence for it; its sole *raison d'être* is to help long-age worldview appear more believable.

And like before, you'll notice, it doesn't answer the problem; it simply pushes it one step further away. Even assuming the existence of the subsurface methane reservoir, it would have run dry eons ago from continually replenishing those fast-evaporating methane lakes, if we insist on a 4.5-billion-year-old solar system.

Saturn's Moon Iapetus

Another of Saturn's moons is Iapetus. Of all Saturn's regular satellites, Iapetus' orbit is the most inclined, relative to Saturn's equator. This means that of all Saturn's regular satellites, the rings would be best seen from Iapetus—since all the others are in the ring plane, the rings would be seen only edge-on, and thus be hard to see. What caused the inclination of Iapetus' orbit is unknown; the Nebular Hypothesis can't explain it.

Image courtesy NASA/Jet Propulsion Laboratory

Saturn's moon Iapetus, with much of its light side and part of its dark side clearly visible.

Iapetus has a light hemisphere and a dark hemisphere, which differ greatly in reflectivity (see photo). The fact that there was a light side and a dark side has been known for centuries: the Italian-born French astronomer Giovanni Cassini (after whom the spacecraft was named) discovered the moon 1671 when its light side was showing. He didn't directly observe the dark side until more than 30 years later, in 1705, when an improved telescope made the dark side visible—it was only 1% as reflective as the bright side. Cassini correctly surmised that there was a bright hemisphere and a dark hemisphere.

The Cassini spacecraft observed Iapetus from up close, and discovered many surprising facts. First, there is an enormous ridge along the equator, which makes Iapetus' shape reminiscent of a walnut (see photo). This ridge is more than 800 miles long (more than a quarter of the way around the moon), 12 miles wide, and 8 miles high. Some peaks in this ridge rise to 12 miles high, making them some of the tallest mountains in the solar system.[276] The ridge is entirely within the dark side of Iapetus, although on the light side, there are occasional mountain peaks along the equator.

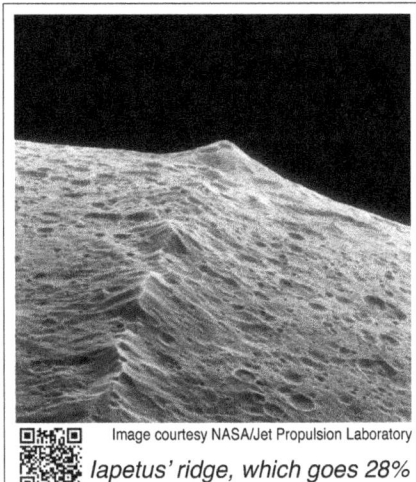

Image courtesy NASA/Jet Propulsion Laboratory

Iapetus' ridge, which goes 28% of the way around the equator.

The second thing that is of interest is the boundary between the light side and the dark side of Iapetus. For centuries after Cassini, astronomers assumed the light-dark boundary was very gradual, going through many shades of gray, as if caused by more black dust or less black dust. Indeed, it looks like that from a great distance. But the Cassini spacecraft took startling closeup photos of a stark white background with pitch-black spots, and the change in spot size and spot count from one location to another causes the boundary to seem gradual from afar.

Image courtesy NASA/Jet Propulsion Laboratory

A closeup of the sharp-edged dark spots on Iapetus, somewhere along the light-dark boundary.

There are tremendous amounts of frozen carbon dioxide (dry ice) on the surface of Iapetus. As you know, carbon dioxide doesn't melt, as water ice does; it **sublimates**—that is, it transitions from a solid directly to a gas, without going through an intermediate liquid phase. Carbon dioxide is volatile,[277]

---

[276] en.wikipedia.org/wiki/Iapetus_(moon); accessed June 8, 2020.

[277] NASA press release on the Cassini mission, October 8, 2007. [Online at NASA.gov/mission_pages/cassini/multimedia/pia10010.html; accessed June 8, 2020.]

which is why it was a surprise to planetary scientists that Iapetus has so much of it. Because it is so volatile, it should have all evaporated away eons ago. Then, as scientists analyzed the data coming in from Cassini, they became even more puzzled. Saturn orbits the Sun in about 29½ Earth years, and according to the calculations of the planetary scientists, Iapetus was losing carbon dioxide at a rate of 2.2 *billion tons* every orbit![278]

Losing that much every orbit around the sun, how long could it keep up that pace? Obviously, not for billions and billions of years. In a JPL press release, the authors stated: "One can see that the long-term stability of CO_2 is problematic." So true. According to their calculations, even if Iapetus started with a 5-kilometer-thick layer covering the whole moon (more than 2½ million square miles), it would be gone in 1.6 billion years, only one third of the imagined age of the solar system.[279]

It looks like deep time needs another rescue device. Astronomers have speculated about comets replenishing the CO_2 in Iapetus, but that makes one wonder why comets didn't also drop CO_2 on the other moons to make them as rich in carbon dioxide as Iapetus. Or perhaps the carbon dioxide is being replenished from beneath the surface of Iapetus, from an underground reservoir. But, like we've seen before, that doesn't answer the question; it just pushes it one step farther away: at the observed rate of depletion, even an internal reservoir wouldn't last for the presumed age of the solar system.[280] Again, this rescue device has not been observed, and there is no evidence for it; it's just that in order to save the notion of deep time, the reservoir's gotta exist—it's just *gotta!*

---

[278] Palmer, E. and R. Brown. May 2008. "The stability and transport of carbon dioxide on Iapetus." *Icarus*. 195 (1): 434–446. DOI: doi:10.1016/j.icarus.2007.11.020. [Online at ScienceDirect.com/science/article/abs/pii/S0019103507005921; accessed June 8, 2020.]

[279] "Iapetus' Equatorial Region." Jet Propulsion Laboratory press release, October 8, 2007. [Online at SolarSystem.NASA.gov/resources/13767/iapetus-equatorial-region; accessed June 8, 2020.]

[280] Coppedge, D. "Enceladus: A Cold, Youthful Moon." *Acts & Facts*. 2006. 35 (11). [Online at ICR.org/article/3112/; accessed June 8, 2020.]

Uranus

Uranus is the seventh planet going out from our Sun. Although it has been called a "gas giant" like Jupiter and Saturn, recent observations have indicated that its core is more likely composed of water ice, ammonia ice, and methane ice. For this reason, Uranus (and Neptune as well) are now referred to as **ice giants**.

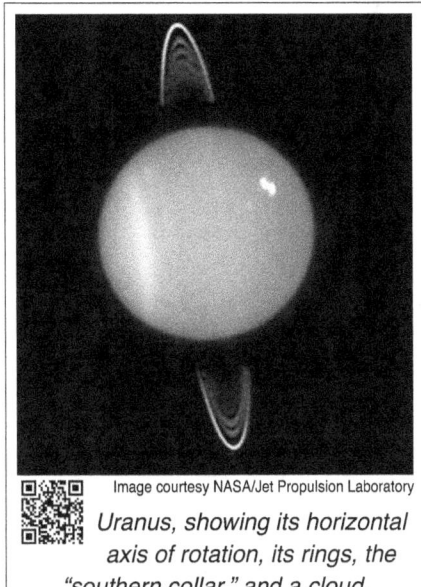

Image courtesy NASA/Jet Propulsion Laboratory

Uranus, showing its horizontal axis of rotation, its rings, the "southern collar," and a cloud.

Uranus orbits the Sun every 84 years, at a distance of 19.22 AU. Its diameter is almost four times the Earth's diameter, so the volume of Uranus is 63 times that of Earth, but its mass is only 14½ times that of the Earth—Uranus is much less dense than the Earth. Like the rest of the outer four planets, Uranus has a ring system, but it's not nearly as spectacular as that of Saturn.

Uranus usually appears as an almost featureless light blue sphere, although around the south pole there is an area of lighter color; this is called the "southern collar." Also, on rare occasions, there are visible clouds, as in the photo (taken by the Hubble Space Telescope).

How does Uranus fit in the creation model better than the evolutionary model? A couple of things. First, its axis of rotation. Like the Sun's initial formation, like the planets' very diverse compositions, like Mercury's volatile materials still being present, like Venus' retrograde axial rotation, like a whole set of Jupiter's moons with oblique orbits, and another set with retrograde orbits, like Europa's lack of craters, like the shiny newness and intricate structure of Saturn's rings—like all of those and many, many more we *didn't* talk about—the Nebular Hypothesis has no explanation for Uranus' axis of rotation; it is almost *90° different* from those of the other planets!

Another way that Uranus fits the creation model better than the evolutionary model is its magnetic field. Most planets have magnetic fields that are pretty well aligned with their axes of rotations, but with Uranus, not so much: it is rotated by 60°! Not only that, but it doesn't go through the center of the planet; it is shifted out from the center of rotation by almost a third of the planet's radius! Not only is Uranus' magnetic field (and Neptune's as well, as we'll see next) "anomalous," to say the least, but if Uranus is really billions and billions of years old, a magnetic field shouldn't even be there anymore; they naturally decay over time.

Neptune

Neptune is the outermost planet of our solar system, and it is often called the Uranus' "sister planet" because of similarities in size and composition. Although it is a bit more massive than Uranus, it is slightly smaller, because its greater gravity compresses its atmosphere more.[281]

Neptune is 30.1 AU from the sun, and it takes almost 165 years to orbit the sun once. Neptune has 14 known moons, as well as a small ring system. Although Neptune is more than 50% farther away from the sun than Uranus, it radiates away much more heat than Uranus. For some unknown reason, Neptune radiates away about 210% of radiation it receives from the Sun—more than twice as much heat going out as coming in. That, in itself, is fascinating, but not problematic, unless you assume it's been doing that for the past 4.5 billion years—the supposed age of the solar system. Then it becomes a problem for the obvious reason of "Where is it getting all that energy to radiate for billions of years?"

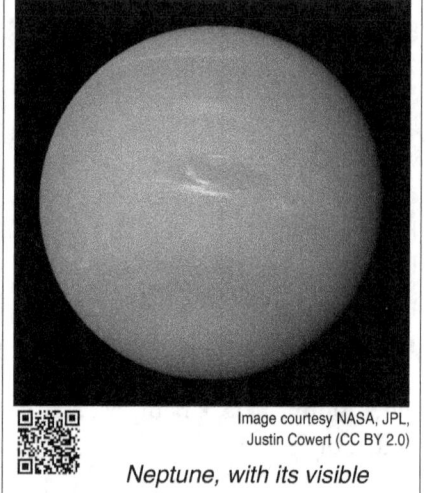

Image courtesy NASA, JPL, Justin Cowert (CC BY 2.0)

Neptune, with its visible weather patterns. The "Great Dark Spot" is the storm in the middle, with a patch of cirrus clouds above, and "Dark Spot Jr." is at the lower left.

Like that of Uranus, Neptune's magnetic field (which is still *way* too strong to be billions and billions of years old) is not aligned with the planet's axis of rotation—not even close. And it doesn't pass through the center of the planet either. Both of these features defy all the dynamo models that have been proposed so far. (Dynamo models are rescue devices that attempt to explain the billions-of-years longevity of the magnetic fields of planets and

---

[281] en.wikipedia.org/wiki/Neptune; accessed June 9, 2020.

moons, when the fields couldn't last that long, because the 4.5-billion-year age of the solar system is settled. Of course it is.)

Neptune's Moon Triton

Triton is Neptune's largest moon; it contains more than 99½% of all the mass in Neptune's orbit, including its rings and remaining moons put together. But Triton is interesting because its orbit is retrograde: it revolves the opposite way of the planet's rotation.

Also, instead of orbiting in the plane of the planet's equator, Triton is off Neptune's equator by 23°.[282] (Yes, these two features again defy the Nebular Hypothesis.)

Triton, which has a retrograde orbit that is also slanted 23° from Neptune's equator.

Planetary scientists, rather partial to deep time, have suggested that Triton's retrograde orbit indicates it was captured at some point after the planets formed. But considering the precise calculations and numerous adjustments needed to put even a small spacecraft in orbit around a planet, that happening so precisely all by itself severely strains credibility.

---

[282] Lisle, Jason. "The Solar System: Neptune." February 28, 2014. [Online at ICR.org/article/solar-system-neptune; accessed June 9, 2020.]

Pluto

For decades after its discovery, Pluto was considered a real planet—the ninth in our solar system. But in 2006, the International Astronomical Union reclassified it to be a "dwarf planet." (See the next section, on Comets, for their reasoning.) The New Horizons spacecraft was launched in 2006 and arrived at Pluto in 2015.[283] What it showed us is truly amazing.

Image courtesy NASA/Johns Hopkins University Applied Physics Laboratory/ Southwest Research Institute/Alex Parker

Pluto, as seen by the New Horizons spacecraft in 2015.

Even though Pluto is no longer classified as an official planet, it is still in the solar system, and it still is a fascinating object that presents numerous problems for the evolutionary theories of solar-system formation.

For one thing, Pluto has way too few impact craters[284] for an object allegedly 4.5 billion years old. There is a large, flat, heart-shaped plain named Tombaugh Regio (named after Pluto's discoverer, Clyde Tombaugh) that should have innumerable impact craters if it is really 4.5 billion years old. It doesn't. And there is an area on the western edge of Tombaugh Regio called Sputnik Planitia. *Zero* craters.[285] Zilch. Zip. Nada. If we say that crater-producing impacts stopped a long time ago, or just started recently, both of those ideas violate the uniformitarianism needed by evolution. The paucity of cratering implies a young planet.

And if someone suggests that geological activity caused eruptions of material that covered up many of the craters that surely "must be" there, that's a problem for evolution too, because such a tiny object should have run out of heat ages ago, if we assume deep time.

---

[283] en.wikipedia.org/wiki/New_Horizons; accessed September 8, 2020.

[284] Feltman, R. "Pluto's surface surprisingly full of mountains and lacking craters." *The Washington Post*. Posted on WashingtonPost.com July 15, 2015. [Online at WashingtonPost.com/national/health-science/plutos-surface-surprisingly-full-of-mountains-and-lacking-craters/2015/07/15/9f9df88e-2b3d-11e5-a250-42bd812efc09_story.html; accessed August 15, 2020.]

[285] Brian Thomas, "Pluto's Craterless Plains Look Young" [Online at ICR.org/article/plutos-craterless-plains-look-young; accessed June 11, 2020.]

In Space.com, one of the authors wrote:

> Scientists had expected that heat to be lost if Pluto was an old object. But New Horizons revealed an active surface on an old planet, and internal heating is the best current guess for what's driving that activity—even if **scientists don't quite know how that heat has lasted over 4 billion years.**[286]

Apparently it's not even considered to be within the realm of possibility that it actually *hasn't* been 4 billion years. Alan Stern, the principal investigator for data returned from the New Horizons spacecraft, states:

> It's a huge finding that small planets can be active on a massive scale, billions of years after their creation.[287]

Note the foregone conclusion of "billions of years;" that appears to be non-negotiable. And that part about their "creation" must have been a slip of the tongue. The surface of Pluto is defying deep-time theories: In one presentation at a planetary conference, the opening line said, "Against many expectations, New Horizons' images of the surface of Pluto and Charon show seemingly young surfaces."[288]

Instead of the multitudes of craters that were expected, *convection* is apparently going on in the crust of Pluto! In this Tombaugh Regio, there is an area of polygonal **convection cells**. Convection cells are caused by the well-known process of hot fluids rising and cooler fluids falling. They can sometimes be seen in a hot cup of

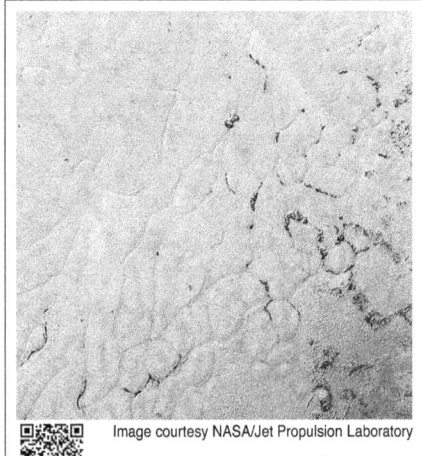

Image courtesy NASA/Jet Propulsion Laboratory

Convection cells in Sputnik Planitia, the western lobe of Pluto's heart-shaped Tombaugh Regio. The dark spots in the boundaries are pits.

---

[286] Redd, N. T. "Part of Pluto's Heart Was 'Born Yesterday.'" *Space.com*. Emphasis added. [Online at Space.com/31080-pluto-heart-born-yesterday.html; accessed June 11, 2020.]

[287] Redd, N. T. "Part of Pluto's Heart Was 'Born Yesterday.'" *Space.com*. [Online at Space.com/31080-pluto-heart-born-yesterday.html; accessed June 11, 2020.]

[288] Trowbridge, A. J., H. J. Melosh, and A. M. Freed. "Vigorous Convection Underlies Pluto's Surface Activity." Presented at the November 2015 47th Annual Meeting, American Astronomical Society, Division of Planetary Sciences, #102.01. [Online at adsabs.harvard.edu/abs/2015DPS....4710201T; accessed June 11, 2020.]

coffee, especially in a ceramic cup, after any leftover turbulence from pouring or stirring has subsided. If there is overhead lighting, you can see slight bumps in the surface of the coffee, where hotter fluid is rising, and the spaces or "cracks" between are where the cooler fluid is falling.

This is what is happening on Pluto. Analysis of the new data indicates some portions of the surface are "less than 10 million years old," and the latest studies indicate an age of ≈180,000 years.[289] *Much* younger than 4.3 billion years old. And if the surface is that young, why not the whole planet? Well, there's (ahem) no need to go that far, is there?

In the words of the New Horizons science team:

> Pluto displays a surprisingly wide variety of geological landforms, including those resulting from glaciological and surface-atmosphere interactions as well as impact, tectonic, possible cryovolcanic, and mass-wasting processes.[290]

Again, how in the world could such processes still be going after 4 billion years?

And by the way, tidal heating, the usual go-to answer for the presence of unexplainable heat, won't work here because Pluto and its largest moon Charon are tidally locked with each other: *both* objects keep the same side always facing the other, so there are no tidal effects at all. And there are no other objects close enough and large enough to cause tidal flexing in Pluto.

---

[289] en.wikipedia.org/wiki/Pluto; accessed June 12, 2020.

[290] Stern, S. A.; et al. (2015). "The Pluto system: Initial results from its exploration by New Horizons". *Science*. 350 (6258): 249–352.

CHAPTER 4: A YOUNG SOLAR SYSTEM

Pluto's Moon Charon

Speaking of Charon, this moon is itself quite a problem to explain within the framework of evolutionary solar system formation. It, too, shows signs of geological activity,[291] and its energy should have run out even before Pluto's.

Charon is even smaller than Pluto, of course, but it has a gargantuan chain of cliffs and troughs that extend almost a third of the way around Charon's circumference. This is visible in the photo to the right, extending from the 8:00 position to the 2:00 position.

Image courtesy NASA/Johns Hopkins University Applied Physics Laboratory/ Southwest Research Institute/Alex Parker

Pluto's largest moon Charon, as seen by the New Horizons spacecraft in 2015.

Charon is half the diameter of Pluto, and one eighth its mass—it is larger than any other moon in the solar system, in comparison to the object it's orbiting. So how in the world did Charon form, evolutionarily speaking? How did such a large object form so close to its primary? One secular model supposes it was a glancing collision between Pluto and some other object, and the resulting fragments attracted each other gravitationally and formed Charon. But such a glancing blow would have caused Pluto to spin so fast, it would have solidified into an oblate spheroid—a squashed sphere. But it isn't; New Horizons found Pluto to be quite spherical.[292]

---

[291] Hebert, J. 2015. "New Horizons, Pluto, and the Age of the Solar System." *Creation Science Update.* [Online at ICR.org/article/8839; accessed June 11, 2020.]

[292] New Horizons, NASA's Mission to Pluto: Media Briefing July 24, 2015. [Online at Pluto.jhuapl.edu/News-Center/Press-Conferences/index.php?page=2015-07-24; accessed June 11, 2020.]

Comets

Comets are small objects with highly elliptical orbits that take them very close to the Sun during the small-and-fast portion of their orbits, and then far away from the Sun on the remaining large-and-slow portion of their orbits.

Comets have been described as "dirty snowballs" because their composition is primarily a mixture of ice, dust, and gravel. During the portions of their orbits where they are close to the Sun, **solar wind** (a constant stream of charged particles emitted from the Sun's corona) and **solar radiation** (electromagnetic radiation of whatever wavelength) evaporates the icy portion and boils it away into space. The roundish cloud of evaporating gases surrounding a comet is called the **coma**, and the trail of gases or dust blown away from the coma in a long stream is called the **tail**.

Image courtesy Philipp Salzgeber (CC BY-SA 2.0)
Comet Hale-Bopp as seen from earth in 1997. Note it has two tails: the bright white dust tail, and the dim upper gas tail.

Pertaining to evolutionary long ages, the problem with comets is that they exist at all! They lose *so much* of their mass during each orbit, they can't last very long; certainly not the billions and billions of years that the solar system is alleged to have existed.

Planetary scientists were in a quandary: we can observe the size of comets, and measure how long it

Image courtesy CC BY-SA 3.0
A closeup of Comet Hale-Bopp's double tail.

would take for them to evaporate completely, and they don't last nearly as long as they "should." So they invented a rescue device—two of them, actually. First, they envisioned a belt of trillions of comet bodies out in the vicinity of Pluto and beyond, and dubbed it the **Kuiper Belt**. There, the comet bodies were in stable orbits for long ages, until something gravitationally knocked a comet body out of its stable orbit, down toward the sun, where it orbited a few dozen times before it evaporated completely. The Kuiper Belt was imagined to be the source of **short-period comets**—those whose orbits take less than 200 years.

The second rescue device scientists invented is called the **Oort Cloud**, and it is imagined to be an enormous spherical reservoir of trillions more comet bodies. The Oort Cloud is too far away to be seen, even with our best instruments. Convenient, that. The Oort Cloud is supposed to be the source of **long-period comets**—those whose orbital periods are longer than 200 years.

The highly questionable existence of the Oort Cloud is made apparent from the clearly un-confident language used to describe it. For example, Wikipedia's article[293] uses these words and phrases in its description: "theoretical," "proposed," "may come from," "may have originated," "conjecture," "no confirmed direct observations," "may be the source," "postulated," and "proposed." And the article is only four paragraphs long!

Even if the Oort Cloud did exist, what are the chances of a random gravitational perturbation dislodging it from its stable Oort-Cloud orbit into a stable solar orbit? *Extremely* small. Much more likely is that any comet dislodged from the Oort Cloud would come flying through the solar system at such a high speed it would fly right through, never to be seen again.[294] Achieving a stable orbit is very tricky. And we're supposed to believe it happens accidentally, for every periodic comet? No, the Oort Cloud is a rescue device, plain and simple, attempting to stem the credibility-hemorrhage deep time is experiencing more every year.

When Voyager 2 flew past Neptune in 1989, objects in the region of the Kuiper Belt (dubbed **Kuiper Belt Objects** or **KBOs**) were observed, confirming the work of astronomers who in 1987 started looking for objects past the orbit of Pluto. However, instead of finding the trillions of small, icy, comet bodies astronomers were expecting, they found far fewer (only in the thousands), relatively large, rocky **planetesimals** (minute planets). Some of these objects are comparable in size to Pluto (Haumea and Makemake), and one object is even larger than Pluto (Eris).

---

[293] en.wikipedia.org/wiki/Oort_cloud; accessed July 3, 2020.
[294] Brian Thomas, "NASA Photographs Young Comet" November 12, 2010 [Online at ICR.org/article/nasa-photographs-young-comet; accessed July 3, 2020.]

It was these larger objects that led the International Astronomical Union in 2006 to define a "planet" as having these four characteristics:

- It orbits a star directly (i.e., it doesn't merely orbit something else that orbits a star, like a moon does),
- It is large enough to be rounded by its own gravity,
- It is not massive enough to cause thermonuclear fusion, and
- It is large enough to clear its immediate vicinity of planetesimals.

It was this new definition of "planet" that caused Pluto to be demoted to the status of a "dwarf planet" in 2006.

The term "Kuiper Belt Objects" (KBOs) seems to be falling into disuse, or at least its use is being limited to only certain kinds of objects, perhaps because it didn't contain what long-age astronomers wanted it to contain. The newer term for such objects is **Trans-Neptunian Objects** (TNOs) because they are beyond the orbit of Neptune. TNOs include KBOs, but also some other kinds of esoteric objects.[295]

In short, the Kuiper Belt did *not* contain the trillions of comet bodies astronomers were expecting. But maybe the Oort Cloud does. Well, maybe. But remember, the Oort Cloud has not been observed, and there is no actual evidence for it. Astronomers who are long-age enthusiasts want it to exist, because it would explain the discrepancy between their presumed age of the solar system (4.5 billion years) with the actual time comets could exist before evaporating completely (usually much less than a million years).

Well, at least the Oort Cloud—if it exists—might help explain the origin of long-term comets, but it would do nothing to explain the presence of magnetic fields in planets and moons that that shouldn't have them anymore, or the volatiles on Mercury, or the young surface of Venus, or the unbalanced atmosphere of Mars, the volcanoes on Io, the existence of the rings of Saturn, the brightness of the rings of Saturn, the geysers on Enceladus, the methane on Titan, the dry ice on Iapetus, the off-axis and off-center magnetic field of Uranus, the outgoing radiation from Neptune being more than twice as much as the incoming, the convection cells on Pluto, and hundreds of other things. Plus numerous things on Earth that we'll cover in later chapters. But the Oort Cloud, if it actually exists, might help explain comets at least a little. . .

But wouldn't it so much simpler to just conclude that the solar system's age actually corresponds to observations, instead of making up all these rescue

---

[295] en.wikipedia.org/wiki/Kuiper_belt; accessed June 11, 2020.

devices to explain why observations *don't* support evolutionary predictions? But no, I guess it wouldn't be simpler, because then we'd have to deal with the reality of the Creator, and that's just too scary.

Chapter 4 Summary

Here are the most important points in this chapter:

- The **Nebular Hypothesis**—the sun, planets, and moons spontaneously forming themselves out of a swirling cloud of dust and gas—fails on account of several things:
    - Thermal expansion
    - Angular momentum
    - Magnetic fields
    - Roche Limit
    - Retrograde axial rotation
    - Retrograde orbits
    - Highly inclined orbits
    - Insufficient mass
    - Widely differing planetary compositions
- **Sun:** The "Young Faint Sun" Paradox: The sun would have been too dim 4.3 billion years ago for life to start on Earth.
- **Mercury:** Has a magnetic field, it is geologically active, and it is still outgassing volatile materials. It shouldn't still be doing any of these things if it formed 4.5 billion years ago.
- **Venus:** It has a retrograde axial rotation (contrary to the Nebular Hypothesis); geological features show similar amounts of erosion, implying recent creation.
- **Earth:** (See the next few chapters.)
- **Mars:** The ^{13}C-to-^{12}C ratio indicates a young atmosphere.
- **Jupiter:** Has retrograde moons. Also, Io still volcanically active and ejecting ultramafic lava, Europa has a youthful appearance, and Ganymede still has a magnetic field, which shouldn't be the case if the solar system is 4.5 billion years old.
- **Saturn:** Rings are too bright and shiny to be billions of years old. Rings shouldn't even exist anymore. Moonlets in rings shouldn't be there anymore. Enceladus shouldn't have enough energy, or material, for geysers

anymore. Titan shouldn't have methane at the equator anymore. Iapetus shouldn't have CO_2 anymore.
- **Uranus:** Has a horizontal axis of rotation. Magnetic field off-axis by 60° and off-center as well. It shouldn't even *have* a magnetic field anymore.
- **Neptune:** Radiates more than twice as much heat as it receives from the sun; that should have stopped a long time ago. Still has a magnetic field, even though it "shouldn't." Triton's orbit is retrograde and tilted 23° from Neptune's equator.
- **Pluto:** Too few craters for 4.5 billion years old, unless geologically active, which it shouldn't be anymore. Convection cells indicate constant heat loss; how could that still be happening? Charon shows signs of geological activity as well.
- **Comets:** They shouldn't exist anymore, since so much of their mass evaporates on every orbit. Kuiper Belt is not a reservoir of comet bodies waiting to be nudged toward the Sun. No evidence for Oort Cloud.

Chapter 5:

A Young Earth

One aspect of a young universe must necessarily be a young Earth: I think you'd agree that, using natural laws only, it would be difficult for the Earth to come into existence before the universe in which it resides.

So let's take a look at the evidence for the age of the Earth and see if evolutionary guesses stand up to scrutiny. While we're at it, let's also look at how the Earth is precisely built and located so that life is possible.

The "Goldilocks" Planet

There are *so* many fortuitous circumstances concerning the structure of the Earth, its composition, its relationship to other planets, and even the laws of physics, that it has been given the nickname of "The Goldilocks Planet." This comes from the child's story The Three Bears, in which the little girl Goldilocks considered items belonging to Papa Bear to be extreme in one direction, items belonging to Mama Bear to be extreme in the opposite direction, but the things belonging to Baby Bear were "just right."

And indeed, there are a great many things about the universe, about the Solar system, and about Earth that are "just right." *So* many, in fact, that it looks decidedly non-random. It looks like Someone made Earth—in fact, the whole *universe*—just for us. Could that be? Let's take a look at some of the items that make the universe, and Earth, "just right" for supporting life.

The Precision of the Universe

Physicist Paul Davies, analyzed the various physical constants of the universe, and observed how, if any of them were even slightly different, life would be impossible.

Marveling at such incredible precision, he commented:

> It is hard to resist the impression that **the present structure of the universe, apparently so sensitive to minor alterations in numbers, has been rather carefully thought out.** . . . The seemingly miraculous concurrence of these numerical values must remain the most compelling evidence for cosmic design.[296]

Although those who are convinced of intelligent design[297] are still in a minority, more and more scientists in all fields are becoming increasingly receptive to the idea of an intentional, non-random universe. The more precise our instrumentation becomes, the less we can write off these seeming coincidences as "accidental."

Walter Bradley, the co-author of *The Mystery of Life's Origin*, says it this way:

> It is quite easy to understand why so many scientists have changed their minds in the past thirty years, agreeing that the universe cannot reasonably be explained as a cosmic accident. **Evidence for an intelligent designer becomes more compelling the more we understand about our carefully crafted habitat.**[298]

---

[296] Paul Davies, *God and the New Physics* (New York: Simon and Schuster, 1983), p. 189. Emphasis added.

[297] The "intelligent design" approach is, in general, an acknowledgment that the universe we live in could *not* have arisen by the random, undirected processes that naturalistic evolutionary theories promote. As such, Intelligent Design proponents differ from Creationists in that the Intelligent Design movement realizes that *some* superintelligence must have created the universe, but they stop short of identifying *Who* that intelligent designer is. Essentially, Creationists acknowledge the God of the Bible as the intelligent designer, but the Intelligent Design movement avoids that final acknowledgment, stopping at the point of understanding that *some* unknown entity had to architect the universe, because randomness is not capable of accomplishing it. Both approaches are good, in that they both expose the numerous shortcomings of evolutionary theories, but only Creationism identifies the Designer. That is, both acknowledge "Evolution couldn't make the universe," but only Creationism completes the thought by adding ". . .but Jesus could."

[298] Walter L. Bradley, "The 'Just So' Universe," in William A. Dembski and James M. Kushiner,

Interesting wording: "carefully crafted habitat."

Astrophysicist Sir Fred Hoyle concurs, specifically in the mechanism by which stars work:

> I do not believe that any scientists who examined the evidence would fail to draw the inference that **the laws of nuclear physics have been deliberately designed** with regard to the consequences they produce inside stars.[299]

Owen Gingerich, a Harvard astronomy professor and the senior astronomer at the Smithsonian Astrophysical Observatory commented on Fred Hoyle's statement above:

> Fred Hoyle and I differ on lots of questions, but on this we agree: a common sense and satisfying interpretation of our world suggests **the designing hand of a superintelligence.**[300]

The above quotes from world-class scientists with impressive credentials—and there are numerous others—are indeed convincing, but some actual examples are in order.

The Strength of Gravity

Suppose you had a ruler that stretched all the way across the observable universe—a ruler more than 90 billion light-years long. This ruler represents the range of possible values for the various forces in the universe: from gravity on the lower end, to the strong nuclear force—which is 10^{40} times stronger than gravity—on the upper end.

In this range of force strengths, the value of the strength of gravity *could* be set anywhere along that 90-billion-light-year-long ruler. What would happen if we could change the setting from where it is? If we moved the setting only *one inch* along this ruler, the force of gravity would change by a factor of one *billion*. Life would be impossible.

Signs of Intelligence, p. 170. Emphasis added.

[299] Quoted from John Barrow and Frank Tipler, *The Anthropic Cosmological Principle* (Oxford: Oxford University Press, 1986), p. 22. Emphasis added.

[300] Owen Gingerich, "Dare a Scientist Believe in Design?" in John M. Templeton, editor, *Evidence of Purpose* (New York: Continuum, 1994), p. 25. Emphasis added.

And another thing: the formula to calculate the gravitational force exerted by two objects on each other is as follows:

$$F \propto m_1 \times m_2 \div r^2$$

In other words, the force of gravity *(F)* is proportional to ($\propto$) the product of the two masses ($m_1 \times m_2$) divided by the square of the distance between them (r^2). Scientists were curious: was the exponent "2" in the denominator really 2, or was it some number that was just very close to 2?

Science News commented on the exponent being exactly two, saying it "has always seemed a little too neat. Is the exponent some fraction near two, which would be messy but might seem more empirical?"[301] And that's a good point: In an evolved, random, chance-driven universe, it seems more likely that it might be a little off; say, 1.993 or 2.0041 or some such. Exactly 2 seems a bit too convenient.

But gravity has been tested with ever-increasing precision, and as far as we can tell, it is *exactly* 2. And by analyzing how gravity interacts with other physical constants, if the exponent were anything else but *exactly* two, orbits could not remain stable. Orbits would decay, so orbiting bodies would spiral into their primaries: the Moon would crash into the Earth, and Earth (and all the other planets) would crash into the Sun, and the Sun (and all the other stars in our galaxy) would crash together in the galactic center. It would be a bad day.

The fact that the "setting" for the force of gravity is what it is, and that the exponent in the denominator is *exactly* two, show an incredible precision that really strains the credibility of "accidental" occurrence.

The Cosmological Constant

One part of Einstein's equations for General Relativity is called the "cosmological constant," and it refers to the energy density of empty space. Fortunately for us, it has a very specific value.

Stephen Weinberg, a Nobel-winning physicist and avowed atheist, is amazed at the way the cosmological constant is "remarkably well-adjusted in our favor."[302] This constant could have any value, and could be either positive or negative, but as Weinberg goes on to say, "but from first principles one would guess that this constant should be very large."

---

[301] "Gravity Very Precisely," *Science News* 118 (July 5, 1980) p. 13.
[302] Stephen Weinberg, "A Designer Universe?" *New York Review of Books* (October 21, 1999).

I mentioned above that it was fortunate for us that the cosmological constant has the value that it does. Weinberg explains:

> If large and positive, the cosmological constant would act as a repulsive force that increases with distance, a force that would prevent matter from clumping together in the early universe, the process that was the first step in forming galaxies and stars and planets and people. If large and negative, the cosmological constant would act as an attractive force increasing with distance, a force that would almost immediately reverse the expansion of the universe and cause it to re-collapse.

Notice that if the cosmological constant was much bigger than it is, or much smaller than it is, life would not be possible. Weinberg continues: "In fact, astronomical observations show that the cosmological constant is quite small, very much smaller than what would have been guessed from first principles." As an atheist, Weinberg is not a creationist, but even so, he sees a delicate and precise balance in the physical constants that points to a Designer, as in "This couldn't have happened by accident."

What are the chances that the cosmological constant would be its precise value to allow life to exist in the universe? About the same as throwing a dart from a distant part of space, at a *particular atom* somewhere on, or in, the Earth, and hitting it on the first attempt. Mathematically speaking, 1 chance in about 10^{53}. Sure. No problem. Happens every day.

Protons and Neutrons

Protons and neutrons, the heaviest particles in the nuclei of atoms, are similar, but not quite identical, in their masses. The neutron is a tiny bit heavier, and so a free neutron decays into a proton, an electron, and an antineutrino.

So what? Is it a big deal that neutrons are slightly heavier than protons? Actually, yes. Suppose protons were more massive than they are, by the tiny amount of 0.2%. What would happen? In that case, the *proton* would become unstable, and decay into a neutron, a positron, and a neutrino.

Now that's a problem, because hydrogen—the most abundant element in the universe—has a nucleus composed of a single proton. If protons decayed, all hydrogen would cease to exist, which means that all water (H_2O, which life absolutely depends on) would vanish. Not only that, but nuclear fusion would cease, and all the stars would go out.[303] It would likely be a problem for us if our sun stopped shining. Of course, all the water in our bodies would have already vanished, so we wouldn't be around to care.

We sure are "lucky" that neutrons are heavier than protons, aren't we?

But it's not just the fact that neutrons are heavier than protons that's important; *neutrons are heavier by just the right amount.* If the mass of a neutron were increased by only 0.14%, the nuclear fusion in stars would stop, and again, all the stars would go out.[304] As mentioned, that might be a problem for life on Earth.

Phase-Space Volume

One of the parameters used to describe the underlying structure of the universe is called "phase space." Without getting too detailed, this parameter too is incredibly fine-tuned—even more so than the ones already discussed.

Roger Penrose, a physicist from Oxford says that one of the universe's parameters, the "original phase space volume," required an accuracy and precision that puts the previously mentioned items to shame. Penrose said that the accuracy required is one part in $(10^{10})^{123}$. This number is so large it's pretty unfathomable: it is more than a trillion trillion trillion trillion times larger than the number of atomic particles in the observable universe. And the phase-space volume had to be calibrated to that precision for the universe to function in such a way as to enable Earth to support life.

As Penrose said, this was "the precision needed to set the universe on its course."[305] This is *extremely* improbable, to say the least. *Discover* magazine wholeheartedly agreed, saying "The universe is unlikely. Very unlikely. Deeply, shockingly unlikely."[306]

---

[303] Donald D. DeYoung, "Design in Nature: The Anthropic Principle," *Acts & Facts*, November 1, 1985. [Online at ICR.org/article/design-nature-anthropic-principle; accessed July 24, 2020.]

[304] Lee Strobel, *The Case for a Creator* (Zondervan, Grand Rapids, Michigan, 2004), p. 134.

[305] Roger Penrose, *The Emperor's New Mind* (New York: Oxford, 1989), p. 344, quoted in Stephen C. Meyer, "Evidence for Design in Physics and Biology" and Michael J. Behe, William A. Dembski, and Stephen C. Meyer, editors, *Science and Evidence for Design in the Universe*, p. 61.

[306] Brad Lemley, "Why Is There Life?" *Discover* (November 2002).

The Precision of the Solar System

The Earth's relationship to the other planets of the solar system is very precise and of immense importance to Earth's ability to support life.

Such precision is no accident, as creationists have been saying all along. But—as mentioned at the beginning of "Chapter 4: A Young Solar System"—evolutionists, unwilling to acknowledge the existence of God, still marvel at how "lucky" we are for everything to have unintentionally, accidentally, and undirectedly "fallen" together so beautifully:

> You might also think that these disparate bodies are scattered across the solar system without rhyme or reason. But move any piece of the solar system today, or try to add anything more, and **the whole construction would be thrown fatally out of kilter.** So **how exactly did this delicate architecture come to be?**[307]

Wow. That's some strong language: add or move any piece of the solar system and the whole thing would be thrown *fatally* out of kilter. And notice what he called the solar system: an "architecture" and a "construction." That kinda makes it sound like Someone architected and constructed it, doesn't it? Was he serious, or just being melodramatic?

Actually, it sounds like he is serious, and his findings are confirmed by Jacques Laskar, a French astrophysicist:

> Jacques Laskar discovered that the orbits of **Jupiter and Saturn keep the earth's orbit from becoming chaotic.** Without the orbital stability produced by Jupiter and Saturn, the earth's orbit would make extreme changes, causing **instability in our climate and making the earth uninhabitable.**[308]

How do evolutionists deal with such strong evidence of precise, intentional design? On the one hand, they "know" there is no God, but on the

---

[307] Webb, R. 2009. "Unknown solar system 1: How was the solar system built?" *New Scientist.* 2693: p. 31. Emphasis added.
[308] Bickel, B. and S. Jantz. 2001. *Creation & Evolution 101: A Guide to Science and the Bible in Plain Language.* Eugene, OR: Harvest House Publishers. Emphasis added.

other hand, they see His handiwork everywhere. Caltech evolutionary astronomer Mike Brown states it this way:

> It really is something that I find deeply weird. . . What does it all mean? I don't know.[309]

The Precision of Earth

What are some of the curiously convenient things about Earth? Glad you asked.

Earth orbits the sun at just the right distance to allow for liquid water, which is essential for life. Also, the atmospheric pressure is just right for liquid water. There is a protective layer of ozone high in Earth's atmosphere, which significantly reduces UV (ultraviolet) radiation from the sun, thus protecting living organisms. Earth has a strong enough magnetic field to deflect cosmic rays, which also protects life on the surface.

And, by the way, Earth's magnetic field is weakening at a rate that is perfectly consistent with the Biblical 6,000-year age of the Earth, but completely incongruous with the imagined 4.5-billion-year age evolutionists promote.

Dating Methods

The theory of evolution, with its required and assumed deep time of billions and billions of years for everything to happen, has a vested interest in proving that the Earth, rocks, fossils, and so forth, really are as old as evolutionary philosophy requires. Calculating the dates of objects uses processes known as *dating methods*.

The Bathtub Illustration

Suppose you arrive home one day and you hear water running. You trace the sound to the bathtub, which (as you quickly measure) is filling at a rate of one gallon per minute. You know that the tub hold 45 gallons, and you notice that it is now two-thirds full. The question is: How long has the water been running?

---

[309] Krulwich, R. "Our Very Normal Solar System Isn't Normal Anymore." *National Public Radio*. Posted on npr.org May 7, 2013. [Online at NPR.org/sections/krulwich/2013/05/06/181613582/our-very-normal-solar-system-isn-t-normal-anymore; accessed June 28, 2020.]

Most people would say that the water's been running for 30 minutes. But is that really true? Yes, *if* a variety of assumptions are valid. For example:

- The water wasn't turned off at any time during the last 30 minutes.
- When the water was on, it was coming in at the same rate the whole time.
- The tub was empty when the water was turned on.
- The plug was in place during the whole process to prevent any water from draining.
- No one added water through any means other than the spout (e.g., a bucket).
- No one removed any water after the tub started being filled.

Clearly, if any of the above assumptions is *not* valid, the answer of 30 minutes is very likely to be wrong, and maybe drastically so. But this illustrates the same problems that all dating methods have: the estimated answers will be roughly accurate only if *all* the implicit assumptions are valid. The problem is that we don't know if *any* of the assumptions are valid. Therefore, the answers given by dating methods are likely to be wrong, and have been shown in numerous cases (as you'll see below) to not even be in the ballpark—the answers can be off by several orders of magnitude. *All* dating methods have this problem.

Back to Dating Methods. . .

There are many dating methods, including circular definitions where the age of rocks is determined by the fossils in them, but the age of the fossils is determined by the rocks they're in. More about this in "The Geologic Column" in "Chapter 7: The Fossil Record."

But other dating methods exist also; for example, various methods of **radiometric dating.** In all such dating methods, the scientific devices used for this purpose do not actually determine ages; they only count atoms of certain kinds. The number of specific atoms found is converted into an age estimate, using an algorithm that makes numerous unverifiable assumptions.

As you remember from the discussion of the "unbalanced" atmosphere of Mars in "Chapter 4: A Young Solar System," **isotopes** are variations in an element's atomic structure, in which different numbers of neutrons reside in the nucleus. Isotopes that are radioactive (not all of them are) decay through the process of **radioactivity** into other isotopes or other elements, and they go about the process in a statistically well-behaved way.

As shown in the illustration, at time zero, there is 100% of the radioisotope in the sample. After one half-life, only ½ of the radioisotope remains; the other half has decayed into something else. After another half-life, only ½ of the previous amount (the quantity present after one half-life) remains, so only ¼ of the original amount remains. After three half-lives, only ⅛ remains, and so forth.

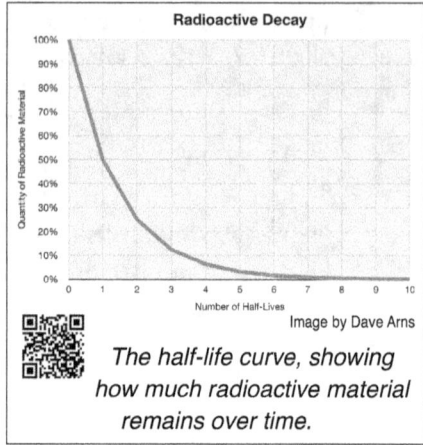

The half-life curve, showing how much radioactive material remains over time.

After only 7 half-lives, the amount of radioisotope remaining is less than 1% of the original quantity, so statistical uncertainty becomes a larger and larger factor as more time elapses.

Three conditions (generalized and boiled down from the six mentioned in the bathtub illustration) must be satisfied for radiometric dating to give even basically accurate answers. They are:

1. We know what the original concentration of the radioisotope was.
2. The rate of radioisotope decay has been constant over all time.
3. No processes other than radioisotope decay affected the amount of radioisotope we see today.

Are these reasonable assumptions? They had better be, if we want to place any credibility in radiometric dating processes. But in actuality, we can't be sure about *any* of the three items. This is not to say that dating methods are completely useless, but the three assumptions upon which they are based need to be addressed much more rigorously, in order for radioactive dating to viewed with any kind of credibility.

Carbon-14

Carbon-14, a radioisotope of carbon, is sometimes called **radiocarbon** for short, and it can be used for estimating the ages of old organic remains. Here's how it works:[310]

1. Radiocarbon, or ^{14}C, is constantly being created in the atmosphere by cosmic rays (high-energy protons from the sun) hitting nitrogen atoms.

---

[310] en.wikipedia.org/wiki/Radiocarbon_dating; accessed June 27, 2020.

The atmospheric ratio of ^{14}C to ^{12}C ("normal" carbon) is about $1:10^{12}$, or 1 in a trillion.

2. The resulting radiocarbon combines with atmospheric oxygen to form radioactive carbon dioxide, or $^{14}CO_2$.
3. The $^{14}CO_2$ circulates in the atmosphere and is absorbed by plants during photosynthesis. So, in live plants, the $^{14}C:^{12}C$ ratio is also roughly $1:10^{12}$—the same as the atmosphere.
4. Sooner or later, plants die, and therefore stop the uptake of atmospheric $^{14}CO_2$ (radiocarbon dioxide).
5. Edible plants are eaten, so the radiocarbon dioxide is ingested into the bodies of humans and animals, so our ratio of $^{14}C:^{12}C$ is also roughly $1:10^{12}$.
6. When the humans or animals die, they too discontinue the uptake of radiocarbon dioxide.

So when the uptake of radiocarbon dioxide stops, because the organism dies, that's when the radioactive decay becomes useful for dating purposes. Radioactive decay had been happening all along, but the ^{14}C was constantly replenished by continual photosynthesis and eating. Once the organism dies, the ^{14}C decays in a statistically well-behaved way that can be used to estimate age (assuming the three aforementioned assumptions are valid).

^{14}C has a half-life of about 5730 years,[311] so its decay rate is as follows:

Number of Half-Lives	Number of Years	Percentage of ^{14}C left
0	0	100%
1	5,730	50%
2	11,460	25%
3	17,190	12.5%
4	22,920	6.25%
5	28,650	3.13%
6	34,380	1.56%
7	40,110	0.78%
8	45,840	0.39%
9	51,570	0.20%
10	57,300	0.10%
50	286,500	89×10^{-15}%
100	573,000	78×10^{-30}%
1000	5,730,000	9.3×10^{-300}%

So the number of years is simply the number of half-lives multiplied by the half-life of ^{14}C, which is 5,730 years. Notice that after only ten half-lives, the quantity of radioisotope remaining is less than one-tenth of 1%, and the statistical uncertainty becomes so large, relative to the quantity of radioisotope remaining, that estimates become very unreliable. Basically, radiocarbon dating is unreliable for ages greater than 50,000 years.[312] That's less than 9 half-lives! True, radiocarbon dating can be "calibrated" to estimate ages greater than 50,000 years, but those calibrations are themselves based on assumptions we can't validate.

And some estimates place the upper limit of ^{14}C dating as even less: only 24,000 years![313]

---

[311] en.wikipedia.org/wiki/Carbon-14; accessed June 27, 2020.
[312] en.wikipedia.org/wiki/Carbon-14; accessed June 27, 2020.
[313] Edouard Bard, "Extending the Calibrated Radiocarbon Record," *Science* 292 (29 June 2001): 2443.

But again, the above table is accurate *only if the three assumptions are valid.* So let's look at some examples to see if they lend credence to the idea that these assumptions are true.

As we saw in the previous chapter, the magnetic fields of planets and moons decay over time, and Earth is no exception. Since this is a repeatedly observable and measurable phenomenon, we know that Earth's magnetic field was stronger in the past. What are the implications of this? These are some of them:

- A stronger magnetic field in the past would deflect more of the cosmic rays (mostly high-energy protons from the sun) away from hitting Earth's atmosphere.
- The resulting decrease in cosmic rays would convert less ^{12}C to ^{14}C, changing the ratio of normal carbon to radiocarbon.
- Atmospheric carbon would continue to combine with oxygen to make carbon dioxide (CO_2), and it would reflect the altered ratio of $^{14}C:^{12}C$.
- Plants and animals that breathe/ingest CO_2 would reflect the altered ratio.
- Carbon-dating plant/animal specimens without taking into account the different $^{14}C:^{12}C$ ratio caused by a stronger magnetic field in the past, would mistakenly indicate a much greater age than is warranted.[314]

So the farther back in time we go, the less ^{14}C would have been produced, so the starting ratio would have been smaller. This overturns Assumption #1 that uniformitarianism depends on. The lower amount of ^{14}C we observe nowadays in an ancient biological specimen would be interpreted as a greater age than what is actually the case. But Earth's decaying magnetic field is rarely taken into account when doing carbon dating. Why? Because it opens up all sorts of "cans of worms" that also militate against deep time.[315] So it's easier just to ignore it. (More about Earth's magnetic field below.)

One of the problems that arises with carbon dating is that evolutionists sometimes claim that they have found ^{14}C in organic specimens that are claimed to be *hundreds of millions* of years old. If that were true, there would be so little ^{14}C left that the margin for error would utterly overwhelm it, ren-

---

[314] Gavin Cox, "Time Fears the Pyramids?", *Creation*, 42(1):18–20, January 2020. [Online at Creation.com/pyramids-of-egypt-age; accessed March 27, 2021.]

[315] D. Russell Humphreys, "The Earth's Magnetic Field Is Young," *Acts & Facts,* August 1, 1993. [Online at ICR.org/article/earths-magnetic-field-young; accessed March 27, 2021.]

dering the measurement ludicrous. ^{14}C would be completely undetectable after only 100,000 years.[316]

So *something* in that radiocarbon dating process was wrong:

- The imagined "hundreds of millions of years" age was wrong,
- The initial amount of ^{14}C was wrong,
- The decay rate was not constant, or
- Other processes affected the amount of ^{14}C in the sample.

Errors probably occurred in all four of the above areas, but at first glance I would say the most likely reason for the error was that the samples in question were "assigned" an age so they would fit with the theory of evolution. Remember, *much* data is interpreted with a presupposition that evolution is unquestionable, so the data *has* to fit, and it is *made* to fit.

In other tests, diamonds that were supposedly *billions* of years old still had detectable levels of ^{14}C!

To illustrate how fast ^{14}C disappears, read what Dr. John Baumgardner of Los Alamos National Laboratories says about it (the "AMS" he refers to stands for "accelerator mass spectrometer," the device used to measure ^{14}C quantities):

> If one started with an amount of pure ^{14}C equal to the mass of the entire observable universe, after 1.5 million years [a tiny fraction of evolutionist time] there should **not be a single atom of ^{14}C remaining!** Routinely finding ^{14}C/^{12}C ratios on the order of 0.1–0.5% of the modern value—a hundred times or more above the AMS detection threshold—in samples supposedly tens to hundreds of millions of years old is therefore a huge anomaly for the uniformitarian framework [the evolutionary time scale].[317]

Baumgardner goes on to say:

> The bottom line of this [^{14}C] research is that the case is now extremely compelling that the fossil record was produced just

---

[316] Jake Hebert, Ph.D., "Rethinking Carbon-14 Dating: What Does It Really Tell Us about the Age of the Earth?" March 29, 2013. [Online at ICR.org/article/rethinking-carbon-14-dating-what-does; accessed June 27, 2020.]

[317] John Baumgardner, "Carbon Dating Undercuts Evolution's Long Ages," *Impact* Article #364 (San Diego: Institute for Creation Research, October 2003): ii. Emphasis added.

CHAPTER 5: A YOUNG EARTH

a few thousand years ago by the global Flood cataclysm. The evidence reveals that **macroevolution as an explanation for the origin of life on earth can therefore no longer be rationally defended.**[318]

Let's validate the above analysis, just to confirm. If every atomic particle in the universe—not just every fully assembled atom, but *every single atomic particle in the universe*—were an atom of carbon-14, how long before they would all be gone? Since the number of atomic particles in the universe is estimated[319] to be about 10^{80}, all we need to do is figure out how many 0.5s you would have to multiply together to arrive at 10^{-80}.

Even without getting into any fancy mathematics, a little trial and error reveals that 265 halves multiplied together, or 0.5^{265}, gives you roughly 10^{-80}. So, after only 265 half-lives of carbon-14, *all* the carbon-14 would be gone, even starting with a whole universe-full. How many years is 265 half-lives of carbon-14? It's 265 × 5,730, or a little more than one and a half million years—a tiny fraction of the speculated age of the universe. This confirms the previous statement from Dr. Baumgardner.

So it is totally ludicrous to think that carbon dating could yield any kind of a reliable answer above a few hundred thousand years, since most carbon-14 analyses start with a sample size smaller than the universe. And, as stated above, statistical uncertainty restricts its upper limit of usability to 50,000 years, or even 24,000 years.

---

[318] John Baumgardner, "Carbon Dating Undercuts Evolution's Long Ages," *Impact* Article #364 (San Diego: Institute for Creation Research, October 2003): iv. [Online at ICR.org/article/carbon-dating-undercuts-evolutions-long-ages.] Emphasis added.

[319] Jay Bennett, *Popular Mechanics,* July 11, 2017. [Online at PopularMechanics.com/space/a27259/how-many-particles-are-in-the-entire-universe; accessed October 28, 2020.]

Potassium-Argon

Unlike radiocarbon dating, in which the substance of interest decays and eventually disappears, potassium-argon (or "K-Ar") dating uses the ratio of ^{40}K to ^{40}Ar as the former decays into the latter. ^{40}K has a very long half-life: 1.248 billion years.

^{40}Ar is an isotope of argon with an atomic weight of 40, and ^{40}K is an isotope of potassium with an atomic weight of 40. Potassium-Argon dating is used to determine the age of rocks, and the idea is that argon, being a gas, readily escapes from molten magma, but when the magma solidifies into solid rock, it's much more difficult to escape. Therefore, as the ^{40}K decays into ^{40}Ar, the argon is trapped, and by measuring how much there is, we can tell how long it has been since the rock solidified. The actual dating uses both the amount of remaining ^{40}K in the sample and the amount of ^{40}Ar (the "daughter" isotope) in its calculations.

The same kinds of assumptions apply here that apply to ^{14}C dating, and again, it is unknown whether or not those assumptions are valid. One of those assumptions is that there is no ^{40}Ar in the rocks at the moment they harden because it all escaped while the rock was molten.

So let's look at some examples[320,321] of potassium-argon dating to see if the assumptions bear out. In the following table, the dates are known because of historical records written by people living at the time of the volcanic events creating the rocks being tested. The older ones are known to within a few

---

[320] G. B. Dalrymple, "^{40}Ar/^{36}Ar Analyses of Historic Lava Flows," *Earth and Planetary Science Letters*, 6 (1969): pp. 47–55.

[321] For the original sources of these data, see the references in A. A. Snelling, "The Cause of Anomalous Potassium-Argon 'Ages' for Recent Andesite Flows at Mt. Ngauruhoe, New Zealand, and the Implications for Potassium-Argon 'Dating'," R. E. Walsh, ed., *Proceedings of the Fourth International Conference on Creationism* (1998, Pittsburgh, PA, Creation Science Fellowship), pp. 503–525. [Online at ICR.org/research/index/researchp_as_r01/; accessed June 27, 2020.]

decades, and the newer ones are known to the exact year and even the exact day. Let's see how K-Ar dating fares:

Source of Sample, Location	Date of Event	K-Ar Age (10^6 yrs)	Error Factor
Mt. Etna basalt, Sicily	122 BC	0.25±0.08	116×
Kilauea basalt, Hawaii	≈1,000	42.9±4.2, 30.3±3.3	42,058×, 29,705×
Sunset Crater basalt, Arizona	1064–1065	0.26±0.09, 0.25±0.15	272×, 261×
Rangitoto basalt, Auckland, NZ	≈1200	0.15±0.47	187×
Medicine Lake Highlands obsidian, Glass Mountains, California	≈1500	12.6±4.5	25,200×
Hualalai basalt, Hawaii	1800–1801	1.6±0.16, 1.41±0.08	7,272×, 6,409×
Hualalai basalt, Hawaii	1800–1801	22.8±16.5	103,636×
Kilauea basalt, Hawaii	≈1800	21±8	105,000×
Mt. Lassen plagioclase, California	1915	0.11±0.03	1,047×
Kilauea Iki basalt, Hawaii	1959	8.5±6.8	139,344×
Mt. Stromboli, Italy, volcanic bomb	1963	2.4±2	42,105×
Mt. Etna basalt, Sicily	1964	0.7±0.01	12,500×
Mt. Etna basalt, Sicily	1972	0.35±0.14	7,291×
Dacite lava flow in Mount St. Helens lava dome	1980	0.35±0.05, 2.8±0.6	8,750×, 70,000×
Anorthoclase in volcanic bomb, Mt. Erebus, Antarctica	1984	0.64±0.03	17,777×

Fascinating. Apparently, at least *some* of the assumptions are not bearing out. . . Look at those error factors! The smallest error was off by a factor of more than 100, and the largest error was off by a factor of more than 130,000! (See also more information on the Mount St. Helens lava dome.[322])

In another experiment in 2000, samples were obtained of lava flows from one of New Zealand's most active volcanoes, Mount Ngauruhoe. These were recent eruptions, seen by living eyewitnesses, so the dates of the eruptions were known to the very day. The eruption dates were: February 11, 1949; June 4, 1954; June 30, 1954; July 14, 1954; and February 19, 1975. Since the study was done in 2000, all samples were between 25 and 51 years old.

---

[322] S. A. Austin, "Excess Argon within Mineral Concentrates from the New Dacite Lava Dome at Mount St Helens Volcano," *Creation Ex Nihilo Technical Journal*, 10 (1996): pp. 335–343. [Online at ICR.org/research/index/researchp_sa_r01/; accessed June 28, 2020.]

Oh, Evolve! (Good Luck With That. . .)

Thirteen samples were sent for potassium-argon dating to one of the most reputable commercial dating labs in the world: Geochron Laboratories in Cambridge, Massachusetts. Here are the results for the 13 samples:

Number of Samples	Age Determined by K-Ar
4	< 270,000 years old
1	< 290,000 years old
1	800,000 years old
3	1 million years old
1	1.2 million years old
1	1.3 million years old
1	1.5 million years old
1	3.5 million years old

And all of these were known to be 51 years old or less![323]

And there are many more. For example, ten diamonds from Zaire were dated as being 6 *billion* years old, which is older than evolutionists' speculation of the age of the Earth itself! Looks to me like some more work needs to be done on dating methods.

Richard G. Klein, a Professor of Biology and Anthropology at Stanford University, described the limitations of potassium-argon dating as follows:

> The very long half life of ^{40}K means that the radiopotassium method has no practical maximum limit (it can be used to estimate the age of the earth) but that in most cases **it cannot be used to date rocks younger than a few hundred thousand years old.** This limitation occurs because they contain too little ^{40}Ar for accurate measurement and because **the statistical error associated with the age estimate therefore may be as large as the estimate itself.**[324]

---

[323] Marvin L. Lubenow, *Bones of Contention* (BakerBooks, Grand Rapids, Michigan), 2004, p. 279.

[324] Klein, Richard G., *Human Career*, 2nd ed., p. 37.

The "Sweet Spot"

You may have noticed in the two previous sections that Carbon-14 dating is unreliable for dates greater than 50,000 (or even 24,000) years, and K-Ar dating is unreliable for any ages less than "a few hundred thousand years." Isn't it convenient that the gap—the "sweet spot"—between these two dating methods is the same area in which most of human evolution is supposed to have occurred?

You see the significance of this, I'm sure. Since both dating methods are unreliable and, as shown above, give very different answers, even for tests of the same kind, *on the same samples,* it's pretty easy to get a dated age that places the fossil or the rock at whatever age is desired, to fit the pre-decided long-age time scale.

Once again, in case you feel I might be exaggerating, read Marvin Lubenow's book *Bones of Contention,*[325] especially Sections V, VI, and VII, to read about many documented cases in which certain fossils' ages were pushed out, and then pulled in, and then pushed out again, because contradictions between the fossils and deep time kept showing up. And remember, deep time must be preserved at all costs.

Basically, the gap between the valid ranges of ^{14}C and K-Ar dating means that there is no such thing as a legitimate fossil sequence leading to modern humans. The fact that evolutionists continue to make absolute age statements on their dating samples only shows that the statements are based on a belief system, not on actual science.

But even more importantly than that, if radiometric dating gives us wrong ages on samples for which we *do* know the actual age, why should we have any confidence that they will give us correct dates on samples that do *not* have any eyewitnesses as to their age? Actually, we shouldn't. If these dating methods give wrong answers about rocks of a known age, it is just foolish to trust them for rocks of *unknown* age.

The above examples were not isolated instances; there are many more. So many, in fact, that there is a growing body of evidence that all radiometric dating methods are completely unreliable.[326]

---

[325] Marvin L. Lubenow, *Bones of Contention* (BakerBooks, Grand Rapids, Michigan), 2004.
[326] For articles that are very enlightening, but not highly technical, see "Radiometric Dating" [Online at AnswersInGenesis.org/home/area/faq/dating.asp; accessed July 24, 2020.]

For a technical discussion, see Larry Vardiman, Andrew Snelling, and Eugene Chaffin, eds., *Radioisotopes and the Age of the Earth,* vol. 1 and 2 (El Cajon, CA: Institute for Creation Re-

Earth's Magnetic Field

You already read other sections talking about the magnetic fields of other planets and moons, and the Earth is no different in that respect. Earth has a magnetic field, and it is losing its strength. Like radioactive isotopes, magnetic fields also have a half-life, and the half-life of Earth's magnetic field is 1465±165 years.[327]

The obvious question that arises at this point is: how could the magnetic field have lasted all this time if the Earth is really 4½ billion years old? And that is an extremely good question. Even when we take into account the evidence that the magnetic field has reversed many times in the past, evolutionary theory doesn't explain how it could have lasted in *any* form until now, if the Earth is really that old.

Using a Biblical timeline, the Earth's magnetic field is much easier to explain—remember Occam's Razor?[328,329] This theory, using a Biblical timeline, matches paleomagnetic data, historic data, and presently measurable data. The researchers' conclusion is that the total energy of the magnetic field has been decaying at least as fast as it is observed to do today, and perhaps even faster in the past. But that also leads to a conclusion evolution is very uncomfortable with: the Earth's magnetic field cannot be more than 20,000 years old.[330]

---

search; Chino Valley, AZ: Creation Research Society, 2000 and 2005).

See also "Half-Life Heresy," *New Scientist,* October 21, 2006, pp. 36–39. [Online at NewScientist.com/channel/fundamentals/mg19225741.100-halflife-heresy-accelerating-radioactive-decay.html; accessed July 24, 2020.]

[327] Humphreys, D. R., "The earth's magnetic field is still losing energy," *Creation Research Society Quarterly,* 39(1):3–13, June 2002. [Online at CreationResearch.org/earths-magnetic-field-still-losing-energy/; accessed June 30, 2020.]

[328] Humphreys, D. R., "Reversals of the earth's magnetic field during the Genesis flood," *Proceedings of the First International Conference on Creationism,* Vol. II, Creation Science Fellowship (1986), Pittsburgh, PA, pp. 113–126. [Online at ICR.org/i/pdf/technical/Reversals-of-Earths-Magnetic-Field-During-the-Genesis-Flood.pdf; accessed June 30, 2020.]

[329] Coe, R. S., M. Prévot, and P. Camps, "New evidence for extraordinarily rapid change of the geomagnetic field during a reversal," *Nature* 374:687–92 (20 April 1995).

[330] Humphreys, D. R., "Physical mechanism for reversals of the earth's magnetic field during the flood," *Proceedings of the Second International Conference on Creationism,* vol. II, Creation Science Fellowship (1991), Pittsburgh, PA, pp. 129–142. [Online at ICR.org/article/reversals-earths-magnetic-field-flood; accessed June 30, 2020.]

That Pesky Erosion...

Erosion, in the geological sense, is the wearing away of rock, dirt, and other sediment, which is then carried away. Wind and especially water are the active agents in this process. The problem with deep time becomes glaringly apparent (again) when comparing the rate of mountain-building with the rate of erosion.

According to the deep-time belief, the earth is about 4.5 billion years old, and the continents formed about 3.5 billion years ago. The Geological Society of America collected data from every continent, converted the data to comparable units, and calculated that continental erosion happens at a rate of about 40 feet every million years.[331] The average thickness of continental crust is 2044 feet above sea level, which means the continents would have eroded away entirely in only about 50 million years. In the 3.5 billion years the continents have supposedly been here, they have had time enough to erode completely away *70 times!* All the continents would have eroded away after less than 1.5% of their alleged age! Why are they still here?

And the above calculations are for outcrops, the high-elevation areas that get less rainfall and less collected water. In continental basins—the lower elevations that get more rainfall, and in which water collects—the erosion rate is much faster: more than 650 feet per million years. At that rate, the continents would have eroded completely away in only three million years.[332] In the ostensible 3.5 billion years since the continents formed, they could have eroded down to sea level more than *1100 times!* After only 0.09% of their alleged age, all the continents would be completely gone—eroded into the oceans!

But perhaps mountain-building happened through tectonic processes, and uplifting from the earth's mantle would have continually replenished the rock being eroded at these alarming rates. That could happen, couldn't it? But the problem with that scenario is that that much continental uplift would have obliterated the very rock layers evolutionists cling to as evidence for evolution.

---

[331] Portenga, E. W. and R. R. Bierman. 2011. "Understanding Earth's Eroding Surface with ^{10}Be." *GSA Today.* 21(8): pp. 4–10.

[332] Brian Thomas, "Continents Should Have Eroded Long Ago" 2011. [Online at ICR.org/article/continents-should-have-eroded-long; accessed December 2, 2021.]

As University of Loma Linda geologist Ariel Roth puts it:

> It has been suggested that mountains still exist because they are constantly being renewed by uplift from below. However, **this process of uplift could not go through even one complete cycle of erosion and uplift without eradicating the layers of the geologic column found in them.** Present erosion rates would tend to rapidly eradicate evidence of older sediments; yet these sediments are still very well-represented, both in mountains and elsewhere.[333]

Oh, dear. It's just not going at all well for the theory of evolution these days. . .

But that's not all. How old are the Hawaiian Islands? According to Wikipedia, the oldest Hawaiian islands are 28 million years old.[334] Is that reasonable? Not if we do the math.

Erosion rates on the Hawaiian Islands has been analyzed for more than a century, and shorelines have been measured to erode at a rate of about 0.11 meters per year, or roughly 0.4 feet per year, on average.[335] Roughly five inches per year. If we do the math, assuming the uniformitarianism that evolutionists swear by, that comes to about 400,000 feet per million years, which is more than 75 miles. And since erosion happens along *all* the shorelines, and not just one side of each island, the diameter of the islands would be shrinking by more than 150 miles per million years.

As you might surmise, this is a problem for a group of islands whose largest—and the only one with active vulcanism to counter the erosion—is the Big Island, which is only about 75 miles at its largest dimension. How could the smaller islands still exist, if they really are tens of millions of years old? Comparable problems exist for all islands worldwide. Yes, that's a really big problem for uniformitarian geology.

---

[333] Roth, A. A. 1986. "Some Questions about Geochronology." *Origins*. 13(2): pp. 64–85. [Online at grisda.org/origins-13064; accessed December 2, 2021.]

[334] wikipedia.org/wiki/Hawaiian_Islands; accessed March 6, 2024.

[335] Fletcher, C.H. et al. National Assessment of Shoreline Change: Historical Shoreline Change in the Hawaiian Islands. U.S. Geological Survey Open-File Report, 2011–1051, 55. [Online at pubs.usgs.gov/of/2011/1051/pdf/ofr2011-1051_report_508_rev052512.pdf; accessed March 6, 2024.]

Earth's Oceans

The land of Earth is not the only aspect of the planet that indicates a young age; the oceans do as well, perhaps even more so than the land.

Ocean-Floor Sediment

Have you ever noticed that rivers are not typically crystal-clear? If you've ever seen a river anywhere, other than mountain streams that are very near to their headwaters, you've noticed that most rivers are a greenish or brownish color because of all the sediment being carried along by the water.

Where does all the sediment go? All rivers eventually flow into the oceans, so the oceans are the destination for all the sediment coming down all the rivers in the world. This immediately brings up a question: if the Earth is really 4.5 billion years old, how much sediment would have washed into the ocean by now?

Wind and water (but mostly water) transport about 20 billion tons of mud and dirt into the ocean each year.[336] All this collects on the ocean floor.

Does such sediment ever go away? Actually, yes, by tectonic plate **subduction**, where one tectonic plate slides underneath another one. Typically, this is an ocean-floor plate sliding under a continental plate, and the built-up layer of sediment goes with ocean-floor plate, where the sediment remelts and becomes part of the mantle. The problem is that this process is not nearly fast enough to keep up with the sediment buildup from rivers. Only about 1 billion tons of sediment is "recycled" by subduction back into the mantle each year.[337]

What happens to the other 19 billion tons per year? It simply builds up on the ocean floor, getting deeper and deeper. If that sedimentation rate has been constant all along (as uniformitarianism requires), all the sediment we currently see would have been built up in only 12 million years! According to evolutionists, the ocean has been in existence for about 3 billion years, so are we to believe that for the first 2,988,000,000 years, no erosion happened?

---

[336] Milliman, John D. and James P. M. Syvitski, "Geomorphic/tectonic control of sediment discharge to the ocean: the importance of small mountainous rivers," *The Journal of Geology*, vol. 100, pp. 525–544 (1992).

[337] Hay, W. W., et al., "Mass/age distribution and composition of sediments on the ocean floor and the global rate of sediment subduction," *Journal of Geophysical Research*, 93(B12):14,933–14,940 (10 December 1988).

If erosion has been happening in the past at the rate at which is happening today, the oceans would be "drowning" (so to speak) in sediment, with layers dozens of miles deep. But we don't see that. Again, uniformitarianism bites evolutionists by completely contradicting observations.

Chemical Imbalances

Sediment is not the only thing that gets washed into the ocean: sodium, nickel, and a host of other chemicals get washed into the oceans by all the rivers of the world eroding the landmasses. Let's take a look at a couple.

Sodium

Sodium (a major constituent of regular salt) gets washed into the ocean constantly: about 450 million tons of sodium each year.[338] This influx is far greater than the 27% of it, or about 122 million tons, that gets removed from the sea each year.[339,340]

Clearly, this will build up sodium to toxic levels after some amount of time. Assuming the oceans started out with no sodium at all, how long would it take to build up the sodium content to the levels we see today? Assuming uniformitarianism, which evolutionists insist on, it would only take 42 million years. That's less than 1½% of the age of the oceans, according to evolutionary doctrine. What happened during the remaining 98½% of the time?

Even being very generous to evolutionary scenarios, they still come up with a maximum age of the oceans as 62 million years. But that's still less than 2.1% of the evolutionists' presumed age of the oceans.

Nickel

Nickel is another chemical clock that discredits the idea of deep time.

---

[338] Meybeck, M., "Concentrations des eaux fluviales en elements majeurs et apports en solution aux oceans," *Revue de Géologie Dynamique et de Géographie Physique* 21(3):215 (1979).

[339] Sayles, F. L. and P. C. Mangelsdorf, "Cation-exchange characteristics of Amazon River suspended sediment and its reaction with seawater," *Geochimica et Cosmochimica Acta* 43:767–779 (1979).

[340] Austin, S. A. and D. R. Humphreys, "The sea's missing salt: a dilemma for evolutionists," *Proceedings of the Second International Conference on Creationism*, vol. II, Creation Science Fellowship (1991), Pittsburgh, PA, pp.17–33. [Online at CreationICC.org/papers.php]

CHAPTER 5: A YOUNG EARTH

Let's look at nickel the same way we looked at sodium, above, but get into a bit more detail.[341] In order to set a maximum limit on the age of the oceans, we need only a few values:

- The amount of water dumped into the oceans by all the world's rivers,
- The amount of nickel per volume unit of river water,
- The amount of nickel currently dissolved in seawater, and
- The total volume of seawater worldwide.

From these four values, we can do the required calculations. The amount of water flowing from the world's rivers into the ocean each year is is 37,288 cubic kilometers, and the ocean holds 1.338 billion cubic kilometers of water.[342] The amount of nickel in river water is about 1–3 micrograms per liter (μg/l)[343], and the concentration of nickel dissolved in the world's oceans is 0.228–0.693 μg/l.[344]

First, we have to convert some units so we're being consistent. In this table, we're using lowest values of nickel concentration in rivers and seawater: 1 μg/l and 0.228 μg/l, respectively. Conveniently, the number of micrograms in a metric ton is exactly the same number as the number of liters in a cubic kilometer, so the conversion is easy (in the table below, see the rows with arrows for step-by-step confirmation to this). As the table shows, calculating the amount of time to bring the oceans up to current nickel concentrations is pretty straightforward:

Flow of all rivers:	37,288 km³/yr
River Ni concentration:	1 μg/l
→m³ (×1000) =	1000 μg/m³
→km³ (×10⁹) =	10¹² μg/km³
→metric tons (÷10¹²) =	1 mt/km³
Ni added to oceans:	37,288 mt/yr
Volume of ocean:	1.338 billion km³

---

[341] David Whyte, "Nickel Concentration Indicates Young Oceans" *Creation* 38(3):54–55, July 2016. [Online at Creation.com/nickel-concentration-indicates-youthful-oceans; accessed May 31, 2022.]

[342] Dai, A. and Trenberth, K.E., "Estimates of Freshwater Discharge From Continents: Latitudinal and Seasonal Variations," *Journal of Hydrometeorology* 3(6):660–687, 2002.

[343] Baralkiewicz, B., and Siepak, J., "Chromium, Nickel and Cobalt in Environmental Samples and Existing Legal Norms," *Polish Journal of Environmental Studies* 8(4):201–208, 1999; pjoes.com.

[344] WHO Europe, *Air Quality Guidelines for Europe*, Second Edition, Chapter 6.10 "Nickel," p. 162, 2000; euro.who.int.

Ni concentration in ocean:	0.228 µg/l
Ni in oceans:	305.1 million mt
Ni accumulation time:	≈ 8181 yr

Well, isn't that interesting? Assuming that oceans originally contained no nickel at all, the observed flow rate and nickel content of rivers, the volume of the oceans, and the nickel concentration observed in seawater today, it would take less that 8200 years to come up to today's observed concentration! Of course, if the oceans had some nickel in them to begin with, it would take even less time to get up to today's observed levels.

Note in the table above, I used the low values of the nickel concentrations in river water and seawater. What happens if I use the high values of each range? It would still only take 8289 years! And what if (just to be generous to the theory of evolution) I used the lowest river water concentration and *highest* seawater concentration? It's still less than 25,000 years to get to where seawater's nickel concentration level is today. All of these values would elicit cries of outrage from those who are believers in deep time. This is happening *way* too fast!

CHAPTER 5: A YOUNG EARTH

The above discussion takes into account how nickel is being *added* to seawater, but it doesn't take into account any nickel that might be *removed* from seawater. And actually, nickel does get removed from seawater, but even so, it's not nearly at fast enough a rate to allow for deep time to be a credible option.

Oceanographers have discovered millions of metallic lumps on the ocean floors all over the world; an example is shown at right. They have been dubbed **manganese nodules** because the most abundant element in them is manganese. Also present in these nodules are iron and nickel, so maybe nickel is being extracted from seawater and stored in the nodules. Is that plausible? Could that make deep time a credible theory? Again, let's do the math.

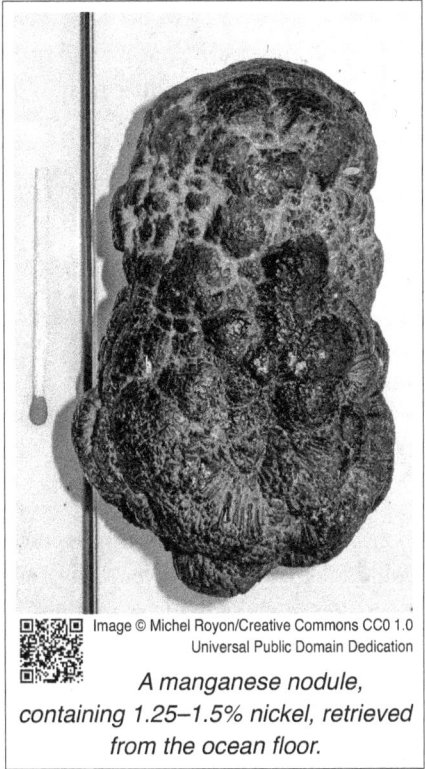

Image © Michel Royon/Creative Commons CC0 1.0 Universal Public Domain Dedication

A manganese nodule, containing 1.25–1.5% nickel, retrieved from the ocean floor.

Judging by the quantity of nodules found on all kinds of seafloors (both shallow and deep, and even in some lakes) and their coverage (obscuring up to 70% of the seafloor in some areas), the total worldwide mass of these nodules is estimated to be 500 billion metric tons. Analysis of these nodules shows them to be composed of 1.25–1.5% nickel.[345] So how long would the rivers take, at their observed rate of adding nickel to the oceans, to supply all the nickel in all these nodules?

---

[345] Parada, J., Feng, X., Hauerhof, E., Suzuki, R., Abubakar, U., "The Deep Sea Energy Park: Harvesting Hydrothermal Energy for Seabed Exploration," *The LRET Collegium 2012 Series*, Vol. 3, University of Southampton, p. 8, 2012.

Glad you asked. In the table below, the rows above the gap are the same as the previous table (but are still needed in these calculations), and the rows below it are the new nodule-related data.

Flow of all rivers:	37,288 km³/yr
River Ni concentration:	1 µg/l
→m³ (×1000) =	1000 µg/m³
→km³ (×10⁹) =	10¹² µg/km³
→metric tons (÷10¹²) =	1 mt/km³
Ni added to oceans:	37,288 mt/yr
Total mass of nodules:	500 billion mt
Ni concentration in nodules:	1.5%
Total Ni in nodules:	7.5 billion mt
Nodule accumulation time:	201,137 yr

So even at the highest observed concentration of nickel in the nodules, it would take the world's rivers only a bit more than 200,000 years to supply all the nickel in these seafloor nodules! And less than 210,000 years to supply all the nickel for the nodules *and* bring the ocean's water up from zero to today's levels! That's only about 0.007% of the ocean's supposed deep-time age; what happened for the other 99.993% of the ocean's lifetime? Are we supposed to believe that no erosion happened? Or that there was no nickel at all in the runoff?

And then, in another desperate attempt to rescue deep time, plate tectonics is considered: maybe seafloor subduction could remove nodules and push them down into the earth's mantle, removing them from consideration. But even this is not enough. There is just *no way*, given current-day observations and measurements, plus uniformitarianism (which evolutionists insist upon), for deep time to be a viable option.

But wait: there's more!

An environmental guideline from the UK states that a nickel concentration greater than 30ppb (30 parts per billion) becomes toxic to sea life. Since nickel is constantly being dumped into the oceans because of normal erosion of the continental landmasses, how long would it take, at the current rate of increase (which uniformitarianism insists upon), for nickel to reach toxic levels in the oceans? *Barely more than a million years!* [346]

---

[346] Paul Price, "The Oceans Show Us a Young Earth," *Creation* 42(2):16–17, April 2020. [Online at Creation.com/oceans.]

Assuming the oceans are three billion years old—and the long-age guesses are often changed as more and more problems are found—that is only 0.03% of the age of the oceans to reach nickel toxicity! So again, what happened during the first 99.97% of the oceans' lifespan? Did no erosion happen at all? Did river sediment contain no nickel at all? Was some nickel-removing process in effect in the past that is no longer happening? All of these suggestions run afoul of uniformitarianism.

And clearly, the oceans are not yet toxic to sea life, so uniformitarianism would require even *more* than 99.97% of the oceans' presumed existence to add no nickel at all to the oceans. Very convenient how it does that.

Other Chemical Clocks

As you might expect, sodium and nickel are not the only things that overturn the whole idea of a 3-billion-year-old ocean.

We can measure the rate at which various chemicals are entering the oceans, and we can measure how much is there now. As demonstrated above, it's not difficult math to calculate the amount of time it would take to get to the point where we are today, if the oceans started with none at all of the chemicals in question. (And if there *were* some of the chemicals in the oceans to begin with, it would even further support the creation model.) *In every case,* uniformitarian estimates yield results that evolutionists claim are impossible, given the presupposition of deep time. And they are correct: it *doesn't* make sense if you assume a 3-billion-year-old ocean.

For much more detail, see the fascinating table in *Acts/Facts/Impacts:*[347] it shows *76* different earth-sciences measurements—*dozens* of which pertain to chemical levels in the oceans that would be much greater, even *toxically* greater, if the oceans really were three billion years old. Things like magnesium, silicon, potassium, copper, gold, silver, mercury, lead, tin, and more. And *all* of these 76 items of measurable quantities indicate the Earth is much, much younger than evolutionists claim.

---

[347] *The Battle for Creation: Acts/Facts/Impacts, Volume 2,* ed. Henry Morris and Duane Gish (Creation Life Publishers: San Diego, CA) 1976, pp. 243–246.

The Moon

Earth's moon is really quite amazing, once you start getting into the details of its construction and specifications.

Two evolutionists, Christopher Knight and Alan Butler, have studied the moon in detail and are flabbergasted by what they discovered. After starting out with the question "Who built the moon?", they continue:

Earth's Moon

> The Moon is 400 times smaller than the star at the centre of our solar system, yet it is also just $1/_{400}$ of the distance between the Earth and the Sun. . . By **some absolutely incomprehensible quirk of nature**, the Moon also manages to precisely imitate the perceived annual movements of the Sun each month.[348]

The Moon's Origin

The question posed in the title of the book quoted above—*Who Built the Moon?*—is a very good question. Many naturalistic hypotheses have been put forward, but all of them have failed. Why? Because of the evolutionists' non-negotiable presupposition of deep time.

Here are some of the ideas that have been put forth over the years:

- Maybe the Moon was formed somewhere else, and was captured by Earth's gravity. Such a near miss is ridiculously unlikely, but conceivable. The problem would be getting rid of the Moon's excess inertia while it's close enough to Earth to be captured. The only thing people thought of was Earth's atmosphere being thick enough to slow down the Moon as it passed by. But no: the Earth doesn't have enough gravity to have an atmosphere that thick. And without such an atmosphere,

---

[348] Knight, Christopher and Alan Butler. 2005. *Who Built the Moon?* London: Watkins Publishing, pp. 4–5. Emphasis added.

if the Moon were caught at all, it would have a highly elliptical orbit, instead of the near-circular one has.

- Or maybe the Earth and the Moon formed at the same time, out of the same primordial dust-and-gas cloud of Nebular Hypothesis fame. But no: if they both condensed from the same dust-and-gas cloud, Earth and Moon would have very similar compositions, and rocks brought back from the Moon have much less metal content than Earth rocks. Also, the Moon's orbital plane would coincide with the Earth's equator—but it doesn't. And we saw above how utterly deficient the Nebular Hypothesis was at explaining pretty much anything in the solar system.

- Or maybe the early Earth was spinning so fast, that a blob flew off and became the Moon. But no: If the Earth was spinning that fast while it was still soft enough to have a part of it fly off, the Earth would not be the almost-perfect sphere that it is; it would have been an oblate spheroid (a squashed sphere). Not only that, the required angular momentum of the Earth-Moon system is not present (and the Conservation of Angular Momentum is one of the most easily proven laws of science).

- Or maybe a rogue planet about twice the mass of Mars flew through the solar system, hitting the early Earth a glancing blow, and launching enough debris into orbit to collect into the Moon. But no: in such a scenario, the impactor would unavoidably contribute some mass to the debris that would form into the Moon, and alter the ratio of rare-earth metals. But the Earth and the Moon have the same ratio —not the same *amount*, but the same *ratio*—of rare-earth metals.[349] (Since the planets in the solar system have very diverse compositions, it is likely that the imaginary planet would too.)

Many ideas of moon formation have come and gone, including the last one mentioned above, which was just proposed (or re-proposed) in 2013. As science journalist Daniel Clery noted:

> As a result, researchers are casting around for new explanations. At a meeting at the Royal Society in London last month—the first devoted to moon formation in 15 years— experts reviewed the evidence. **They ended the meeting in**

---

[349] Samec, R. 2013. "Lunar formation—collision theory fails." *Journal of Creation*. 27 (2): pp. 11–12. [Online at Creation.com/lunar-formation-collision; accessed July 28, 2020.]

an even deeper impasse than before, as several proposed solutions to the moon puzzle were found wanting.[350]

The Moon's genesis is indeed a problem for evolutionists, as Robert C. Humes states in his book on elementary astronomy:

> The whole subject of the origin of the moon must be regarded as highly speculative.[351]

In other words, "We won't accept the Bible's explanation, even though we have nothing to offer in its place."

Geophysics professor Dr. Louis B. Slichter of MIT has put much study into the problem as well, and he concurs: ". . .the time scale of the Earth-moon system still presents a major problem."[352]

The Moon's Construction

The Moon is really enormous, as compared to the planet it is orbiting. No other planet in the solar system has a moon so large in comparison, as does the Earth and its Moon:[353] It is over 25% the diameter of the Earth.

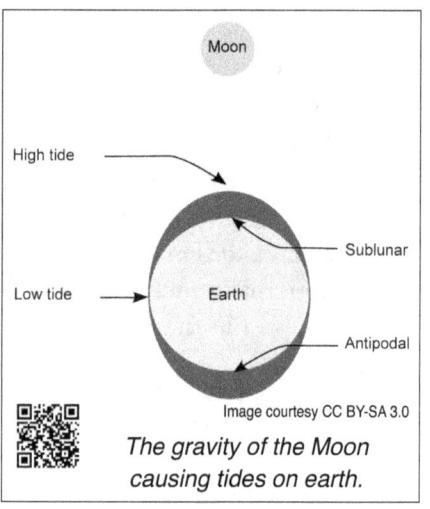

The gravity of the Moon causing tides on earth.

Because the moon is so large, it causes tides to happen in Earth's oceans, as shown here. The sun also contributes to the tides, but the effect of the Moon is much larger. These tides keep the oceans fresh; without them, the oceans would stagnate. It

---

[350] Clery, Daniel. "Impact Theory Gets Whacked." *Science*. 342 (6155): October 11, 2013, pp. 183–185. [Online at Science.ScienceMag.org/content/342/6155/183; accessed July 28, 2020.] Emphasis added.

[351] Robert C. Humes, *Introduction to Space Science* (John Wiley, 1971).

[352] Slichter, Louis B., "Secular Effects of Tidal Friction Upon the Earth's Rotation," *Journal of Geophysical Research*, 1964, Vol. 8, No. 14, pp. 4281–4288.

[353] Charon is fully half the diameter of Pluto, so the Charon/Pluto ratio is bigger that the Moon/Earth system, but Pluto is no longer considered a planet, so technically, this statement is still correct.

provides light at night,[354] brighter than any other moon in the solar system, as seen from its planet.

So how did the moon come to be? As we saw above, evolutionists are at a loss to explain it, because their doctrine insists on the 4.5-billion-year age of the solar system, as well as uniformitarianism, and these two requirements invariably cause contradictions and impossible situations to arise.[355]

Because the moon and sun appear exactly the same size in Earth's sky, precise solar eclipses are possible, and therefore, we are able to study the sun's chromosphere and other aspects of the sun that would otherwise be impossible to see because of the sun's glare. No other known moon in the solar system covers the sun so precisely, when viewed from its planet.

Another evolutionist commented:

> **The Moon's orbit is fiendishly difficult to explain,** moving as it does around a rotating Earth, which together form [essentially] a 'double-planet' system that orbits around the Sun. It is a classic example of a three-dimensional, gravitational three-body problem.[356]

The Moon's Magnetic Field

Planetary scientists have been trying for years to figure out why the Moon still has a magnetic field. Like so many of the moons of other planets, Earth's Moon is way too small to maintain a magnetic field over its hypothetical 4.5-billion-year age.

According to the National Academy of Sciences, these three facts are germane to the discussion:

- The moon has a magnetic field today. Granted, it is weak, but it shouldn't be there at all if evolution's deep time is real.
- The moon rocks that the Apollo missions returned still exhibit remanent magnetism that indicates the Moon's magnetic field was once as strong as the Earth's is today.

---

[354] Genesis 1:16 (NIV): God made two great lights—the greater light to govern the day and **the lesser light to govern the night.** He also made the stars.

[355] Whitcomb, John C. and Donald B. DeYoung, *The Moon, Its Creation, Form and Significance,* BMH Books, Winona Lake, Indiana, p. 41.

[356] Dumé, B. "Moon's bulge linked to early orbit." *PhysicsWeb.* Posted on physicsworld.com August 3, 2006, accessed September 21, 2015. [Online at PhysicsWorld.com/a/moons-bulge-linked-to-early-orbit; accessed June 28, 2020.] Emphasis added.

- The favored theory held by secular astronomers today is that there is a dynamo effect caused by molten fluids in the Moon's core, rotating at different speeds. Such a dynamo is plausible, but it would have worn down *millions* of years before the time evolutionists assign to the period of high field strength.

So the "fluid dynamo" model was abandoned because it would have lost its strength way too early.

Another idea is that asteroids, crashing into the moon, would slosh the aforementioned fluids in the Moon's core, restarting their movement and reenergizing the magnetic field. Kinda like throwing rocks at a generator to make it start. Even if this were plausible, the resulting magnetic field would have lasted only about 10,000 years, which is much too short for the long ages needed by uniformitarian evolutionists. Not only that, but the Moon rocks would have indicated a magnetic field that got stronger and weaker and stronger and weaker. Which they didn't.

A final proposal was that the moon's axis of rotation was **precessing** (wobbling), and that may have caused the dynamo effect mentioned earlier.[357] But more analysis revealed that it wouldn't have been strong enough to account for the strength of the magnetic field, as indicated by the moon rocks.

According to the study's authors, they admit:

> . . .the high paleointensities of [three moon rocks] still present a major challenge. . .[358]

And this is not surprising, given the laws of physics. Friction is going to slow down any dynamo in the moon's core that generates a magnetic field. This magnetic field would be too weak, and die out billions of years too soon, to satisfy evolutionists' requirement of deep time, even if it was assisted by imaginary asteroid strikes and precessions.

---

[357] Thomas, B. "What Magnetized the Moon?" Creation Science Update. December 8, 2011. [Online at ICR.org/article/what-magnetized-moon; accessed June 28, 2020.]

[358] Sauvet, C. et al. 2013. "Persistence and origin of the lunar core dynamo." *Proceedings of the National Academy of Sciences.* 110 (21): 8453–8458. [Online at PNAS.org/content/110/21/8453.short; accessed June 28, 2020.]

The Moon's Recession

As mentioned above, the Moon's gravity causes tides on the Earth. But water experiences friction with the ocean floor as the tides ebb and flow, and this friction slows down the Earth's rotation.

Not only that, but the Moon's gravity also causes tidal effects on the Earth's crust. Since the crust is solid, the tidal effects are nowhere near as pronounced as they are on the oceans, but that effect, too, tends to slow the Earth's rotation down. A secondary effect, via the law of Conservation of Angular Momentum, is this: since the Earth and the moon gravitationally affect each other, the moon is always moving farther away from the Earth, by about 38mm per year.

During three of the Apollo missions to the Moon,[359] the astronauts left **corner reflectors**,[360] which bounce light back exactly in the direction from which it came. Since then, astronomers have been able to fire a laser at the Moon and measure the time it takes for the light to return. These extremely precise measurements have shown us that the moon is receding from the Earth 38mm per year.

That's all well and good, and even interesting, but now let's apply it to the deep time required by evolution. The moon is about 239,000 miles away from Earth (center to center), so how close would it have been 4.5 billion years ago? Keep in mind that this recession rate is non-linear: the closer the Moon is to the Earth, the greater the tidal effects on Earth's oceans and crust would have been, which means the Moon would have receded faster in the past. Taking this into consideration, the Moon would have been *touching* the Earth only about 1.4 billion years ago.[361,362] Think of it: you could reach up and touch the Moon as it went by, orbiting only inches above your head.

Of course, the moon would have broken up because of tidal forces long before that point—or, technically, the moon would never have formed. About 11,500 miles out is the Roche limit, discussed earlier. Which means the Moon never would have coalesced in the first place; Earth would instead have Sat-

---

[359] en.wikipedia.org/wiki/Lunar_Laser_Ranging_experiment; accessed June 28, 2020.

[360] en.wikipedia.org/wiki/Corner_reflector; accessed June 28, 2020.

[361] Jason Lisle, "The Solar System: Earth and Moon" September 30, 2013. [Online at ICR.org/article/solar-system-earth-moon; accessed June 28, 2020.

[362] Green, J. A. M., M. Huber, D. Waltham, J. Buzan, and M. Wells. 2017. "Explicitly modelled deep-time tidal dissipation and its implication for Lunar history." *Earth and Planetary Science Letters*. 461: pp. 46–53. [Online at ScienceDirect.com/science/article/pii/S0012821X16307518; accessed July 9, 2020.]

urn-like rings. Then again, if we use the Biblical time scale of 6,000 years, the Moon would only have been 730 feet closer to Earth than it is now, and the whole problem goes away.

Because the receding moon is such a problem for evolutionary explanations of the Earth/Moon system, naturally uniformitarians have objected strenuously to these calculations. Claims have been made that perhaps the recession rate may have been smaller in the past, so the Moon would not be catastrophically close to the Earth at any time in its history.

And granted: technically, that is possible. *If* several conditions were satisfied, the recession rate of the Moon from the Earth may have been less a couple billion years ago (assuming the Earth and Moon existed a couple billion years ago). Attempts to infer the recession rate from geological and paleontological data give mixed results: in one study, three out of four scenarios still had the Moon catastrophically close to Earth at some recent point in its alleged 4½-billion-year history.[363]

And, if you think about it, it doesn't much matter exactly *when* the Moon would have collided with the Earth. Taking its measurable recession from Earth back into the past, sooner or later, there's a problem. In the study mentioned in the previous paragraph, it only did the simulation for 250 million years back—less than 6% of the alleged history of the Earth/Moon system.

And even supposing that the Earth/Moon system *were* stable over the entirety of its alleged 4½-billion-year age—that all the conditions were just right to prevent the Moon from breaking up from tidal stresses—there are still many other factors that cast great doubt on the assertion that the Earth, the Moon, and the rest of the solar system are 4½ billion years old.

---

[363] Williams, G. E. 2000. "Geological Constraints on the Precambrian History of the Earth's Rotation and the Moon's Orbit." *Reviews of Geophysics.* 38 (1): pp. 37–59. [Online at AGUPubs.onlinelibrary.wiley.com/doi/abs/10.1029/1999rg900016; accessed August 15, 2020.]

Chapter 5 Summary

Here are the most important points in this chapter:

- Other planets in the solar system are **configured to keep Earth's orbit stable.**
- Earth is just the **right distance from the sun** to support liquid water. And we have the **right atmospheric composition and pressure.** And we have an **ozone layer** to reduce harmful UV (ultraviolet light). And we have a **magnetic field** to deflect cosmic rays, which are also harmful.
- Dating methods don't have a stellar track record for accuracy, probably because **at least three assumptions have to be valid** in order for it to work, and we don't know if any of them *are* valid.
- **Carbon-14 (radiocarbon) is unreliable** (even if all three assumptions are valid) **beyond about 50,000 years.** Or beyond 24,000 years.
- K-Ar dating is **unreliable for any ages shorter than "a few hundred thousand years."**
- The "sweet spot," between Carbon-14's usable range and K-Ar's usable range (assuming all the assumptions are valid) is where virtually all human evolution is supposed to have taken place. This means that **all "sequences" of fossils that lead to modern humans are mere wishful thinking.** (And these wishes change often.)
- Earth's **magnetic field shouldn't still be here** if it is really 4½ billion years old.
- With the observed rate of continental erosion at high elevations, all **the continents would have eroded completely away more than 70 times**, if the continents were as old as evolutionists believe. And, if you use the erosion rate observed at lower elevations, they would have eroded away 1100 times! And if even *one* cycle of mountain-building happened, **all the geologic layers and fossils would be gone.**
- **Sediment depth on the bottoms of the oceans is way too small** for a 4½-billion-year-old Earth.
- There is **way too little sodium in the oceans** for a 4½-billion-year-old Earth. The same goes for levels of nickel, magnesium, silicon, potassium, copper, gold, silver, mercury, lead, tin, and more.
- **The size of Earth's Moon is** unique to any planet in the whole solar system: **larger, compared to its primary, than any other.**

- There is **no known naturalistic process** whereby the moon could have come to be.
- If **the Moon** is 4½ billion years old, it **should not have a magnetic field anymore. But it does.**
- Because of tidal friction and the conservation of angular momentum, the Moon is receding from Earth, at a current rate of 38mm/year. So **1.4 billion years ago, the moon would have been** *scraping* **Earth** as it orbited. But of course, the Roche Limit would have broken it up long before that.

Chapter 6:

Mutations

In evolution, enormous timeframes are absolutely essential, because they make the theory so much harder to falsify. Experiments can, and often are, done on processes that take seconds, days, months, or years, and the results are in soon enough that any necessary alterations can be made to the hypotheses that were used to make the predictions tested.

But if a proposed process—say, evolution, for example—takes 10,000 times longer than a human lifespan, or more, it becomes much harder to prove it wrong, simply because nobody lives that long. Since the Second Law of Thermodynamics has so far proven to be valid in every observable place and time, there is strong motivation to move evolution to some *unobservable* place and time. This is convenient for the theory of evolution, be that evolution of the universe, the solar system, geological formations, life on Earth, or whatever, because it is impossible to test something in an unobservable time or place.

However, even though it is impossible to test things in an unobservable time or place, we *can* test things in *observable* time and place—if those processes are claimed to have happened in said *un*observed time and place—and realistically extrapolate them into the dim, dark past.

And here is another situation where the uniformitarianism that evolutionists swear by actually works against them. If the laws of physics were the same back then as they are today—which is a very reasonable assumption—we can do experiments using the same laws of physics and see if they yield results which, when extended into long timeframes, would result in the outcomes evolutionists desire.

In short, the theory of evolution seems more plausible in the minds of many people if we just have enough time. It would be silly to think that organisms just built themselves, or "fell together accidentally" in the blink of an eye—that would be just magic, and as such, is clearly impossible. However, if we allow millions or billions of years, set aside that pesky Second Law of Thermodynamics as well as numerous other laws of physics, and then claim that organisms just built themselves, or "fell together accidentally" over that whole long duration—then that would be science, you see.

In other words, *quick* spontaneous generation is silly and ridiculous, but *slow* spontaneous generation is the source of every living thing. So they say. But what, exactly, *is* spontaneous generation?

Spontaneous Generation

Spontaneous generation is a theory that claims that organisms can arise without being descended from similar organisms; that is, without a reproductive process having been used.

This theory came about by watching, for example, meat decay over a period of several days: first, there weren't any maggots, but then later, there were. The conclusion was that maggots spontaneously arose from the rotting meat. Later on, it was discovered that the flies that ate the rotting meat laid eggs, and *that's* where the maggots came from.

Let's look at some well-respected information sources and see what the common thoughts are, concerning spontaneous generation.

Wikipedia

Wikipedia describes spontaneous generation as a "**disproven theory** of life arising from nonliving matter," and it states further that it is "an **obsolete** body of thought."[364]

In describing this theoretical process, Wikipedia goes on to state:

> "**Crucial** to this doctrine are the ideas that life comes from non-life and that **no causal agent, such as a parent, is needed**. The hypothetical processes by which life routinely

---

[364] en.wikipedia.org/wiki/Spontaneous_generation; accessed May 23, 2020.

emerges from nonliving matter on a time scale of minutes, weeks, or years. . . are sometimes referred to as abiogenesis. Such ideas have **no operative principles in common with the modern hypothesis of abiogenesis,** which asserts that life emerged in the early ages of the planet, **over a time span of at least millions of years,** and subsequently diversified, and that there is **no evidence of any subsequent repetition** of the event."

Several fascinating points, just in this one article:

- Wikipedia calls spontaneous generation a "doctrine"—that almost makes it sound like a religion, doesn't it? And indeed it is: spontaneous generation, along with anything that depends on it (like evolution) requires faith. And it takes more faith to believe that the complex and interdependent biosphere we observe fell together accidentally than it does to believe that God created it.
- The event of life arising from non-life has "no causal agent." In other words, the beginning of life on Earth is an effect without a cause. Never mind that the definition of "effect" is "that which is produced by a cause." An "uncaused effect"—say, for example, evolution—is an oxymoron, a logical absurdity.
- Spontaneous generation (life arising *quickly* from nonlife), is disproven and obsolete. *Why* is it disproven and obsolete? Because it's so easy to disprove over such short timeframes. But, abiogenesis (life arising *slowly* from nonlife) is absolutely essential for the doctrine of evolution. Are things likely to get *more* organized, complex, and interdependent, the more time they have? (Remember the discussion of "The Second *Law* of Thermodynamics" above.)
- Note the underlying statement: Even though the two theories are fundamentally identical—life arising from nonlife—they are described as having "no operative principles in common." It almost leads one to wonder if the evolutionists' position could be more clearly stated as "Even though we *absolutely* require spontaneous generation, let's call it something different, so people are less likely to notice our self-contradiction."
- After life "arose" on this planet, the article states that "there is no evidence of any subsequent repetition of the event." If it did not repeat, then it happened only once, and if it happened only once, why did this single event require "at least millions of years?"

Dictionary.com

As pointed out above, another name for spontaneous generation is "abiogenesis," which comes from a Greek phrase that means life being "generated from not-life."

It's fascinating to research this word; Dictionary.com has two definitions for "abiogenesis:"

1. The now-discredited theory that living organisms can arise spontaneously from inanimate matter; spontaneous generation.
2. The theory that life forms on Earth developed from nonliving matter.

Look at that! Abiogenesis (a.k.a. spontaneous generation) is "now discredited" but it is also the theory of how life came to be on Earth! Evolution is completely dependent on spontaneous generation/abiogenesis, in spite of the fact that it is "now discredited!" Do people actually not notice the self-contradictions here?

Encyclopedia Britannica

Encyclopedia Britannica describes spontaneous generation as the "**hypothetical process** by which living organisms develop from nonliving matter," and adds "By the 18th century **it had become obvious that higher organisms could not be produced by nonliving material.**"[365] So the Encyclopedia Britannica states that spontaneous generation is a "hypothetical" process, but what, exactly, does "hypothetical" mean? It means "supposed, but not necessarily real or true."[366] Also, more than two centuries ago, it had become "obvious" that higher organisms could not be produced from nonliving material.

In a different article, Britannica.com states that: "**Life ultimately is a material process that arose from a nonliving material system spontaneously**—and at least once in the remote past. How life originated is discussed below. Yet **no evidence for spontaneous generation now can be cited.** Spontaneous generation, also called abiogenesis, **the hypothetical process by which living organisms develop from nonliving matter, must be rejected.**"[367]

---

[365] Britannica.com/science/spontaneous-generation; accessed May 23, 2020.
[366] *New Oxford American Dictionary*, Copyright © 2010, 2019 by Oxford University Press, Inc. All rights reserved.
[367] Britannica.com/science/life/Evolution-and-the-history-of-life-on-Earth#ref1013858; accessed May 23, 2020.

Did you catch that? Britannica bluntly states that "life arose from non-living material," and did so "spontaneously." Then, two sentences later, "no evidence" for such a process has been found (you know that they would have cited such evidence if it could be found). And then, *in the very next sentence,* Britannica states that this process "must be rejected."

This shows the confusion that results when people are so determined to exclude God from their thinking: this hypothetical process of spontaneous generation—even though it "had become obvious" that it was not valid, and there is "no evidence" for it, and that it "must be rejected"—is *still* bluntly stated as the process through which life arose! (If you just give it enough time.)

Note the similarity between the following two quotes:

Image courtesy Creative Commons CC0 1.0 Universal Public Domain Dedication

One has only to contemplate the magnitude of this task to concede that **the spontaneous generation of a living organism is impossible.** Yet we are here—as a result, I believe, of spontaneous generation.[368]

—George Wald, Nobel Laureate

Image in the Public Domain

Alice laughed. 'There's no use trying,' she said. 'One can't believe impossible things.'

'I daresay you haven't had much practice,' said the Queen. 'When I was your age, I always did it for half an hour a day. Why, **sometimes I've believed as many as six impossible things before breakfast.'**

—Lewis Carroll, *Through the Looking Glass*

---

[368] George Wald [biochemist and winner of Noble Prize in Physiology or Medicine, 1967], "The Origin of Life," *Scientific American* 191 no. 48 (1954): p. 46.

Note that in the George Wald quote, he states that he believes spontaneous generation is impossible, yet possible. That is, spontaneous generation is both false and true at the same time, thus violating the Law of Non-Contradiction, the most fundamental law of rational thought: "A statement cannot be both true and false at the same time and in the same sense." Enough said.

Fascination with Mutations

Once life has spontaneously generated—once life has arisen from non-life, via unknown processes that defy the laws of physics—evolution's self-improvement program kicks in: mutations.

What are mutations? There seems to be a fascination with mutations, and the wondrous enhancements to people's abilities that they supposedly cause. Movies and television abound with mutant stories: from movies like the *X-Men* series, *The Hulk*, and *The Fantastic Four*, through television shows such as *Alphas, The Gifted,* and *The Tomorrow People*, all the way down to the *Teenage Mutant Ninja Turtles*.

Why the fascination? People know that there ought to be something more to life than they currently experience—something more exciting, meaningful, and significant for them to be involved in. And indeed there is. But this much-more-meaningful and exciting life has the "disadvantage" of requiring one to arrange one's life according to God's priorities, which is something that the outspoken proponents of evolution strenuously resist.

So, a Plan B was needed—a Plan B that allows for design without a Designer and creation without a Creator.

It didn't take much observation at all to realize there is an awful lot of different kinds of plants and animals all over the world. They are unquestionably here, so the question arises: "How did they get here?" Yes, there are those religious fuddy-duddies who believe that God created them, but that is apparently unacceptable. In the foreword to the anniversary edition of Darwin's book, *The Origin of Species*, Sir Arthur Keith states, "Evolution is unproved and unprovable. We believe it only because the only alternative is special creation, and that is unthinkable." Hence the need for a Plan B. (Notice the unmistakably religious thinking: "We believe this because we don't want to believe that.")

Another requirement for Plan B is that everything couldn't have come about at the same time, or even within a short period of time, because that sounds perilously close to Creation, and we can't have that. So let's say they arose over enormous periods of time. Millions and millions of years! Even billions and billions! Surely that will be enough time for all sorts of unexplainable things to happen.

But how did such a wide variety of things evolve? Well, we know that random mutations happen, so let's go with that. True, we haven't ever seen a random mutation that was actually beneficial, but theoretically it's possible, so that will have to do.

So, Plan B was emphasized, promoted, popularized, and advocated, although couched in much more scientific-sounding language than used above. Since random-but-beneficial mutations are supposedly the cause of the vast increases in complexity of organisms as they evolved, the proponents of evolution set about to discover some. What was actually discovered was disappointingly unhelpful to the cause of the theory of evolution, and alarmingly frequent. We'll cover some of those discoveries next.

What Are Mutations?

The "stuff of life" is the DNA. DNA stands for **deoxyribonucleic acid**, and it is the very long molecule that contains the genetic information needed to describe and define the being of which it is a part.

DNA is quite remarkable in the amount of genetic information it contains. The very fact that DNA exists is a whole *bunch* of nails in the coffin of naturalistic evolution, because random processes cannot result in information; information can *only* come from intelligence, as a molecular biologist we'll hear from later points out.

Most of a person's cells have 46 **chromosomes** (DNA molecules); only the reproductive cells—sperm and eggs—have 23, so each child has a mixture of the parents' attributes. If all 46 chromosomes' worth of DNA in one of your non-reproductive cells were stretched out and connected end to end, it would be about 7 feet long. This strand would be so thin that even an electron microscope could not see its structure. If all the information from a single cell of a human was written in book form, it would fill a library of about 4000 books. If all the DNA in your whole body were placed end to end, it would stretch from here to the moon and back more than 250,000 times. If one cell's worth of DNA from every person who ever lived on this planet were placed in a pile, it would weigh less than a single aspirin![369]

---

[369] Walt Brown, *In the Beginning: Compelling Evidence for Creation and the Flood,*, Eighth edition, 2008 (Center for Scientific Creation, Phoenix, Arizona, CreationScience.com), p. 3.

As the illustration at the right shows, DNA is shaped like a twisted ladder, usually called a **double helix**. Of course, the double helix of DNA is enormously longer than the tiny snippet in the illustration. Each "rung" of this ladder is composed of a pair of chemical groups, or **bases**. These bases are the same for each "side rail" of the ladder, allowing the ladder to be split and copied, so the DNA can be duplicated during cell reproduction.

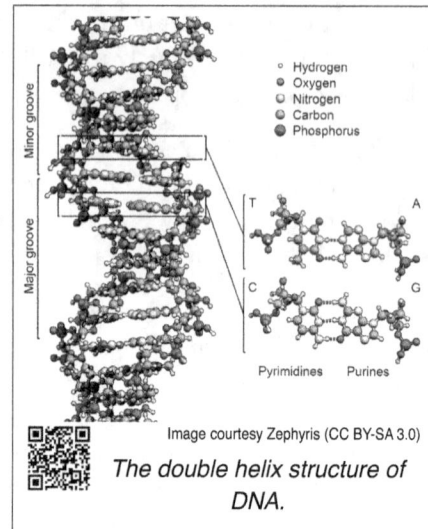

The double helix structure of DNA.

When the DNA "ladder" is split during cell reproduction, and then a duplicate base is added to each ladder-half, the DNA is duplicated. The problem is that, despite ingenious error-correction mechanisms discussed later in the chapter, errors are still made in the copying process—"DNA typos," if you will. Not only that, but DNA can be damaged by radiation that is constantly bombarding us. Such radiation may come from space, from radioactive minerals in the soil, radon in our basements, X-rays from our CT scans and dentist appointments, and so forth. But here's the rub: when DNA is damaged by radiation or whatever else, that damage—that mutation—if it occurred in our reproductive cells, can become part of our genetic definition, and therefore is propagated to our children. And that happens far more often than you might imagine.

So our children inherit the genetic changes we received, but then their DNA is damaged throughout life as well, so even more radiation damage or copying mistakes take place, on top of the ones they inherited from us. When they have children of their own, the process repeats. So the number of genetic differences (mutations) increases with each generation. This ever-increasing number of genetic mutations is called the **genetic load**.

A Linguistic Illustration of Evolution

As an illustration of the problems involved with random mutations having useful results, let's consider "evolving" one English word into another. Remember, one of evolution's most fundamental tenets of faith is that *it is random;* nothing can be done with intention, foresight, or awareness.

And, oh, the problems that causes! To make it simple, we'll use words of only ten letters and, since there are only 20 amino acids used in living organisms,[370] we'll use words composed of only the first 20 letters of the alphabet.

Let's think about randomly evolving the word "trademarks" into "headstrong:"

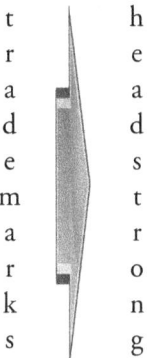

We might be momentarily pleased that positions 3 and 4 of the destination word contain "a" and "d", respectively; they already match the desired outcome before we even get started. But then we remember that evolution has no foresight; it has no "awareness" that certain parts already have the right values, so they should stop mutating. It is just as likely that the correct ones will be mutated into incorrectness, as the other way around. In fact, it is much *more* likely that things will go wrong than right. For each position, there is only one correct letter, but there are nineteen incorrect letters. So for any random change of one letter into another, wrong answers will outnumber right ones by factor of 19 to 1.

Someone might say, "But there are many different ten-letter words; we don't need that *particular* one!" Actually, we do. Suppose you wanted to say, "My brother-in-law is very headstrong," but instead, you said, "My brother-in-law is very trademarks." Or, suppose you wanted to say, "We need to obtain trademarks for our new product lines," but instead, you said "We need to obtain headstrong for our new product lines." Would it make any difference to the meanings of the sentences? Yes, clearly. Why? Because "headstrong" and

---

[370] There are certain kinds of bacteria that use two more kinds of amino acids—selenocysteine and pyrrolysine—so technically, there are 22 amino acids known to be used in living organisms. But since these two are used only in a few kinds of bacteria, there are not included here. For technical details, see Atkins, J.F. and Gesteland, R., "The 22nd amino acid," *Science* 296(5572):1409–10, 24 May 2002; commentary on technical papers on pp. 1459–62 and 1462–66.

"trademarks" mean different things; they do different things in the sentences. And, you'll notice, in each case, one of them is meaningless nonsense. Just like it would be if you picked the wrong amino acid.

But even assuming it *didn't* matter which word you picked, how many valid words are there, compared to the number of nonsensical letter combinations? With 20 possible values for each of ten positions, there are 20^{10} possible combinations. How many of them are actual words? From my English wordlist of more than 267,000 words, there are only 15,918 valid ten-letter words that use only the first 20 letters of the alphabet. Therefore, less than one in every 643 million possible letter combinations is an actual word. And that's to arrive at *any* of the valid ten-letter words! How much less likely are we to arrive at the *desired* ten-letter word?

In the same way, amino acids can't be put together just any ol' way and expect them to do anything useful. Or even anything at all. Proteins are made of specific numbers of amino acids in specific sequences, and they perform specific functions. And they have to work compatibly with thousands of other proteins, also made of specific numbers of amino acids in specific sequences to perform specific functions.

Our linguistic illustration above contained only ten units; real proteins contain anywhere from 50 to *30,000* amino acids that have to be the correct ones in the correct order! And non-functional combinations, which would result in the death of the organism (or, actually, the prevention of life ever starting), outnumber the functional ones by trillions of trillions to one. To randomly make even the smallest protein, of only 50 amino acids, there is exactly *one* way of assembling it correctly, and about 1.13×10^{65} wrong combinations! At a thousand attempts per second, it would take much more than quadrillions of times longer than even the supposed deep-time age of the universe to randomly assemble it correctly. And that's just *one* protein! The *simplest* one! And that also assumes that the constituent amino acids had accidentally fallen together already and are still present (because amino acids are quickly destroyed by normal environmental conditions).

And it's actually much worse than that, because this assumes all the amino acids to choose from are left-handed—even *one* right-handed amino acid in the mix would cause the protein to fail. And random processes cause 50% of the molecules to be left-handed and 50% to be right-handed, so at every step, there are *twice* as many things than can go wrong. (More about left-handed and right-handed amino acids later.) So this makes the likelihood of even the simplest amino acid forming by chance trillions of times less likely than even the impossibility mentioned above.

And we're supposed to believe that all of earth's life forms accidentally fell together in only 14 billion years? Mathematically, there's no way.

Let's talk a bit more about the transformation concept above, but instead of letters in a word, let's talk about amino acids in a protein, because that's where the failures of evolution really become apparent. (Well, again, it's another *one* of the places the failures of evolution become apparent.) Finally, biology has a quantitative measurement process to determine the genetic "distance" between any two organisms. By comparing the amino-acid sequences of two proteins, we can arrive at an objective value that represents their degree of similarity: for any given position, the amino acids are either the same or they aren't; there's little room for personal bias or opinion to muddy the results.

For example, **cytochrome c**—a molecule that is involved with energy production in cells—varies from one kind of organism to another. By comparing the amino-acid sequences of cytochrome c from two types of organisms, evolutionists hoped to determine if Organism A is the descendent of Organism B, if B is the descendent of A, or if A and B are both descendents of some third Organism C.

What evolutionists found was disheartening. Rather than the orderly increase of complexity that evolution assumes—and depends on—what biologists discovered showed no sequence at all! For example, in comparing the cytochrome c from a bacterium to the cytochrome c of a horse, a pigeon, a tuna, a silkmoth, wheat, and yeast, it was discovered that they were *all* almost exactly the same genetic "distance" from the bacterium's cytochrome c![371] The genetic distances—the percentages of difference in their respective cytochrome c proteins—were 64%, 64%, 65%, 65%, 66%, and 69%, respectively.

That is shocking for someone who believes in evolution! To imagine that, in terms of their cytochrome c, a horse is the same genetic distance away from the bacterium as is a pigeon! And a tuna is the same distance as a silkmoth! And one-celled yeast is even farther away from the one-celled bacterium than wheat is!

---

[371] Michael Denton, *Evolution: A Theory in Crisis* (Bethesda, MD: Adler and Adler, 1985), p. 281.

And this was not a unique feature of cytochrome c; many other biochemicals were analyzed in a similar way:

> Thousands of different sequences, protein and nucleic acid, have now been compared in hundreds of different species but **never** has any sequence been found to be *in any sense* **the lineal descendent or ancestor of** *any* **other sequence.**[372]

Interesting: there was no evolutionary sequence to be found. *Anywhere.*

These are some of the astonishing realizations that came from the comparative biochemistry described above:

> At a molecular level there is no trace of the evolutionary transition from fish→amphibian→reptile→mammal. So **amphibia,** always traditionally considered intermediate between fish and the other terrestrial vertebrates, **are in molecular terms as far from fish as any group of reptiles or mammals!** To those well acquainted with the traditional picture of vertebrate evolution the result is truly astonishing.[373]

So is evolution being portrayed as having actual explanations for how life came to exist on this planet, when in actuality, it has none? Yes:

> Despite the fact that **no convincing explanation of how random evolutionary processes could have resulted in such an ordered pattern of diversity,** the idea of uniform rates of evolution is presented in the literature as if it were an empirical discovery. The hold of the evolutionary paradigm is so powerful that **an idea which is more like a principle of medieval astrology** than a serious 20th-century scientific discovery **has become a reality for evolutionary biologists.**[374]

The confounding evidence of comparative biochemistry and related fields of study has led many biologists to make statements like the following:

> **No species can be considered ancestral** to any other.[375]

---

[372] Ibid., p. 290. Emphasis added.
[373] Ibid., p. 285. Emphasis added.
[374] Michael Denton, *Evolution: A Theory in Crisis* (Bethesda, MD: Adler and Adler, 1985), p. 306. Emphasis added.
[375] Beverly Halstead (1981) "Halstead's Defence Against Irrelevancy", *Nature,* 292: p. 403. Emphasis added.

Much of **today's explanation of nature,** in terms of neo-Darwinism, or any synthetic theory, **may be empty rhetoric.**[376]

Referring to evolution's presumed "ancestral relationships" of some species to others,

> . . .**they exist not in nature** but in the mind of the taxonomist, as abstractions. . . **they are always discussed as if they have some reality. . .**[377]

Very enlightening, isn't it?

Beneficial Mutations

We've talked about mutations, but are all mutations created equal? Oops—I'd better stay away from the word "created" here—so, are all mutations *evolved* equal? In other words, do they all do the same things? Do they all do comparable things? Do some mutations cancel the effects of others?

When talking about beneficial mutations, there are two possible ways to interpret that, so let me clarify exactly what I'm talking about here. Conceptually, a beneficial mutation could be:

- A mutation that results in loss of genetic information, but which causes a serendipitously—and temporarily—beneficial outcome, in spite of the loss of genetic information. Also called "equivocally beneficial."
- A mutation that results in the *addition* of functional genetic information, which is beneficial to the organism that mutated, as well as its descendents. Also called "unequivocally beneficial."

Let's look at both of these in more detail.

Equivocally Beneficial Mutations

As mentioned, an "equivocally beneficial mutation" is a mutation that actually loses genetic information, but in a specific environment, has a convenient side effect.

---

[376] Colin Patterson (1980) "Cladistics", *Biologist,*27: p. 238. Emphasis added.

[377] Colin Patterson (1976) "The Contribution of Paleontology to Teleostean Phylogeny" in *Major Patterns of Vertebrate Evolution,* ed M.K. Hecht, Plenum Press, London, p. 623. Emphasis added.

As an example of this type of beneficial mutation, suppose some kind of flying insect is a favorite meal for the local predators, and whenever the insect lands on a tree branch, the nearby predators notice, and eat him. Now suppose this flying insect experiences a mutation that deforms his wings to the point that he can no longer fly. Since he stays on the ground now, he is less often noticed and consumed by his predators.

Such a mutation could be considered serendipitously beneficial because of the current predatory circumstances: for the moment, it is eaten less often. However, the insect's DNA was damaged and it is now less "fit" to survive in the long run. (And evolutionists *love* long runs.) This is *not* the kind of mutation I am primarily discussing in this book. And technically, it is inaccurate to call this kind of mutation a "beneficial mutation" because it *isn't* beneficial; it's much more accurate to call an equivocally beneficial mutation a "harmful mutation with a convenient side effect."

Here's another example of an equivocally beneficial mutation: The mutation that causes sickle-cell disease could be considered beneficial, if you live in a region with a high incidence of malaria, because sickle-cell disease protects against malaria.[378] But sickle-cell disease also causes pain, anemia, swelling in the hands and feet, bacterial infections, and stroke.[379] Not only that, but the sickle-shaped blood cells clog small blood vessels, starving the associated organs of oxygen, resulting in tissue- and organ death. So is it really a beneficial mutation? No; this equivocally beneficial mutation is a harmful mutation that serendipitously has a convenient side effect in one specific area and, as usual, *many* detrimental consequences.

Antibiotic Resistance

Yet another example of equivocally beneficial mutation is the oft-touted antibiotic-resistant bacteria. This example is commonly offered by evolutionists as living proof of evolution, and indeed it could look that way, as long as you constrain the environment and don't look too closely at the mechanisms involved.

---

[378] Ken Ham and Bodie Hodge, editors, *The New Answers Book 1* (Master Books, Green Forest, Arkansas), 2006, p. 200.

[379] en.wikipedia.org/wiki/Sickle_cell_disease; accessed June 17, 2020.

There are three different mechanisms whereby bacteria have been seen to acquire antibiotic resistance. They are:[380]

1. Modification of the antibiotic's target (where the "target" is the part of the bacterium that the antibiotic acts upon to accomplish its purpose),
2. Restriction of the antibiotic's access to the target, and
3. Deactivation of the antibiotic.

As an example of Mechanism #1 above, the first step in killing Gram-negative bacteria is binding a positively-charged molecule (e.g., **colistin**) to the negatively-charged molecules of **lipopolysaccharides** (LPS) in the cell membranes. One particular mutation destroys the bacteria's ability to create LPS, so the binding doesn't occur, and therefore the bacteria survives the antibiotic. However, LPS contributes greatly to the structural integrity of the cell membrane, and its absence weakens the membrane considerably. The cell wall, now weakened from the lack of LPS, more easily tears open, spilling the contents of the cell, which then dies. Thus, even though the bacteria are more resistant to antibiotics now, they are less robust in general—less "fit to survive."

Another kind of mutation pertains to the production of positively-charged magnesium ions that keep the cell membrane stable. When the bacteria's control system to limit the production of magnesium ions suffers a mutation, the system runs amok and over-produces the positively-charged magnesium ions, which in turn reduces the LPS in the cell membrane. The result, again, is that colistin has too few binding sites to be effective, and the bacteria survive the presence of the antibiotic. But here too, the antibiotic resistance comes at the cost of structural integrity of the cell membrane.

You'll notice in both cases above, it wasn't newly-created genetic information that provided the antibiotic resistance; that is, no actual evolution happened. Rather, the mutations caused *damage* to the existing genetic information, and although this damage happened to have a beneficial side-effect in an antibiotic-rich environment, it was detrimental in general because some of the bacteria's robustness was lost.

As an example of Mechanism #2 above, many kinds of antibiotics—**fosfomycin**, for example—must be taken *into* a bacterium to kill it. Cells have tiny **flux pumps** (i.e., "transport pumps") that pull certain kinds of molecules

---

[380] Don Batten, "Antibiotic Resistance: Evolution in Action?" *Creation* 39(4):46–48, October 2017. [Online at Creation.com/antibiotic-resistance-not-evolution-in-action; accessed May 30, 2022.]

into the cells. These pump mechanisms are selective about what they pull into the cell, but fosfomycin is similar enough to some of the molecules the cell wants, that it is pulled in along with other ones. Mutations to the genes containing the instructions for the assembly of these transport pumps cause the pumps to degrade in their efficiency, or perhaps fail completely. Since some of these pumps would, but now don't, pull in molecules that are shaped like fosfomycin, fosfomycin itself doesn't get into the cell to kill it.

But again, it wasn't newly-created genetic information that provided the fosfomycin resistance; it was damage to the existing genetic information that created those pumps. This damage kept out the fosfomycin, but also kept out some desirable molecules as well. Thus, the beneficial side-effect was resistance to fosfomycin, but this very mutation is detrimental in general because some of the bacteria's viability was lost.

As a variation on Mechanism #2, many kinds of bacteria have **efflux pumps**—pumps that push things *out* of the cell. In a healthy cell, these efflux pumps are created at a certain rate, but if the control mechanism breaks (e.g., because a mutation happens), many more efflux pumps than needed are created, because the shutoff switch is broken. This can be convenient if one of the things being pumped out happens to be an antibiotic, but in a normal environment, so much energy is consumed building unnecessary efflux pumps that fewer resources remain to enable other necessary tasks. Therefore, the bacterium experiences a net loss in its fitness to survive.

In Mechanism #3, the antibiotic itself is neutralized. This is the one that looks most like actual evolution, but alas, this one isn't either. For example, certain bacteria create enzymes called **β-lactamases**, which eat penicillin, neutralizing its antibiotic effect. This enzyme-creating ability is encoded in small DNA loops called **plasmids**. The penicillin-eating ability is nice, but these penicillin-resistant bacteria, through a process known as **gene transfer,** can give *other* bacteria the ability to make β-lactamases, and become penicillin-resistant themselves. The plasmid containing the genetic instructions of how to make β-lactamases is physically transferred into the non-resistant bacterium, via a tiny tube called a **pilus**, after which the newly-injected bacterium knows how to neutralize penicillin.

It is basically the cellular equivalent of downloading a new app to your phone, giving it new capabilities: the app, with all its programming (which, by the way, had to be created by an intelligent designer), already existed on the server. No new information "evolved," you simply downloaded it, using a protocol that already existed on both the server and your phone. Your phone doesn't "evolve" in a Darwinian sense simply because you download an app; it is simply doing what a large team of very intelligent, goal-directed, decision-making people programmed it to do. This is nothing at all like evolution: random, unintentional changes with no goal in mind.

But like the others, Mechanism #3 is not a case of new genetic information being created, so no actual evolution happened. It was simply a case of existing genetic code being copied to a new location, via a mechanism whose description and specifications were already existent in the respective cells' DNA.

So, in the first two cases above, the genetic damage caused by the mutations happened to have a side-effect that is convenient in an antibiotic-rich environment. But it's still damage; it is *devolution*, not evolution. And in the third case, again, it was not evolution, because no new information was created; it was simply copied to a place it was needed, using the copying process and apparatus that already existed in the genetic code of the donor. So while antibiotic resistance is often touted as an example of "evolution before our very eyes," it is nothing of the sort: it is mutational damage, which *decreases* the organism's overall viability and fitness to survive, or simply copying already-existing genetic information from one place to another.

Unequivocally Beneficial Mutations

As an example of the second type of beneficial mutation, suppose an organism is happily going through life, and because of a random, undirected bit of radiation, his DNA is changed in such a way as to *add* genetic information, and a new feature, characteristic, or ability appears because of the added information. Basically, a staggeringly complex new subroutine is written, *by accident*, and it correctly performs a useful function, and it interfaces properly with the operating system, and has no detrimental side effects. How often do you think *that* happens?

Yet this is the kind of beneficial mutation that the theory of evolution absolutely depends on happening trillions of times since the Earth "formed." And this is the kind of beneficial mutation that has never actually been seen.[381,382]

---

[381] Henry Morris III, "Science Does Not Observe Evolution Happening Today" podcast, June 8, 2016. [Online at ICR.org/article/science-not-observe-evolution; accessed June 17, 2020.]

[382] Barney Maddox, "Mutations: The Raw Material for Evolution?" *Acts & Facts*, September 1, 2007. [Online at ICR.org/article/mutations-raw-material-for-evolution; accessed June 17, 2020.]

It is this latter kind of beneficial mutation that I refer to in this book, because that is the Big Problem. Looking at the DNA of any living organism, we see *scads* of organized information—programming instructions for the assembly of proteins necessary to carry on the processes of life. And a *very* important question is: where did the information come from? We'll address this excellent question below.

"Hello, World!"

Speaking of programming instructions, let's illustrate—with actual programming instructions—one of the problems that random evolution encounters. When beginning to learn any computer programming language, one of the most common exercises is to write a program that simply prints the words "Hello, world!"

An example of such a program is below, written in the PHP computer language:

```
print("Hello, world!");
```

How long would such a program take to be assembled at random, like evolution requires? There are only 23 characters in the entire program, and the number of possible characters with which to write a program is only 95 (the standard ASCII character set).[383]

Since there are 95 possibilities for each position, and 23 characters in the whole program, the number of combinations is 95^{23}, or 3.07×10^{45}. The age of the universe, according to evolutionists, is about 13.8 billion years, which is a paltry 4.35×10^{17} seconds. Attempting to write the "Hello, world!" program at the rate of a billion attempts per second, the universe would have to be on the order of a billion billion times older than even the deep-time hopes of the age of the universe to randomly assemble this tiny program even once! Just to make a simple 23-character program!

Considering that the operating system of a smartphone or a laptop (which required teams of intelligent, cooperating engineers working toward the same goal for years) is orders of magnitude more complex than "Hello, world!", and that even the tiniest virus is many orders of magnitude more complex than a computer's operating system, we can see than random processes are utterly, completely, totally, absolutely inadequate to get the job done. Just in case I hadn't mentioned that before.

---

[383] "ASCII" stands for "American Standard Code for Information Interchange." It has 128 possible characters, only 95 of which are used for computer programming: 26 uppercase letters, 26 lowercase letters, the digits 0–9, and a variety of punctuation characters.

Back to Mutations...

So, back to mutations. There are three possible kinds of mutations:

- Beneficial mutations: These are additions to the genetic information in an organism's DNA that increase an organism's complexity or improve its likelihood of survival: increased speed, agility, intelligence, resistance to disease, and so forth. As mentioned, this kind has never been seen.[384]

- Neutral mutations: These are changes to the genetic information that have no significant impact on the organism's ability to survive and thrive. There are two conceivable kinds of neutral mutations:
    - First, a mutation could actually make a negligible change in the offspring; the resulting genetic change offers no detectable difference in viability of the children than that of the parents. Therefore, natural selection would not notice, and the mutation would not be weeded out of the population.
    - Second, a mutation that results in a combination of noticeably beneficial results and noticeably detrimental results in the offspring, and these effects cancel each other; the beneficial changes in one area are offset by the detrimental changes in some other area. These are exceedingly rare, because no one has ever seen a mutation that was actually beneficial, but it is conceivable.

- Harmful mutations: These are changes to the genetic information—damage to, or loss of, genetic information—that are harmful to the organism or decrease its ability to survive: fewer defensive capabilities, increased susceptibility to disease, decreased mental capacity or ability to reproduce, and so forth. Occasionally, these harmful mutations have a convenient side-effect, but there is invariably a net loss of fitness.

By definition, neutral mutations are genetically undetectable, and they don't cause any noticeable change in the organism, so they cannot influence natural selection. For this reason, they are useless from an evolutionary standpoint. However, they still degrade the DNA, so a better term would be "near-neutral" mutations. And even these near-neutral mutations are piling up in the DNA so quickly that organisms would have gone extinct long ago if they had existed for the millions of years evolution demands.[385]

---

[384] John D. Morris "Can the Small Changes We See Add Up to the Big Changes Needed for Evolution?" May 1, 2002. [Online at ICR.org/article/can-small-we-see-add-up-big-changes-needed-for-evo; accessed June 21, 2020.]

[385] Sanford, J. 2005. *Genetic Entropy and the Mystery of the Genome.* Lima, NY: Elim Publish-

But clearly, for evolution to have taken place, the beneficial mutations must *far* outnumber the harmful ones. The DNA of humans is *so* complex, and contains *so* much essential information, that beneficial mutations must be the order of the day.

The problem is, that's not what we observe. In experiment after experiment, observation after observation, harmful mutations far outnumber the beneficial ones. And neutral mutations? Well, the jury's still out on whether there *is* such a thing: what used to be called "neutral mutations" were called that only because we hadn't discovered their negative effects yet. The more we learn about DNA, the more we realize that what we used to call "neutral" mutations are actually harmful ones.

And actual beneficial mutations? These are astonishingly rare. As a matter of fact, *no one has ever seen a single verifiable case.*[386] Yet evolution rises or falls on the existence—indeed, the *predominance*—of beneficial mutations. *Trillions* of them would be required to evolve a single-celled organism into a human.

Evolutionists often sidestep this showstopping problem (of adding genetic information) by employing a hand-waving, dismissive statement such as "Beneficial mutations accumulate." There are several significant problems with this "explanation:"

- It's highly debatable whether beneficial mutations—ones that add functional genetic information—actually exist, since, as mentioned above, no one has ever seen one. Not to mention the fact that no one has yet come up with a plausible explanation of how it even *could* happen, using only unguided laws of physics.
- By definition, evolution is blind and purposeless, so if a beneficial mutation ever actually happened, evolution couldn't recognize that fact so as to preferentially protect it from the detrimental mutations, which evolutionists admit outnumber beneficial ones by 10,000 to 1.
- In all living organisms, there are *so* many complex systems that have numerous necessary parts, that no individual part would be beneficial until *all* the parts were present. This means that natural selection would filter out any mutations that could eventually be beneficial long before all the other necessary parts could accidentally fall into place.

---

ing, 33–41, 150.

[386] Morris, J. 2008. "The Secrets of Evolution." *Acts & Facts.* 37 (3): 13.

But it's very useful to claim that random processes can create information, and do it so often, and so reliably; it avoids *so* many knotty problems like, say, those pesky laws of physics.

Lip Service

Evolutionists vociferously proclaim that life increased in ability, complexity, and viability because of beneficial mutations caused by random processes, but they don't really believe that. And the public doesn't believe that either. Not really.

How can I say such a thing? Because if people actually believed that random mutations were beneficial in the numbers that are claimed by evolutionists, people would be *flocking* to Chernobyl in Ukraine, Three Mile Island in the United States, Fukushima Daiichi in Japan, Bikini Atoll in the Pacific, and other sites of nuclear powerplant meltdowns and nuclear bomb tests, to live there. Just *think* of all that wonderful radiation! Just *imagine* the amazing mutations that would occur: invisibility, super strength, X-ray vision, teleportation, telekinesis, instant healing of injuries, and more!

No. Everybody knows that radiation causes mutations of the harmful kind, evolutionary orthodoxy notwithstanding. Everybody knows what happens when you get a large dose of radiation: you get sick and die.

And this is precisely why, when you get your teeth x-rayed at the dentist's office, they drape you in a lead apron, and step out of the room to push the button that activates the x-ray machine. They don't want you, or themselves, to receive any more radiation than absolutely necessary to get the job done. They know, as does pretty much everybody in technologically advanced nations, that radiation causes mutations, and mutations will kill you. Maybe by a slow and lingering disease, maybe by a quick and aggressive one, and maybe by the inexorable decrepitude of advancing age, but barring God's direct intervention, sooner or later, *mutations will kill you.*

Fixing Typos

In the process of copying DNA, errors sometimes happen. Although the copying process is very reliable, it is still subject to the Second Law of Thermodynamics, which says that disorder (for example, copying errors) is always increasing.

Consider copying a file from one computing device to another, or downloading a file from the internet. In a perfect world, the information would come across perfectly accurately 100% of the time. However, in this post-Fall world we now live in, in which the Second Law of Thermodynamics reigns supreme (at least for the present), transmitted data occasionally is corrupted in transit, and the copy of the data is not the same as the original data.

Computer hardware/software designers realize that a certain level of unreliability is inevitable. Even though hardware is getting more reliable all the time, it will never be 100% reliable, so how can we trust the files we copy or the data we download? Because of error correction. Built in to the hardware or software is a process that compares the original data and the copied data, and if it doesn't match, the original data is sent again—repeatedly, if necessary—until the original data and the copied data match each other. In this way, we can be confident that the copied data is identical to the original data. Unless, of course, the error-correcting mechanism is itself defective, or the data was corrupted after the copying process was completed.

When DNA is copied in the process of cell division, occasional errors creep in here as well. And just like the error-correcting algorithms in computer hardware and software, there are error-correcting mechanisms resident in the cellular operations.

As might be expected, the error-correcting mechanism for DNA is much more elaborate than that of a computer. If a DNA base is copied incorrectly (mutated), the damage can be fixed by referring to backup copies within the cell. Or, the mutation could be corrected to the matching base (the matching "half-rung") on the other "upright" of the DNA strand. Or, the alternate backup on the sister chromatid. And there are "repair enzymes" that fabricate entire "patches" for the DNA when necessary. And, several hundred other enzymes also detect and correct copying errors.[387]

DNA's error-correcting mechanism is elaborate and awe-inspiring. It is also a tremendous problem for evolution. Here's why.

Evolution, as we've seen numerous times in this book, absolutely depends on mutations for its very existence. Assuming life somehow got started in the first place (which evolution is still at a loss to explain), *trillions* of mutations would be required to convert a single-celled organism into a person. But these error-correcting mechanisms—these mutation-protection mechanisms—prevent that very thing! So in order for evolution to happen, mutation protection

---

[387] Thomas, Brian. "Can Evolution Hurdle the 'Mutation Protection Paradox'?" May 3, 2011. [Online at ICR.org/article/can-evolution-hurdle-mutation-protection; accessed July 10, 2020.]

must be turned off. But if it is turned off, errors build up so fast, the organism quickly dies.

Evolutionists acknowledge that 99.99% of mutations are harmful,[388] and even with the mutation-protection mechanisms in place, more than 60 additional mutations *per generation* are observed in living humans.[389] So mutation protection would have to be *on* in order for the organism to survive, but they would have to be *off* to allow enough mutations to happen that the organism could evolve. Evolution seems to be between the proverbial rock and the hard place.

This is the evolutionary paradox called the "mutation protection paradox:" How can the mechanisms be on and off at the same time? They can't. Here is yet another deal-breaking self-contradiction in the evolutionary belief system.

Not only that, but why would a mutation-hindering process—especially with so many redundant processes to ensure success in preventing mutations—evolve in the first place? Why would random mutations create a process that resists, cancels, reverses, and repairs mutations? That seems rather self-defeating, doesn't it? Yep, it does.

Do Mutations Happen in People?

Human beings, the biological organisms that we are, have DNA, live in the universe, and are subject to environmental radiation and other mutagenic factors, such as less-than-perfect copying of genetic material as cells divide and reproduction happens.

According to the doctrine of evolution, life "arose" from non-life, spontaneously grew more complex as information was somehow added to the genome—things led to things, and hominids and humans eventually appeared on the scene. (**Hominid** originally meant modern man and his closest extinct relatives, but eventually it came to mean primates, including chimpanzees, gorillas, orangutans, and humans. So a new term was needed: now, **hominin** means modern man and his closest extinct relatives. Of course, both these terms assume that man actually *has* extinct relatives from which he evolved.)

---

[388] Ernst Mayr, in *Mathematical Challenges to the Neo-Darwinian Interpretation of Evolution*, ed. P. S. Moorhead and M. M. Kaplan (Philadelphia: Wistar Institute Press, 1967), p. 50.
[389] Kong, A. et al. 2012. "Rate of de novo mutations and the importance of father's age to disease risk." *Nature.* 488 (7412): 471–475.

But, back to mutations: the subjects of radiation and other mutagenic factors were mentioned. Well, that sounds bad, doesn't it? We saw in the previous section that harmful mutations outnumber beneficial mutations, countless to one. And remember that these mutations that evolution relies upon are *undirected:* they are random and purposeless. Therefore, there is no option of "helping" the beneficial ones happen more often, or the harmful ones less often. Likewise, there is no option of helping the beneficial ones be much more effectual (to cause big steps forward) than the harmful ones (to cause small steps back).

Now it's sounding even worse. But how often do mutations happen in humans? This is an exceedingly important question, and has been the subject of numerous studies. Since heritable mutations are, by definition, aspects of the genetic material we pass on to our offspring, mutations build up, for good or ill, generation after generation. Each generation is more of a mutant than any previous generation in its ancestry, regardless of species.

Evolutionists have known for a long time that if the mutation rate was even *one* mutation per generation, the species would eventually go extinct.[390] So how many mutations per generation actually happen?

With recent advances in DNA-analysis tools, geneticists have been able to analyze DNA with unprecedented speed and accuracy. In one study in 2012,[391] geneticists actually counted the mutations in a group of Icelandic families, including 78 sets of grandparent-parent-child trios. Their findings? There was an average of 63.2 new mutations appearing *every* generation!

And it's actually worse than that. John Sanford, Courtesy Associate Professor at Cornell University, noted:

> All these studies **just look at genic** regions. The most mutable parts of the genome are the tandem repeats and satellite DNA. If these parts of the genome were included, I am confident **it would at least double the total mutation rate.**[392]

With evolutionists' acknowledgement that even *one* harmful mutation per generation would eventually lead to species extinction, and we find much

---

[390] John D. Morris, Ph.D. 2006. "How Old Is Life?" *Acts & Facts.* 35 (6). [Online at ICR .org/article/how-old-life; accessed June 17, 2020.]

[391] Kong, A. et al. 2012. "Rate of de novo mutations and the importance of father's age to disease risk." *Nature.* 488 (7412): 471–475.

[392] Thomas, Brian. "The Human Mutation Clock Is Ticking" July 7, 2011 [Online at ICR .org/article/human-mutation-clock-ticking; accessed July 10, 2020.] Emphasis added.

CHAPTER 6: MUTATIONS

more than that, human beings are unswervingly headed for certain extinction, barring divine intervention.[393]

Like so many other natural processes we've discussed earlier, a mutation rate of at least 60 additional mutations every generation indicates that the entire human history is just thousands, not millions, of years.[394]

As of this writing, the most accurate way to simulate genetic change over many generations is to use the population genetics-modeling computer program called "Mendel's Accountant,"[395] which is named in honor of Gregor Mendel, a meteorologist, mathematician, biologist, Augustinian friar, abbot of St. Thomas' Abbey, and now recognized as the founder of the field of genetics.[396] This software was developed by a team of scientists, including Cornell University plant geneticist John Sanford, and it calculates the cumulative effects on the average survivability—the "fitness to survive"—of organisms that inherit mutations over multiple generations.

Mendel's Accountant can simulate, with accuracy never before possible, the genetic results of the accumulation of mutations. Given an initial population size of 2,000 people, and given that each mother has six children, they assumed a rate of 60 mutations per generation in the algorithms (note that this is *less* than the number of mutations per generation actually observed in the Icelandic study cited above). What do you think they found? The simulation shows that the human race would go extinct *after only 350 generations*.[397] This simulation also assumed that natural selection would have been effective at removing the "least fit" from the population every generation, so each succeeding generation would use the "most fit" individuals. And still: 350 generations.

If 350 generations is how long the human race could survive before the genetic load caused extinction, how in the world could we still even *be* here if we assume deep time? That is an excellent question, and I am not the first to realize the contradiction (see the Alexey Kondrashov quote in the next section). It is interesting to note that the 350-generation extinction estimate matches

---

[393] Which, by the way, will happen; see I Thessalonians 4:15–17 for details.
[394] Conrad, D. F. et al. 2011. "Variation in genome-wide mutation rates within and between human families." *Nature Genetics.* 43 (7): p. 712–714.
[395] Genetic simulation software package "Mendel's Accountant: Simulating Genetic Change Over Time." [Online at mendelsaccount.sourceforge.net]
[396] "Gregor Mendel" [Online at en.wikipedia.org/wiki/Gregor_Mendel]
[397] Brian Thomas, "New Genomes Project Data Indicate a Young Human Race" [Online at ICR.org/article/new-genomes-project-data-indicate-young; accessed June 3, 2022.]

quite well with the Creation model, since the ≈6000 years since creation would result in approximately 300 generations (assuming 20 years per generation).

Mutations that we observe in nature, or are induced in a lab, are always harmful. Even evolutionists admit it is doubtful that, with all the mutations that have been observed, even *one* of them can definitely be said to have increased the viability of the mutated plant or animal.[398] However, knowing that the viability of the evolutionary doctrine absolutely depends on the existence of beneficial mutations, and trying to reconcile the repeatably observable evidence that they are vanishingly rare, evolutionists say that a very small fraction, on the order of 1 in 10,000 mutations, is beneficial.[399]

Now, seriously: even if 1 in 10,000 mutations was beneficial (and added genetic information, as evolution assumes), *how in the world* could anyone claim, with a straight face, that taking one step forward and 9,999 steps backward—or at best, sideways—would *ever* result in progress? Such is the magical world of evolution.

And remember, these are *random* mutations, which means we cannot make the beneficial mutations (if one ever occurs) highly effectual and helpful, and keep the harmful mutations tiny and negligible.

And, by the way, have I mentioned yet that *no* beneficial mutations have ever been seen? That means *all* of them are harmful, even the so-called "neutral" ones, which are too small to influence natural selection, but they still accumulate damage to the programming instructions encoded in the DNA.

So just in case there is still any question about whether repeatedly doing "little" bits of damage can improve an organism, let's read comments from some people well-versed in the field:

> Accordingly, mutations are more than just sudden changes in heredity; they also affect viability, and, to the best of our knowledge, **invariably affect it adversely.** . . . Mutation does produce hereditary changes, but the massive evidence shows that **all, or almost all, known mutations are unmistakably pathological and the few remaining ones are highly suspect.**[400]

---

[398] C. P. Martin, *American Scientist* 41:100 (1953).

[399] Ernst Mayr, in *Mathematical Challenges to the Neo-Darwinian Interpretation of Evolution*, ed. P. S. Moorhead and M. M. Kaplan (Philadelphia: Wistar Institute Press, 1967), p. 50.

[400] C. P. Martin "A Non-Geneticist Looks at Evolution," *American Scientist*, January 1953, p. 102. Emphasis added.

Lethal mutations outnumber visibles by about 20 to 1. Mutations that have small harmful effects, the **detrimental mutations, are even more frequent than the lethal ones.**[401]

I took a little trouble to find whether a single amino acid change in a hemoglobin mutation is known that doesn't affect seriously the function of that hemoglobin. **One is hard put to find such an instance.**[402]

The one systematic effect of mutation seems to be **a tendency towards degeneration. . .**[403]

Such is the legacy of random mutations.

This concept of randomness is *so* important, it would be good to consider another illustration. Evolutionists insist upon randomness for their belief system's mutations, but it is that very randomness that continues to bite them. Suppose you wanted to raise money for a charity, and as a fundraiser, you decided to walk from Redding, California to Times Square in New York City. *But*—and here's the catch—every step is going to be straight east (toward Times Square) or straight west (away from Times Square), and *the direction will be picked at random.*

It's about 2,882 miles, so how long would it take to get to Times Square? You'd never get there. If the direction were truly chosen at random at each step, you would die of old age before you got 100 yards from your starting point. Assuming you started in the center of Redding, you'd never reach the city limits, let alone cross the country.

Yet that is exactly what evolutionists insist upon, even though all of empirical science denies it. Richard Dawkins, passionate advocate of evolution, stated that evolution has been observed, but then added, "It's just that it hasn't been observed while it's happening."[404]

---

[401] A. M. Winchester, *Genetics,* Fifth Edition (Boston: Houghton Mifflin Company, 1977), p 356. Emphasis added.

[402] George Wald, as quoted by Murray Eden, "Inadequacies of Neo-Darwinian Evolution as a Scientific Theory," *Mathematical Challenges to the Neo-Darwinian Interpretation of Evolution,* editors Paul S. Moorhead and Martin M. Kaplan, pp. 18–19. Emphasis added.

[403] Sewell Wright, "The Statistical Consequences of Mendelian Heredity in Relation to Speciation," *The New Systematics,* editor Julian Huxley (London: Oxford University Press, 1949), p. 174. Emphasis added.

[404] Bill Moyers and David Brancaccio interviewing Richard Dawkins on NOW, the weekly newsmagazine from PBS. [Online at PBS.org/now/transcript/transcript349_full.html#dawkins; accessed July 24, 2020.]

In other words, like Carl Sagan, Dawkins *assumes* evolution happened, and his evidence is the fact that we exist.

But does it make sense to conclude that the mere existence of something proves that it evolved? Clearly not. If it did make sense, we could just as reliably conclude that commercial jets evolved (with no intentional actions whatsoever, remember) from gliders, which themselves evolved from paper airplanes. And that the cell phone evolved, all by itself, from two tin cans and a string. And that the computer evolved by accident from the abacus, with no decisions having been made to go that direction. And that—well, you get the picture.

The Effects of Mutation

So what are the effects of harmful mutations? The researchers in the Icelandic study came to the conclusion that such mutation-accumulation is very likely conducive to a greater incidence of diseases, including autism and schizophrenia. Alexey Kondrashov, an evolutionary geneticist and expert in mutation effects, reviewed the results of the Icelandic study and concurred. Other research corroborates these conclusions as well.[405]

Kondrashov adds:[406]

> "Because deleterious mutations are much more common than beneficial ones, **evolution** under this relaxed selection **will inevitably lead to a decline in the mean fitness** of the population."[407]

Isn't that interesting? Notice what evolution will do: it "will inevitably lead to a decline in the mean fitness of the population." Undirected evolution, which supposedly goes uphill, and increases order, complexity, and fitness, *ac-*

---

[405] Tennessen, J. A. et al. 2012. "Evolution and Functional Impact of Rare Coding Variation from Deep Sequencing of Human Exomes." *Science.* 337 (6090): 64–69.

[406] Kondrashov, A. 2012. "The rate of human mutation." *Nature.* 488 (7412): 467–468.

[407] Even "tight selection," wherein the four least fit out of six hypothetical children die and only the two most fit survive, could never reverse mutational buildup, not only because mutations are overwhelmingly non-beneficial, but for other reasons as well. See Sanford, J. 2008. *Genetic Entropy and the Mystery of the Human Genome*, 3rd ed. Waterloo, NY: FMS Publications.

But of course, such "tight selection" is an intentional, directed choosing, which evolution disallows. This means the results of random mutations are actually even worse than the "overwhelmingly non-beneficial" results Sanford describes.

tually goes downhill. This is obvious, but it's still rare to find an evolutionist who admits it.

What is evolution? What happens during this process of evolution? "Deleterious mutations" happen. And these deleterious mutations are "much more common than beneficial ones" because of the Second Law of Thermodynamics. Random mutations causing an "inevitable decline in fitness" hardly describes the ever-upward version of evolution we constantly hear about in the media and in textbooks, does it?

So what would a "decline in fitness" look like? Sicknesses and diseases. The most severe diseases the medical profession deals with are mutation-based diseases, because these diseases are not caused by the presence of pathogens (which often can be dealt with by medications) or harmful chemicals (which often can be flushed out), but they are malfunctioning protein machines, built using mutated/damaged programming instructions in the DNA, *in every cell of the body.* Such diseases are caused by the random mutations that evolution lauds so warmly. In 1994, we knew of nearly 4,000 diseases that were caused by mutations to DNA,[408] and the list is growing constantly.

One of the discoveries made by geneticists working on the Human Genome Project is described as follows:

> The human genome contains a complete set of instructions for the production of a human being. . . Genome research has already exposed **errors** [mutations] in these instructions **that lead to heart disease, cancer, and neurological degeneration.**[409]

---

[408] Nora, J. et al. 1994. *Medical Genetics: Principles and Practice.* Philadelphia: Lea and Feliger, 3.
[409] The Human Genome Project. Announcement from the University of Texas Southwestern Medical School, May 6, 1993.

OH, EVOLVE! (GOOD LUCK WITH THAT. . .)

As an example of this, peruse the image[410] below, paying special attention to the labels on each chromosome:

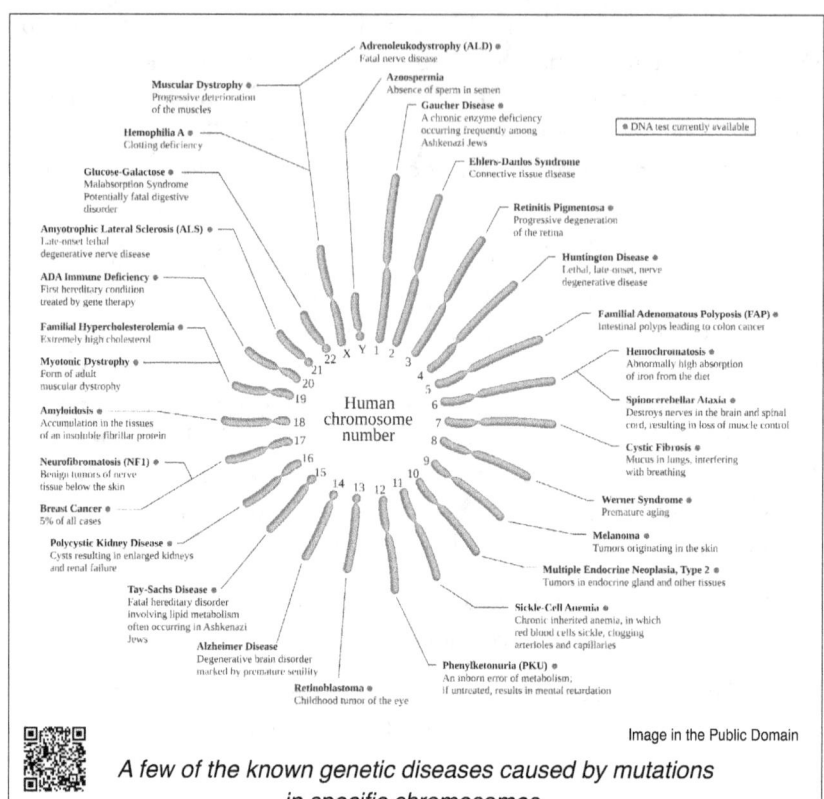

A few of the known genetic diseases caused by mutations in specific chromosomes.

And note this quote from the same article:

> There are **well over 6,000 known genetic disorders, and new genetic disorders are constantly being described** in medical literature. More than 600 genetic disorders are treatable. Around 1 in 50 people are affected by a known single-gene disorder, while around 1 in 263 are affected by a chromosomal disorder. **Around 65% of people have some kind of health problem as a result of congenital genetic mutations.**[411]

---

[410] "Genetic disorder" [Online at en.wikipedia.org/wiki/Genetic_disorder; accessed January 14, 2022.]

[411] "Genetic disorder" [Online at en.wikipedia.org/wiki/Genetic_disorder; accessed January 14, 2022.] Emphasis added.

Look at that list! Enzyme deficiency; connective tissue disease; degeneration of the retina, nerves, and muscle; colon cancer and breast cancer; loss of muscle control; excess mucus in lungs; premature aging; tumors in skin, eyes, nerves, and elsewhere; sickle-cell anemia; mental retardation; premature senility; inability to properly metabolize iron, phenylalanine, lipids, cholesterol, and sugars; renal failure; fiber buildup; problems with muscles, immune system, and blood clotting; nerve disease; and an inability to reproduce. Many of these conditions are fatal. Such is the legacy of random mutations.

Now think about it: if *random* mutations caused these previously smooth-running chromosomes to malfunction and cause all these diseases, how in the world did they become smooth-running in the first place? Evolution demands *random* mutations, but we see in the diagram above the results of randomness in mutations. Are we supposed to believe that randomness *assembles* working machines as easily as it damages and destroys them? In what universe?

In an analysis of known mutations, the goal of which was to identify the ratio of beneficial mutations to harmful ones, a leading geneticist discovered even more evidence to confirm that evolution's central dogma—that mutations are the mechanism of evolutionary progress—is simply wishful thinking. After analyzing almost a half a million known human mutations, he identified 186 beneficial mutations, and 453,732 harmful mutations. And even the 186 "beneficial" mutations also had detrimental effects, so they fall into the category of "equivocally beneficial" mutations—those with a beneficial outcome, but a downside as well.[412]

So the mutations causing only harmful results outnumbered the mutations that had any beneficial results by a factor of more than 2,439 to 1! So again I ask: How in the world could anyone claim, with a straight face, that one step sideways (no steps forward at all) and more than 2,400 steps backward could *ever* make any forward progress? One random improvement (plus its associated downside) for every 2,400 burdensome, crippling, or fatal mutations could result in the life forms we see today? It is completely ludicrous.

Just in case someone suspects me of exaggerating the harmfulness of mutations, let me quote Dr. Lee Spetner, a scientist and professor who taught in-

---

[412] Sanford, J. 2005. *Genetic Entropy and the Mystery of the Genome*. Lima, NY: Elim Publishing, 26.

formation and communication theory at Johns Hopkins University. He wrote the following in his book *Not By Chance:*

> In this chapter I'll bring several examples of evolution, particularly mutations, and show that **information is not increased**... in all the reading I've done in the life-sciences literature, **I've never found a mutation that added information.**
>
> **All point mutations** that have been studied on the molecular level turn out to **reduce the genetic information and not to increase it.**
>
> The NDT [neo-Darwinian theory] is supposed to explain how information of life has been built up by evolution. The essential biological difference between a human and a bacterium is in the information they contain. All other biological differences follow from that. The human genome has much more information than does the bacterial genome. **Information cannot be built up by mutations that lose it. A business can't make money by losing it a little at a time.**[413]

At this point, someone might object by saying that almost any combination of atoms, resulting from random mutations, could represent information. Is this a valid objection?

Actually, no, by the very nature of information:

> No matter how many "bits" of possible combinations it has, **there is no reason to call it "information" if it doesn't at least have the potential of producing something useful.** What kind of information produces function? In computer science, we call it a "program." Another name for a computer software is an "algorithm." **No man-made program comes close to the technical brilliance of even the Mycoplasmal genetic algorithms.** Mycoplasmas are the simplest known organisms with the smallest known genome, to date. How is its genome and other living organisms' genomes programmed?[414]

---

[413] Lee Spetner, *Not By Chance,* The Judaica Press, Brooklyn, New York, 1997, pp. 131–132, 138, 143. Emphasis added.

[414] David L. Abel and Jack T. Trevors, "Three Subsets of Sequence Complexity and Their Rel-

Now it's true that we have a lot of genetic information in our DNA, so how long would it take for 60 mutations per generation (the observed rate for protein-coding genes) to cause extinction? According to the theory of evolution, man has been evolving for about 2.4 million years. Assuming 20 years per generation, that works out to 120,000 generations, which would result in 7,200,000 mutations. There are about three billion "letters" with which human DNA is written, so more than seven million mutations in three billion letters far surpasses the mutational tolerance of the human genome.[415] In other words, we'd already be extinct. Last I checked, we weren't yet. This, by itself, is fatal to the theory of evolution.

Kondrashov understands the import of this ever-increasing genetic load, and asks, "Why aren't we dead 100 times over?"[416] We would be, if deep time were actually a thing.

But on the other hand, if we consider the Bible to be an accurate description of the age of humanity, we've been around for about 300 generations since Adam and Eve. This works out to 12,000 non-lethal mutations so far. This is not enough to cause outright extinction, but it is enough to cause more diseases with every generation.[417] And this is exactly what I've seen just in my own lifetime. Now in my sixties, I see much more widespread disease these days, even in young people, and many more *types* of disease, than I did when I was in college, medical advances notwithstanding.

Geneticist John Sanford, whose work was quoted above, describes the situation as follows:

> One of my reviewers told me that the message of this book is both terrifying and depressing. He suggested that perhaps I am a little like a sadistic steward on board the Titanic, gleefully spreading the news that the ship is sinking. But that is not correct. I hate the consequences of entropy (degeneration). I hate to see it in my own body, in the failing health

---

evance to Biopolymeric Information," *Theoretical Biology & Medical Modeling*, Volume 2, August 11, 2005, Page 8. [Online at TBioMed.com/content/2/1/29; accessed November 9, 2020.] Emphasis added.

[415] One study estimated human genome collapse after 32,800 years in a population of 10¹¹ and a 1.5% fitness decline per generation. See Williams, A. 2008. "Mutations: Evolution's Engine Becomes Evolution's End!" *Journal of Creation*. 22 (2): 60–66.

[416] Kondrashov, A. 1995. "Contamination of the genome by very slightly deleterious mutations: why have we not died 100 times over?" *Journal of Theoretical Biology*. 175 (4): 583–594.

[417] Thomas, B. 2012. "Human Mutation Clock Confirms Creation." *Acts & Facts*. 41 (11): 17.

of loved ones, or in the deformity of a newborn baby. I find it all absolutely ghastly, but also absolutely undeniable. **Surely a real steward on the Titanic would have a responsibility to let people know that the ship was sinking, even if some people might hate him for it. I feel I am in that position.** Responsible people should be grateful to know the *bad news,* so they can constructively respond to it. If we have been putting all our hope in a sinking ship, would it not be expedient to recognize this and abandon the false hope? It is only in this light that we can appreciate bad news. Only in the light of the *bad news* can we really appreciate the *good news* that there is a lifeboat.[418, 419]

Sanford goes on to say:

I believe the Author of Life has the power to defeat death and degeneration. . . He gave us life in the first place, so He can give us new life today. . . I humbly put before you this alternative paradigm for your consideration—Jesus is our one true hope.[420]

How Likely is Evolution?

According to the proponents of evolution, there are only two requirements for evolution to work. The first is beneficial mutations, covered already, and the second is **natural selection**, or "survival of the fittest." According to the theory, natural selection acts like a filter that allows "fit" specimens to continue propagating the species, but causes "unfit" specimens to die out, stopping the propagation of whatever made them unfit.

Because evolution requires random mutations (any *deliberate* mutations would require an intelligent Mind, which is verboten), it is subject to statistical analysis. To help put the above discussion of random mutations in perspective, let's consider a hypothetical organism of only 200 working parts.[421]

---

[418] John Sanford, *Genetic Entropy and the Mystery of the Genome,* 3rd edition (FMS Publications, New York, 2008), p. 158. Emphasis added.

[419] Royal Truman, *From Ape to Man Via Genetic Meltdown: A Theory in Crisis, Journal of Creation* 21(1):43–47, April 2007. [Online at Creation.com/from-ape-to-man-via-genetic-meltdown-a-theory-in-crisis; accessed June 9, 2022.]

[420] John Sanford, *Genetic Entropy and the Mystery of the Genome,* 3rd edition (FMS Publications, New York, 2008), p. 159.

[421] My appreciation goes to Dr. Henry Morris for the basis of this discussion. See Henry M.

Starting from nothing, we have none of the parts that will ostensibly evolve into the required 200-part organism. So we will clearly need 200 random-but-beneficial mutations to build it up into the 200-part organism we're hypothesizing. These mutations must be **heritable**—that is, they must be able to be passed on genetically to offspring—so in order to confirm that they are heritable, these mutations must show up in every generation after the one in which they first appeared.

These 200 beneficial mutations could be continuous—200 beneficial mutations in a row—but we don't actually need that. All we need is for the beneficial mutations to outnumber the harmful ones by 200. (Let's apply some suspension of disbelief and pretend for a moment that random beneficial mutations—those that add genetic information—actually exist.)

So, for example, suppose you got 100 beneficial mutations in a row, and then had a harmful one, which negates the improvement of one of the beneficial ones: you'd be back down to 99 beneficial mutations. But if you had a beneficial mutation that cancelled the harmful one you just experienced, you'd be back up to 100 and could go on from there. So again, all you need for this *gedanken* **experiment** ("thought experiment"), is for the beneficial ones to outnumber the bad ones by 200. (Let's apply some more suspension of disbelief and pretend for a moment that the benefit of the beneficial mutations averages out to be the same magnitude as the harm of the harmful ones.)

Let's further assume the best of all possible scenarios where *every* mutation is a beneficial one, so we can get all 200 *in a row*. This will minimize the time required. We could allow harmful mutations and subsequent beneficial ones that cancel out the harmful ones, but that would make the whole process take even longer. We're looking for the fastest possible time here.

Let's further assume a wildly optimistic probability of beneficial mutations: that fully 50% of the mutations are beneficial, rather than the observed rate of an almost infinitesimal fraction of 1%.

Doing the Math

Before we get to the actual math part, here is a refresher on what is called **scientific notation.** Scientific notation is basically a method of expressing very large or very small numbers in a more manageable way.

Morris, Ph.D. 2003. "The Mathematical Impossibility of Evolution." *Acts & Facts.* 32 (11). [Online at ICR.org/article/mathematical-impossibility-evolution; accessed June 17, 2020.]

For example, the Earth weighs about 6,600,000,000,000,000,000,000 tons (6 sextillion, 600 quintillion tons). That's a rather large number, and we can much more easily make mistakes if we make calculations with long strings of zeroes like that. The commas certainly help, but still, it's easy to miscount. It's a lot more convenient to express the above number as 6.6×10^{21}, which means 6.6 times 21 tens all multiplied together. Calculating with numbers expressed this way is faster and less error-prone than writing them out the long way.

One way of using scientific notation is a variation called **engineering notation**. Engineering notation differs from scientific notation only in that the exponent (the little superscripted number) is always a multiple of three, so groupings like million, billion, trillion, and so forth—which are more easily understood by humans—can be used.

For example, suppose you wanted to represent the number 6.8×10^{13}. This number is in scientific notation, and is a perfectly valid number. In human terms, this could be read as "six point eight ten-trillions" (since 10^{13} is ten trillion), but that may be awkward for a human to grasp. So if we convert the representation to engineering notation by multiplying the number on the left by 10, and subtracting 1 from the exponent, we have 68×10^{12}, or "sixty-eight trillion." The two representations express exactly the same number, but I think you'd agree that "sixty-eight trillion" is more easily understood and less cumbersome than "six point eight ten-trillions."

For this reason, engineering notation will be predominantly used in the discussions below.

Our Hypothetical Organism

Okay, so back to our hypothetical 200-part organism: assuming that beneficial mutations have a 50% success rate, and that all 200 of them happen *in a row*—in successive generations, to minimize the time required—the likelihood (let's call it P, for the **probability** of success for any given attempt) of that 200-part organism actually coming into existence is:

$$P = \frac{1}{2} \times \frac{1}{2} \times \ldots \times \frac{1}{2} \times \frac{1}{2} \times \frac{1}{2} \text{ (200 times)}$$

In other words, ½ multiplied by itself 200 times, or one half to the 200th power. This could more simply be expressed as:

$$P = 0.5^{200} \approx 622 \times 10^{-63} \text{ chance of success}$$

So this means that for each attempt at getting 200 beneficial mutations in a row, you have less than 1 chance in 10^{60} to assemble all 200 parts the correct way to make a viable organism. Notice that we're not talking about arriving at a *particular* 200-part organism, but *any* viable 200-part organism. (And we haven't even addressed whether we even *have* all the raw materials yet. Or where they came from.)

According to accepted laws of probability, if the chance of something occurring is less than 10^{-50} (less than 1 chance in 10^{50}), it can safely be considered impossible.[422] If a friend started flipping a coin and got ten heads in a row, you would already be thinking something fishy—something decidedly non-random—was going on. But the probability of ten heads in a row is only about 10^{-3}, which is enormously more likely than our poor little organism's chances!

Since the likelihood of success for our little organism is *so* small, the number of attempts required to arrive at our 200-part organism needs to be huge. *So* huge, in fact, it's hard to grasp its magnitude. But let's tackle it anyway.

The Surface of the Earth

Unless you're used to dealing with very large numbers, it's difficult to wrap your brain around a quantity like 10^{60}, so let's bring it more down to earth. I encourage you to grab a calculator (there's probably one on your phone) and follow along, to satisfy yourself that my calculations are correct. The Earth is close to being a perfect sphere, so spherical calculations will be close enough.

The radius of Earth, at the equator, is approximately 3,963 miles:

$r \approx 3963$ mi

The formula for calculating the surface area of a sphere is:

$A = 4 \pi r^2$

Therefore, the surface area of Earth, in square miles, is:

$A \approx 4 \times \pi \times 3963^2$
$\approx 197{,}359{,}487$ sq mi

---

[422] Probability expert Emile Borel wrote, "Events whose probabilities are extremely small never occur… We may be led to set at 10^{-50} the value of negligible probabilities on the cosmic scale." (Emile Borel, *Probabilities and Life*, (New York: Dover Publications, 1962), p. 28.)

But it will be more convenient for our purposes if we convert the units from square miles to square millimeters. There are 5,280 feet in a mile, 12 inches in a foot, and 25.4 millimeters in an inch; therefore:

$$\begin{aligned} 1 \text{ sq mi} &= (5280 \times 12 \times 25.4)^2 \text{ sq mm} \\ &= 2{,}589{,}988{,}110{,}336 \text{ sq mm} \\ &\approx 2.6 \times 10^{12} \text{ sq mm (2.6 trillion)} \end{aligned}$$

Therefore:

$$\begin{aligned} A &\approx 197{,}359{,}487 \text{ sq mi} \times 2.6 \times 10^{12} \text{ sq mm/sq mi} \\ &\approx 511 \times 10^{18} \text{ sq mm} \end{aligned}$$

So the Earth has about 511 billion billion square millimeters of surface area, counting both land and water.

How Much Time?

So now we have the surface area of the Earth. And, as mentioned above, the theory of evolution is of the opinion that Earth has been in existence for roughly 4.5 billion years:

$$\begin{aligned} T &\approx 4{,}500{,}000{,}000 \text{ yr} \\ &= 4.5 \times 10^9 \text{ yr} \end{aligned}$$

Let's convert that duration to seconds, just for fun. Since there are about 365¼ days in a year, 24 hours in a day, 60 minutes in an hour, and 60 seconds in a minute:

$$\begin{aligned} 1 \text{ yr} &\approx 365¼ \times 24 \times 60 \times 60 \text{ sec} \\ &\approx 31{,}557{,}600 \text{ sec} \\ &\approx 31.6 \times 10^6 \text{ sec (31.6 million seconds)} \end{aligned}$$

Therefore:

$$\begin{aligned} T &\approx (4.5 \times 10^9 \text{ yr}) \times (31.6 \times 10^6 \text{ sec/yr}) \\ &\approx 142 \times 10^{15} \text{ sec (142 million billion seconds)} \end{aligned}$$

So, according to the theory of evolution, Earth is about 142×10^{15} seconds old.

Back to Our Hypothetical Organism

Now let's apply those calculations to our hypothetical 200-part organism. Let's assume that all 200 beneficial mutations happened in only 200 consecutive generations, and all 200 generations of this life-form could take place in a *single* second. This means that a new attempt at getting 200 random-but-beneficial mutations in a row could begin *every second* of every day for 4½ billion years. Let's also assume that it takes no time at all to recognize a failure, clean everything up, and start over.

Further assume that these one-second-long attempts at getting 200 beneficial mutations in a row are happening *in every square millimeter* of the Earth's surface, and they have been occurring for the entirety of the alleged 4½-billion-year duration of the planet's existence.

And *further* assume that the probability of a random-but-beneficial mutation is (a preposterously generous) 50%. So with all of these assumptions, how long would it take for our 200-part organism to evolve?

We just need to multiply some things together to get P, the probability that we would get 200 more beneficial mutations than harmful ones (and assuming no time wasted by having a harmful mutation needing to be subsequently cancelled by a beneficial one):

$$P \approx 511 \times 10^{18} \text{ (sq mm on Earth)} \times$$
$$142 \times 10^{15} \text{ (Earth's assumed age)} \times$$
$$622 \times 10^{-63} \text{ (chance of evolving correctly)}$$
$$\approx 45 \times 10^{-24}$$

So, we could expect such experiments to yield desirable results—that is, one single viable 200-part life form in our 4½-billion-year experiment—less than one billionth of one billionth of one percent of the time. Or, roughly once every 22 billion trillion times this 4½-billion-year experiment was performed, where every square millimeter on Earth was doing the whole 200-generation attempt *every second*. To get *one* successful outcome of our 200-part organism. Seriously, how much more ridiculous could random evolution get?

According to the National Geographic magazine,[423] the odds of dying from various causes are as follows:

Lightning:	1 in 135,000
Meteorite:	1 in 1,600,000
Shark:	1 in 8,000,000

---

[423] Howard, Brian Clark, "What Are the Odds a Meteorite Could Kill You?" *National Geographic* February 9, 2016. [Online at NationalGeographic.com/news/2016/02/160209-meteorite-death-india-probability-odds/#close; accessed on December 1, 2019.]

The odds of all of these happening to you at the same time—that you experience *all* of these accidents seriously enough to die from them—is pretty small. The odds of you getting fatally attacked by a shark at the same time you are fatally struck by lightning and a meteorite is about 578×10^{-21}, or about 1 chance in a billion billion. *This is still more than ten thousand times more likely* than our little 200-part organism evolving even *once* in 4½ billion years, with every square millimeter on Earth making another attempt every single second.

Or, if you play cards, here's another way of looking at it. In poker, the highest-ranked hand possible is the royal flush: it is composed of a ten, jack, queen, king, and ace, all in the same suit. What are the odds of being dealt such a hand?

Assuming that the player is being dealt five cards from a well shuffled standard 52-card deck, keeps all five cards, and no cards are wild, we need just two quantities in order to figure out the probability of being dealt a royal flush. First, we need to determine the total number of possible poker hands; that is, the number of different combinations of five cards out of the 52 we started with.

This is calculated by:

$$N_{ph} = C(52, 5)$$
$$= 52! \div (5! \times (52-5)!)$$
$$= 2{,}598{,}960$$

So there are more than 2½ million possible poker hands. How many of them can make a royal flush? A royal flush needs 10, J, Q, K, and A, but it doesn't matter which suit; any one will do, as long as all cards are of the same suit. So there are only four ways, out of the 2½ million, to get a royal flush. This works out, on average, to 1 chance in 649,740 deals.

The chance of our little organism evolving successfully even once, after making a new attempt every second for 4½ billion years, in every square millimeter of the Earth's surface, is still *80,000 times less likely* than being dealt three royal flushes *in a row* from a well-shuffled deck.

CHAPTER 6: MUTATIONS

And keep in mind, in our hypothetical scenario, we were ridiculously generous and optimistic in giving our little organism a fighting chance to evolve by lucky random mutations. Specifically:

- We allowed all 200 required generations to happen *in a single second* for each attempt. In actuality, each generation would take *much* longer than 5 milliseconds for any newborn organism to mature enough to be ready to reproduce. Even single-celled organisms take much longer to reproduce: the reproduction cycle in baker's yeast, for example, can be "as short as one hour."[424] We permitted our organism to reproduce 720,000 times faster than this.
- We allowed the start-over time to be exactly zero, which means after each unsuccessful attempt at evolving a 200-part organism, it could start over with a "clean slate" exactly one second after the previous one-second attempt was started.
- We allowed the likelihood of a random-but-beneficial mutation to be a *full 50%*. In actuality, the likelihood of a random-but-beneficial mutation is an extremely tiny fraction of 1%. In fact, as you remember, a random, heritable, beneficial mutation has not yet been observed anywhere on Earth.[425] But still, evolution absolutely depends on it happening, and in *astronomical* numbers.
- We allowed *every single mutation* to be beneficial, so there were no harmful mutations at all to impede progress, and which would have to subsequently be "cancelled" by another beneficial one. In other words, we'll pretend that the *theory* of evolution actually overrides the Second *Law* of Thermodynamics.
- We allowed *every square millimeter* on Earth—both land and water—to host these one-second evolutionary attempts simultaneously.
- We allowed the *entirety* of the alleged age of the Earth—4½ billion years—for this process to take place 24/7, even though evolutionists themselves claim that the surface of the Earth was molten lava for a long time.

And still, even with all the ridiculously generous allowances noted above, the chances of our little 200-part organism evolving successfully, even once,

---

[424] Article "Baker's Yeast and Its Life Cycle." [Online at phys.ksu.edu/gene/a1.html, accessed December 7, 2019.]

[425] Henry M. Morris, Ph.D. 2003. "The Mathematical Impossibility Of Evolution." *Acts & Facts*. 32 (11). [Online at ICR.org/article/mathematical-impossibility-evolution; accessed June 17, 2020.]

in our 4½-billion-year experiment, is less than one billionth of one billionth of one percent! And, by the way, even a single-celled plant or animal has *millions* of molecular "parts," and we can't even grow a measly 200-part organism, in spite of granting laughably optimistic suspensions of the laws of physics.

The essence of naturalistic evolution is randomness—the non-directed nature of change. But if you just do the math, the likelihood of the evolution of *anything* vanishes before our eyes.[426] And even evolutionists themselves have acknowledged that this is an enormous monkey wrench in the believability of their theory.[427]

And did you notice the other show-stopping monkey wrench that was not even mentioned in the above discussion of our little organism? Even with the absurdly generous conditions we allowed, there is another factor that would take the likelihood of success from 1 chance in 10^{63} down even further. Remembering that anything less than 1 chance in 10^{50} can safely be considered impossible, this new monkey wrench that we just glossed over takes it down to *exactly zero*. Yes, this other factor takes it all the way down to *no way*—completely, absolutely impossible. Can you guess what it is? Think about it for a while, and then when you get to the section called "Reproduction," later in this chapter, it will become clear.

But this is a terrifying thought to proponents of evolution: the only alternative to a random process is a process that is *not* random. And if it is not random, it is, by definition, directed. And if it is directed, there is intention. And if there is intention, there is a Mind behind it all. And if there is a Mind behind all of Creation, we have to acknowledge a very real possibility that we have a purpose in life, and we are accountable for what we do regarding that purpose.

Haldane's Dilemma

J.B.S. Haldane (1892–1964) was an evolutionary geneticist, and was one of the founders of the field of **population genetics**—the study of how genetic changes propagate through an entire population. This process would, of course, be essential if any beneficial mutations that popped up were to persist, and the species, as a whole, evolved into something better.

---

[426] James Coppedge, *Evolution: Possible or Impossible* (Zondervan, 1973, 276 pp.)

[427] Wistar Institute Symposium, *Mathematical Challenges to the Neo-Darwinian Interpretation of Evolution*, 1967, 140 pp.

So in 1957, Haldane did some math on this process, and analyzed how long it would take for, say, an ape to evolve into a human being. To guarantee that any given beneficial mutation would be passed on to the next generation, it must propagate through the *entire* population. Otherwise, the effect of the beneficial mutation would be so diluted from interbreeding with the "unimproved" versions of the species, that the improvement would be overwhelmed by the sheer numbers of the old version, and would vanish from the population. So, to avoid this scenario, the unimproved version of the species must be *completely* replaced by the improved version.

Let's run some numbers and see how likely this actually is. We'll start with the following suppositions:

- We'll assume we start with a population of 100,000 apes. Large primates that they are, the length of a generation would be about 20 years.[428]
- We'll also assume that an unequivocally beneficial mutation happens in one of the males. That is, a mind-bogglingly complex DNA-based, protein-implemented algorithm, more intricate than anything mankind has yet developed, appears all by itself, out of nothing. (Never mind that an unequivocally beneficial mutation has never been found.)
- We'll also assume that the same mind-bogglingly complex algorithm appears in a female of the species, within the lifetime of the male. We need the same mutation to appear in the female because it takes a male *and* a female to create a baby.
- We'll also assume that this improved ape, with the new mutation, is *so* superior to the unimproved versions, that the entire population is replaced, in a single generation, by the improved version. That is, the two improved apes do away with all the others, and quickly have 99,998 babies to replace the old version entirely. (Yes, I know this is preposterous to the highest degree, but work with me here.)
- We'll also assume that a male and a female of the new-and-improved type also simultaneously experience a new random mutation that is so superior to their parents that they, too, do away with all of the now-old version and have 99,998 babies of their own. So the entire popu-

---

[428] Gibbons, Ann, "Generation Gaps Suggest Ancient Human-Ape Split," *Science*, August 13, 2012. [Online at science.org/content/article/generation-gaps-suggest-ancient-human-ape-split; accessed November 24, 2023.] The empirical science in this article is good—actual observations and repeatable tests and measurements. But the historical science woven throughout the article—ideas about how random accidents could have created the information and intricate complexities of life—are merely speculations. Beware.

lation of apes is completely replaced by a new and improved model *every generation.*

- Because evolutionary guesses place the Most Recent Common Ancestor (MRCA) of apes and man at 10 million years ago, that works out to be only half a million beneficial mutations: 10 million years divided by 20 years per generation equals half a million generations. At one beneficial mutation per generation, this is half a million beneficial mutations.

Even with the above ridiculously optimistic assumptions, half a million beneficial mutations is only about 0.02% of the human genome. This is still 50 times smaller than the value given by evolutionist Richard Dawkins in his book *The Blind Watchmaker,* which states that the difference between chimp (closer to humans than apes are) and human genomes was less than 1%.[429] One might wonder how Dawkins came up with this value, since he wrote his book in 1986, and the human and chimp genomes weren't even published until 2001 and 2005, respectively. The 1% value was based on a 1975 estimate that used only tiny bits of selected sections of the respective DNA strands, and evaluated by a process laden with assumptions. But since the genome data required to even attempt to make such a statement wasn't published until almost 20 years *after* he made the statement, it is clear that Dawkins' claim was little more than wishful thinking.

Statements that the difference between chimp and human DNA was "only 2%" were bandied about for years. But even with the laughably optimistic concessions given above, the maximum number of beneficial mutations is still 100 times smaller than the difference between the DNA of chimps and humans that used to be matter-of-factly stated as only 2%. But then further research revealed in 2002 that it was more like 5%[430] because they used some genome data wasn't previously considered, and didn't ignore some differences that previously had been.

Then further research, done by teams taking into consideration larger and larger fractions of the entire DNA strands (instead of short, hand-picked segments), came to the conclusion that the similarities between human and

---

[429] Richard Dawkins, *The Blind Watchmaker: Why the Evidence of Evolution Reveals a Universe Without Design,* [W.W. Norton, New York, 1986], p. 263.

[430] David DeWitt, "Greater than 98% Chimp/Human DNA Similarity? Not Any More." [Online at creation.com/greater-than-98-chimp-human-dna-similarity-not-any-more; accessed November 24, 2023.]

chimp DNA is somewhere between 81% and 87%.[431] So the more of the DNA that is actually compared, the larger the differences discovered. And Haldane's Dilemma quietly stands, pointing out the impossibility, even under the most optimistic conditions, of evolution ever happening.

One might think that it would help to start with a smaller population, because one couple having almost 100,000 babies is a bit of a stretch. But that doesn't help, because that would also make it that much less likely that a beneficial mutation could randomly happen (like it's possible to get less likely than 0%). Similarly, increasing the population to increase the likelihood of getting a random-but-beneficial mutation would also make the task of replacing the entire population more problematic. Something about "a rock and a hard place" comes to mind. . .

So Haldane's Dilemma has not been solved,[432] even though it was originally published more than 65 years ago. Recent articles presented for publication on the subject are quietly rejected. It seems no one wants to talk about Haldane's Dilemma, because it is yet another nail in the coffin of the theory of evolution, yet another devastating showstopper for a theory that is holding on by sheer desperation.

"Primitive" Man

Evolution states that all life came from non-life, and that would necessarily include man. So naturally, evolutionists are always on the lookout for evidence that would support the contention that man evolved from "lower forms" of primates—apes, chimpanzees, or some such.

Proponents of evolution are *very* eager to find support for their opinions—sometimes perhaps a little too eager. All of the alleged "primitive" forms of man mentioned below were, at one time or another, lauded as "the missing link" that genetically connected apes to modern man. So eager were evolutionists to jump at the chance to prove their theory that they've had to retract their discoveries on more than one occasion, acknowledging, to their chagrin, that maybe the discoveries weren't missing links after all.

---

[431] Jeffrey Tompkins and Jerry Bergman, "Genomic monkey business—estimates of nearly identical human–chimp DNA similarity re-evaluated using omitted data" *Journal of Creation* 26(1):94–100, April 2012. [Online at Creation.com/human-chimp-dna-similarity-re-evaluated; accessed November 24, 2023.]

[432] Don Batten, "Haldane's dilemma has not been solved," *Journal of Creation* 19(1):20–21, April 2005. [Online at Creation.com/haldanes-dilemma-has-not-been-solved; accessed November 24, 2023.]

If you dig into further research on "early man," you will find much disagreement and mutual contradiction among evolutionists. You'll also notice that answers that were considered "right" a while back are now considered wrong. It is interesting to note that in many cases, when a previously accepted answer is discovered to be wrong, the new idea is introduced with "Now we know that. . ." It's not "Now we think," "Now we believe," or "Now it looks like," but "Now we *know.*" After a certain number of retractions, you'd think they would become a bit more humble in their claims.

There's another confusion factor as well, mentioned earlier: Sometimes a fossil's alleged age determines the category/species to which a fossil is assigned, but other times the category/species in which it has been placed determines the fossil's alleged age.[433] As you research, be on the lookout for this.

Neanderthal Man

The Neanderthal man—*Homo neanderthalensis*—is named after the Neander Valley in Germany, where it was found in 1856. Much debate ensued about the validity of the finds, what they really represented, the intelligence of the species found, and so forth. The conclusion at the time was that the Neanderthals were considered to be slouching brutes, ape-like primitives.

Le Moustier, artist Charles Knight's impression of Neanderthals in 1920.

Image is in the public domain.

The idea of the Neanderthal man being a primitive version of man, somewhere between apes and modern man in the evolutionary continuum, received a lot of impetus when it was promoted by Marcellin Boule in the *Annales de Paleontologie.* He wrote a series of articles from 1911 through 1913, championing his ideas.[434]

Although Boule's ideas were later refuted, his notions remain in textbooks and museum displays to this day, proselytizing the average person to believe that Neanderthal is one of the "missing links" in the evolutionary progression from ape to man.

---

[433] Marvin L. Lubenow, *Bones of Contention* (BakerBooks, Grand Rapids, Michigan), 2004, p. 18.

[434] Lubenow, M. 2004. *Bones of Contention.* Grand Rapids, MI: Baker Books, p. 53.

It's actually amazing how little evidence there is for these fossils being those of "early man." The discoverers of fossils are extremely secretive about them and it is rare they are ever seen by others, which means that any claims made are difficult, if not impossible, to verify or falsify. This secretiveness was present with the original Neandertal fossils unearthed in 1856 by Herr von Beckersdorf, the owner of the land. He gave the fossils to J. K. von Fuhlrott, who was the president of the local natural history society.

What did Fuhlrott do with them?

> The skull and the bones were Fuhlrott's private property, and he did not show them to many. Only very few scholars in Britain and on the Continent had seen the skull or obtained a cast. [Even Rudolf Virchow, the greatest medical man of his time] **could only study the remains in Fuhlrott's house after gaining access from his wife when Fuhlrott was away.**[435]

William King never saw the original fossils, but he pronounced them to be a different species than modern humans, and named them *Homo neanderthalensis* in 1864. Darwin never saw these fossils, or any other fossil humans, but still, in 1871 he published a book entirely devoted to human evolution. Thomas Huxley, "Darwin's bulldog," never saw the original fossils either, although he described them in his 1863 book, *Man's Place in Nature*.[436]

Hopefully, this clears up any romanticized notion that theories of human evolution are actually based on fossil evidence. In actuality, paleoanthropology is in the odd state of being the science in which most of its workers do not have access to the material upon which their science is based.

Because evolutionists *assume* evolution is valid, and nothing else is even considered, human, animal, and plant fossils are plugged into a pre-existing narrative wherever they seem to fit best at the moment (and this changes whenever deemed necessary). This narrative can look very different, depending on the particular evolutionist's biases. As anthropologist Ian Tattersall said, this process is "both political and subjective" to the point where he sug-

---

[435] David Van Reybrouck, "Imaging and imagining the Neanderthal: the role of technical drawings in archaeology," *Antiquity* 72 (March 1998): 62. Emphasis added.

[436] David W. Frayer, "Naming Our Ancestors," review of *Naming Our Ancestors*, edited by W. Eric Mielke and Sue Taytlor Parker, *American Journal of Physical Anthropology* 98, no. 2 (October 1995): 235.

gested that "paleoanthropology has *the form but not the substance of a science*."[437]

But what has been discovered about Neanderthals since? They were actually quite intelligent. Neanderthal technology includes stone tools, fire and cave hearths, making birch bark tar for an adhesive, weaving, making clothing such as blankets and ponchos, seafaring through the Mediterranean, making use of medicinal plants, treating severe injuries, storing food, using various cooking techniques (e.g., roasting, boiling, smoking). Neanderthals ate a wide variety of foods, including hoofed mammals and other big game, plants, small mammals, birds, and fish.[438]

Neanderthals also were able to treat each other medicinally: for simple pain relief, they would chew the bark of poplar trees. Bark from the poplar tree (as well as some other trees) contains salicylic acid, and a modified version of this substance—acetylsalicylic acid—is still used as a pain reliever. You may know it as aspirin. Neanderthals also ate a certain kind of mold called *Penicillium* which, as you might have guessed, produces the antibiotic penicillin.[439]

Neanderthals are also known to have treated things more serious than minor pain relief and bacterial infections. Study of Neanderthal remains indicate that they treated tooth abscesses, broken limbs, and even fractured skulls. Bone growth after treatment shows that the patients lived and recovered from the injuries. In fact, one researcher notes:

> The high level of injury and recovery from serious conditions, such as a broken leg, suggests that others must have collaborated in their care and helped not only to ease pain, but to fight for their survival in such a way that they could regain health and actively participate in the group again.[440]

---

[437] See: Ian Tattersall, "Paleoanthropology and Preconception," in: W. Eric Meikle, F. Clark Howell, and Nina G. Jablonsky, editors, *Contemporary Issues in Human Evolution,* Memoir 21 (San Francisco: California Academy of the Sciences, 1996); Jeffrey A. Clark, "Through a Glass Darkly: Conceptual Issues in Modern Human Origins Research," in G. A. Clark and C. M. Willermet, editors, *Conceptual Ideas in Modern Human Origins Research* (New York: Aldine de Gruyter, 1997), Quoted in: Jonathan Wells, *Icons of Evolution*, p. 223. Emphasis added.

[438] en.wikipedia.org/wiki/Neanderthal; accessed June 12, 2020.

[439] Lita Cosner and Robert Carter, "The sophisticated Neandertal", *Creation* 42(1):12–13, January 2020. [Online at Creation.com/sophisticated-neandertal; accessed March 27, 2021.]

[440] University of York, "Neanderthal Healthcare Practices Crucial to Survival", Oct 4, 2018. [Online at ScienceDaily.com/releases/2018/10/181004110042.htm; accessed March 27, 2021.]

The same article states that evidence indicates that Neanderthals were knowledgable about medicinal botanicals, how to bring down fevers, care for wounds, and use midwives to assist in birthing.

When Neanderthals did die, they buried their dead ritualistically, and included tools and other artifacts with the body,[441] indicating that they distinguished life and death, honored one another, and mourned their dead.

As a matter of fact, continuing discoveries of the Neanderthal has caused him to be promoted. He is now considered fully human—a real member of the *Homo sapiens* species. That fact may be a shock to many people nowadays, who were raised with the "common knowledge" that Neanderthals were primitive half-apes. But Neanderthals had larger brains than modern humans, they had a well developed language center,[442] they cared for their ailing and aged, they created art, practiced religious rites, and created music.[443] Their culture was indeed unsophisticated compared to our twenty-first-century technological world, but no more unsophisticated than many people groups living around the world today.

A fossil elbow and fossilized Laetoli footprints (Laetoli is an archaeological site in Tanzania) cannot be distinguished from modern man, but they have been dated by evolutionary scientists as 4 million years old, which is earlier than the earliest Neanderthals.[444] Using dates that were published by evolutionists themselves:

> . . .anatomically modern *Homo sapiens,* Neandertal, archaic *Homo sapiens,* and *Homo erectus* [as well as Lucy-like *Australopithecines*] all lived as contemporaries.[445]

So, *why* are Neanderthals still considered by many to be evolutionary stepping-stones to our modernness?

---

[441] Brian Thomas, "Neanderthal Babies Were Human Babies," [Online at ICR.org/article/neanderthal-were-human-babies; accessed June 12, 2020.]

[442] Phillips, D. 2000. "Neanderthals Are Still Human!" *Acts & Facts.* 29 (5). [Online at ICR.org/article/468/; accessed June 12, 2020.]

[443] Morris, J. 1997. "Is Neanderthal in Our Family Tree?" *Acts & Facts.* 26 (9). [Online at ICR.org/article/1170; accessed June 12, 2020.]

[444] Tuttle, R. 1990. "The Pitted Pattern of Laetoli Feet." *Natural History.* March Issue, 60–65. Quoted in Lubenow, *Bones of Contention,* 170.

[445] Lubenow, M. 1992. *Bones of Contention.* Grand Rapids, MI: Baker, 178.

Cro-Magnon Man

The Cro-Magnon Man—*Homo sapiens fossilis* or *Homo sapiens cromagnonensis*—another supposed intermediate form in the long, upward, evolutionary climb toward humanity, is so named because the first such skeletons were found in the Cro-Magnon rock shelter in France in 1868.

Image courtesy Wellcome Images and Creative Commons (CC 4.0). Photo cropped.

1918 bust of Cro-Magnon Man, presumed age to be "at least 25,000 years."

So just how primitive was the Cro-Magnon man? In Cro-Magnon culture,[446] they made weapons out of bone, antlers, and flint; they pierced bones, shells, and teeth to make body ornamentation and jewelry. They hunted big game, including mammoth, cave bears, horses, and reindeer, hunting with spears and spear-throwers, and built huts out of mammoth bones. They spun, dyed, and knotted flax fibers into cloth for garments. They played music using flutes made from bone.

Cro-Magnon Man also buried their dead with jewelry and tools, indicating an understanding of ritual. They took care of wounded or ill members of their population. In fact, the term "Cro-Magnon Man" is no longer the preferred term for this type of person; the preferred term is now "European early modern humans," or "EEMH." So if the Cro-Magnon man is a "modern human," what, exactly, is the difference? Apparently, none.

---

[446] en.wikipedia.org/wiki/European_early_modern_humans; accessed June 13, 2020.

Java Man

The Java man—*Homo erectus* (formerly known as *Pithecanthropus erectus* or *Anthropopithecus erectus*)—is described as an "early human" fossil, and it is so named because it was first discovered in 1891 on the island of Java in the Dutch East Indies, which is now part of Indonesia. It is stated to be between 700 thousand and a million years old. The *Homo erectus* part is there because he was believed to be the first species of "early man" to walk erect on his feet, as opposed to walking on all fours in a stooped, knucklewalking manner characteristic of certain species of monkeys and apes.

Image is in the public domain.

J. H. McGregor's 1918 conception of the "ape-man of Java."

The Java Man was conceptualized (imagined, invented, concocted, made up) by Dutch anatomist Eugène Dubois on the basis of finding a molar and a skullcap in 1891. The next year he found a femur 50 feet away and assumed it was from the same individual as the skullcap and the tooth, and from there, he concluded that this was a transitional form between apes and man.[447]

Did Dubois seek input from colleagues on the nature of what he found? Hardly. Describing Dubois, G. H. R. von Koenigswald wrote:

> On this point he was as unaccountable as a jealous lover. Anyone who disagreed with his interpretation of *Pithecanthropus* was his personal enemy.[448]

There was much disagreement about the Java Man; some considered him an ape, while others considered him a human. And even *when* he lived is a subject of much debate. In 2010, *ScienceNews* reported that the Java Man's remains were dated as being 30,000 to 50,000 years old, using a dating method that measured "radioactive elements in the fossil-bearing sediment."[449]

---

[447] en.wikipedia.org/wiki/Java_Man; accessed June 14, 2020.

[448] Von Koenigswald, *Meeting Prehistoric Man*, p. 32.

[449] Bower, B. "'Java Man' takes age to extremes." *ScienceNews*. [Online at ScienceNews.org/article/java-man-takes-age-extremes; accessed June 14, 2020.]

But then a team dated the rocks that contained the Java Man's remains, and the age of the rocks was measured at 550,000 years! Interesting how the rock layers, both above and below the Java Man, waited half a million years for the Java Man to nestle in between them. This result, presented at the 2010 meeting of the American Association of Physical Anthropologists, used radioactive argon measurements to "determine" the age of the rocks. Susan Antón, a researcher at New York University, acknowledged to *ScienceNews* that it wasn't clear why the estimates were so drastically different.

All such radioactive-decay dating methods make several assumptions, the trustworthiness of which we don't know. As we saw in the "Dating Methods" section in the previous chapter, we do know that something can and often does go wrong with such dating methods.[450] For another example, in 1996, the lava dome that had formed when Mount St. Helens erupted ten years earlier, was dated with the radioactive-argon method. The result? The ten-year-old lava dome measured 350,000 years old![451] Off by 349,990 years in a 350,000-year date determination—more than 99.997% wrong—seems like a bit of an inaccuracy, doesn't it?

But back to Java Man. A team of researchers analyzed 166 freshwater mussel shells that were found on Java at Trinil, the same place where Java Man was found. These had been in a museum in the Netherlands since 1900, but were retrieved and examined again. The researchers discovered that there were intentional markings, shaped like a "W" or an "N," etched in the shells, probably with a shark's tooth. Not only that, but many of the shells had precisely drilled holes in them, and the location of the holes was exactly the spot that would disable the mussels' shell-closing muscles.

The researchers described the holes this way:

> The absence of holes similar to those at Trinil in natural shell assemblages, the difference between the holes observed in the Trinil assemblage and those produced by non-human animals and abiotic factors, and the similarity between our ex-

---

[450] Marvin Lubenow, "Alleged Evolutionary Ancestors Coexisted with Modern Humans," *Acts & Facts,* April 1, 1997. [Online at ICR.org/article/alleged-evolutionary-ancestors-coexisted-with-mode; accessed June 14, 2020.]

[451] Austin, S. A. 1996. "Excess Argon within Mineral Concentrates from the New Dacite Lava Dome at Mount St. Helens Volcano." *Creation Ex Nihilo Technical Journal.* 10 (3): 335–343. [Online at ICR.org/article/argon-mount-st-helens; accessed June 14, 2020.]

perimental holes, human-made *Lobatus* holes and Trinil holes all suggest that *H. erectus* was the responsible agent.[452]

Translating the jargon, this means: "We noticed three things. First, we don't see holes like this in natural habitats. Second, the holes here don't look like holes caused by animals or natural damage. And third, the holes we drilled in mussel shells, and the holes humans drilled in other shells, both looked like what was dug up with Java Man. These three reasons tell us that Java Man did the drilling." (*Lobatus* is a kind of conch shell in the Caribbean, known to have had holes drilled in them by humans.)

In fact, the more time goes on, the more some paleoanthropologists are concluding that the Java Man—*Homo erectus*—was pretty smart. A scientific literature survey of the Java Man/*Homo erectus* was published in 2017; it contained the following examples of intelligent human behavior:

- Boat construction and navigating at sea
- Language skills
- Making jewelry
- Creating rope and understanding knot-making
- Creating and using stone and bone tools
- Cooking and controlled use of fire
- Catching fish and processing them
- Building of organized living spaces and work spaces
- Creating art, such as petroglyphs, figurines, and using red ochre paint
- Woodworking skills
- Coordinated hunting and processing of big game
- Creating clothing from animal skins
- Use of fibers and resins
- Family and social structure
- Care for the elderly and weak

So once again, this ostensible "missing link" shows the understanding, skill, artistry, and strength present in modern man, to drill such precise holes in precise locations, and do all the other things in the above list. And, by the way, the skull cap turned out to be that of a regular ape, and the femur was human.[453]

---

[452] Joordens, J. C. A. et al. "Homo erectus at Trinil on Java used shells for tool production and engraving." *Nature*. [Online at hesp.irmacs.sfu.ca/sites/hesp.irmacs.sfu.ca/files/joordens_et_al._2014_homo_erectus_at_trinil_used_shell_for_tool_production_and_engraving.pdf; accessed June 14, 2020.]

[453] John Morris, "Did the Evolutionists Present a Good Case at the Scopes Trial?" *Acts & Facts*.

In an assignment a professor gave his university students, in which they had to research various fossils of "early man," one student asked him: "Why are many recent Australian fossils assigned to *Homo sapiens* when the evolutionists state that they are almost identical to the older *Homo erectus* fossils from Java?" After this question and many others equally probing, the students began to realize: In the mind of the evolutionist, *fossils exist to serve evolution, rather than to serve objective science.*[454]

Piltdown Man

The story of the Piltdown man—*Eoanthropus dawsoni*—begins in February of 1912, when Charles Dawson, an amateur archaeologist, contacted the Keeper of Geology at the Natural History Museum, a man named Arthur Smith Woodward, claiming to have found a human-like skull in the gravel beds near Piltdown, in East Sussex, England.

Image is in the public domain.

This illustration of the Piltdown Man, the newly-discovered "missing link," was published in 1913 in the Popular Science Monthly, Volume 82.

That same summer, both men discovered more bones at the site, including more skull fragments, a jawbone and set of teeth, which they connected to the previous find. They also found primitive tools. Smith Woodward reconstructed the fragments of skull and came to the conclusion that it was from a human ancestor from about half a million years ago. The discovery was announced at a meeting of the Geological Society, and hailed as a "missing link" between apes and humans. The species was given the name *Eoanthropus dawsoni*, which means "Dawson's Dawn-Man."[455]

August 1, 1995. [Online at ICR.org/article/did-evolutionists-present-good-case-at-scopestria; accessed June 14, 2020.]

[454] Marvin L. Lubenow, *Bones of Contention* (BakerBooks, Grand Rapids, Michigan), 2004, p. 19.

[455] en.wikipedia.org/wiki/Piltdown_Man; accessed June 12, 2020.

Then, in 1953, it was shown to be a forgery: the jawbone and the teeth were from an orangutan, and the teeth had been filed down to make them look old. These were combined with the skull fragments of a small modern human. What is most astonishing is not that someone tried to pull off a hoax, but that it took more than *40 years* for it to be discovered!

The success of this hoax, and the amazingly long time before it was discovered, even though the best evolutionary authorities in the world had examined it, prompted the evolutionist Sir Solly Zuckerman to remark:

> **It is doubtful if there is any science at all** in the search for man's fossil ancestry.[456]

It is often said by evolutionists that creationists make too big of a deal out of the Piltdown fraud. And really, creationists don't *need* to make a big deal out of it—it isn't the frauds like Piltdown that reveal the flimsiness of human evolutionary theory; what falsifies human evolution much more effectively are the *legitimate* fossils.

Dr. W. R. Thompson, a Ph.D. in both Zoology and Philosophy, had this to say about the Piltdown Man, shortly after the hoax was exposed:

> **The success of Darwinism was accompanied by a decline in scientific integrity.** . . . A striking example, which has only recently come to light, is the alteration of the Piltdown skull so that it could be used as evidence for the descent of man from the apes; but even before this **a similar instance of tinkering with evidence** was finally revealed by the discoverer of *Pithecanthropus [Java man]*, who admitted, many years after his sensational report, that **he had found in the same deposits bones that are definitely human.**[457]

It's interesting to note that the success of Darwinism was accompanied by a decline in scientific integrity, but if you think about it, it's not surprising at all. Since evolution begins with a foundational belief that there is no God, it immediately follows that we will not stand before Him on Judgment Day to give an account of our lives. Therefore there is no reason to adhere to any moral standards, like honesty. After all, your opinion of morality is no better

---

[456] Solly Zuckerman, *Beyond the Ivory Tower* (London: Weidenfeld & Nicolson, 1970), p. 64.

[457] W. R. Thompson, "Introduction to *The Origin of Species*," Everyman Library, Number 811 (New York: E. P. Dutton & Sons, 1956; reprint, Sussex, England: J. M. Dent and Sons, Ltd., 1967), p. 17. Emphasis added.

than mine, and I can decide for myself what is acceptable behavior. It's very reasonable to think that, in this evolutionary, survival-of-the-fittest world, anything I can do to get ahead of everybody else is acceptable. So it's no wonder that the success of Darwinism was accompanied by a decline in scientific integrity; it is indeed inevitable.

Nebraska Man

The Nebraska Man—*Hesperopithecus haroldcookii*—was trumpeted as the first "higher primate" (a supposed evolutionary ancestor of man) found in America. Henry Fairfield Osborn gave the Nebraska Man that moniker in 1922, after having received a tooth from Harold Cook, who was a rancher and geologist.

This drawing, by Amédée Forestier in 1922, illustrated his impression of the Nebraska Man.

Cook had found this tooth in 1917, and when he gave it to Osborn, and told him where he found it, Osborn and another paleontologist named William Matthew came to the conclusion that it was the tooth of an anthropoid ape "more closely related to humans than other apes." Shortly thereafter, *Science* published an article trumpeting the discovery of a manlike ape right here in North America. The *Illustrated London News* quickly jumped on board, showing this illustration.

Further research determined that the tooth was not that of a "missing link" in the evolutionary climb to become man, but was the tooth of an extinct pig—a peccary, to be exact.[458] They misidentified it because it was "weathered." Oops.

---

[458] "Nebraska Man" [Online at en.wikipedia.org/wiki/Nebraska_Man; accessed December 5, 2021.]

CHAPTER 6: MUTATIONS

Australopithecus

The *Australopithecus,* which means "southern ape," once hailed as a primitive ancestor of modern man, is in trouble.

A skull of Australopithecus sediba, dated as being younger than its descendents.

One of the flavors of Australopithecus, named *Australopithecus sediba,* was discovered in 2010 and dated as being 1.9 million years old. The problem is that 1.9 million years old is at least a million years younger than some other remains that were classified as modern humans.[459] Do you see the problem here? If *Australopithecus sediba* were really an ancestor to modern humans, why did it show up *later* than modern humans? Why does it appear after its descendents? To believe that *A. sediba* is really an ancestor of modern man is simply a matter of faith. Who said evolutionists weren't religious?

Did *A. sediba* walk upright like a human, or on all fours? Lee Berger, a paleoanthropologist claims that it could have walked upright.[460] This can be determined by analyzing the shape of the hip bones. The problem is that the hip bones weren't found.[461]

There are several different subcategories of *Australopithecus* (e.g., *A. garhi, A. africanus, A. afarensis, A. anamensis,* and so forth), and Dr. Richard Leakey, the discoverer of several of them, began to suggest that it was possible that all the variations of *Australopithecus* were not in the direct line leading to modern humans, but rather, that is was a sterile sidebranch—an extinct ape.[462]

---

[459] Lubenow, M. L. 2004. *Bones of Contention,* Revised and Updated. Grand Rapids, MI: Baker Books, p. 337.

[460] Berger, L. R. et al. 2010. "*Australopithecus sediba:* A New Species of Homo-like Australopith from South Africa." *Science.* 328 (5975): pp. 195–204.

[461] Thomas, B. "A New Evolutionary Link? *Australopithecus sediba* Has All the Wrong Signs" *Acts & Facts,* April 15, 2010. [Online at ICR.org/article/new-evolutionary-link-australopithecus; accessed June 17, 2020.]

[462] Duane Gish, "Man... Apes... *Australopithecines...* Each Uniquely Different" *Acts & Facts,* November 1, 1975. [Online at ICR.org/article/manapesaustralopithecineseach-uniquely-different; accessed June 18, 2020.]

Further research of *Australopithecus* by Dr. Charles Oxnard, a professor in the Departments of Anatomy and Anthropology at the University of Chicago, casts even more doubt on its role as a primitive form of modern man. Oxnard's study was a computerized multivariate analysis of thousands of data points, from corresponding places on the respective skeletons, and numerous measurements and comparisons were made. This allows an objective assessment of much data, rather than a subjective evaluation of just a small amount of data, as had often been done before. He compared numerous datapoints of *Australopithecus,* apes, and modern humans, and his conclusion was that *Australopithecus* was "uniquely different" from both apes and man; genetically, it was as far from both apes and man, as apes and man are from each other.[463]

Peking Man

The Peking Man—*Homo erectus pekinensis*—was unearthed in a group of fossils discovered near Peking, China (since renamed Beijing) during excavations between 1929 and 1937.[464]

The Peking Man is said to have been alive between 300,000 and 750,000 years ago. All the original fossils mysteriously disappeared in 1941, but we still have descriptions and plaster casts.[465] But of course, you can't do fossil dating on descriptions and plaster casts. (Assuming, of course, that fossil dating is actually reliable. See "Dating Methods" for numerous examples of how unreliable dating methods really are.)

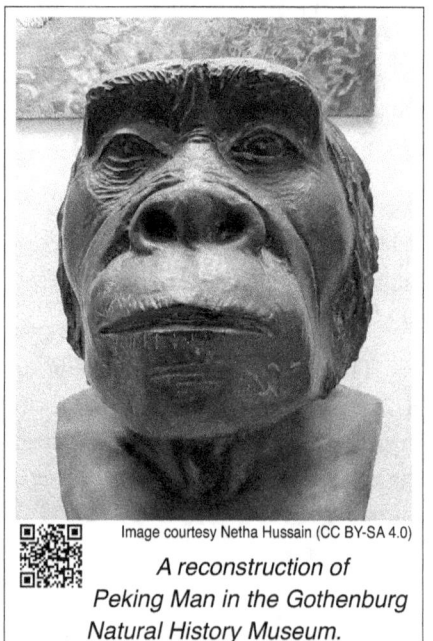

Image courtesy Netha Hussain (CC BY-SA 4.0)
A reconstruction of Peking Man in the Gothenburg Natural History Museum.

Later analysis of the Peking Man and the Java Man has resulted in them being considered the same species: *Homo habilis,* "handy man," because he

---

[463] Duane Gish, "Man... Apes... *Australopithecines...* Each Uniquely Different" *Acts & Facts,* November 1, 1975. [Online at ICR.org/article/manapesaustralopithecineseach-uniquely-different; accessed June 18, 2020.]

[464] en.wikipedia.org/wiki/Peking_Man; accessed June 18, 2020.

[465] Ibid.

used tools, or sometimes *Homo erectus,* because it is believed that they walked upright. This is disputed as well. Why is there so much confusion and disagreement? Part of it, to be sure, is what was mentioned earlier: That sometimes the fossil's alleged age determines the category to which a fossil is assigned, but other times the category in which it has been placed determines the fossil's alleged age. And sometimes both. No wonder there's confusion.

And then Richard Leakey claims that one of his finds, a skull labelled KNMR 1470, is more modern in its morphology (form) than the 500,000-year-old Peking Man, dating KNMR 1470 at 3 million years![466, 467] So how can a three-million-year-old fossil of primitive man be more modern than a half-million-year-old fossil of primitive man? Now that is an excellent question.

But if Leakey's 3-million-year age of KNMR 1470 is accepted, that throws out the Peking Man as being an ancestor of modern man.

Zinjanthropus

In 1959, Dr. Louis Leakey and his wife Mary unearthed a fossil skull that they designated East Africa Man—*Zinjanthropus boisei*. They claimed that this was a previously unknown missing link in man's evolution.

Later, paleoanthropologists realized that *Zinjanthropus boisei* (or "bosei") was an *Australopithecus,* so it was renamed *Australopithecus boisei.* Then, a growing consensus stated that there were two species of *Australopithecus*: one was *A. robustus* (equated to *A. boisei*) and *A. africanus.* Thereafter, it was believed that *A. robustus* went extinct, while *A. africanus* evolved into man—at least for a while. Then, when they were found

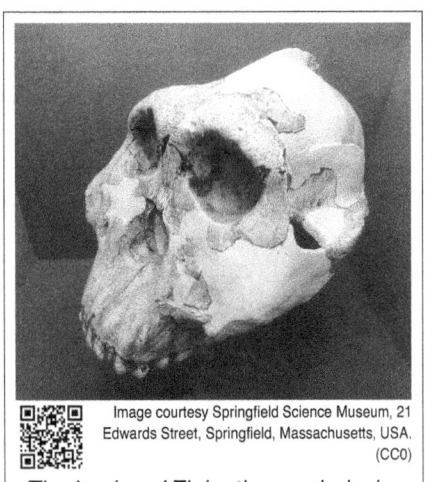

Image courtesy Springfield Science Museum, 21 Edwards Street, Springfield, Massachusetts, USA. (CC0)

The Leakeys' Zinjanthropus boisei, a "missing link" that was eventually reclassified as just another Australopithecus.

---

[466] *Science News,* Vol. 102, p. 324 (1972).

[467] R. E. F. Leakey, *National Geographic,* Vol. 143, p. 819 (1973).

together, it was realized that one was the male and the other was the female.[468, 469]

So *Zinjanthropus* was realized to be an *Australopithecus*, which was determined to be different both from apes *and* man.

And On and On. . .

There are many other lesser-known paleontological finds that have been triumphantly heralded as no-longer-missing links between apes and man, but all of them are so vague, so disputed, or so disproven, that the missing links are just as missing as they've ever been.

Perhaps G.K. Chesterton was right when he said: "Evolutionists seem to know everything about the missing link except the fact that it is missing."[470]

The main problem paleoanthropology has—well, *one* of them—is described well by Marvin Lubenow:

> Since the original fossils are virtually beyond access even to most who teach and write in the field of paleoanthropology, and only a few of the fossils are available as reproductions, and reproductions are not recommended in the preparation of scientific papers, and those scientific papers themselves cannot adequately convey differences between fossils, the "science" of paleoanthropology seems to have a problem.
>
> The myth in the minds of the public is that the human fossil material is readily available and is thoroughly studied by all who teach and write on the subject. The truth is that **paleoanthropology is in the awkward position of being a science that is several steps removed from the very evidence upon which it claims to base its findings.**[471]

An unshakable devotion to deep time, combined with uncooperative observational results, unsurprisingly causes perplexity in evolutionists' minds. As in, "Why aren't observations lining up as they should? After all, we *know* evolution is true!"

---

[468] R. E. F. Leakey, *Nature,* Vol. 231, p. 241 (1971).
[469] *Science News,* Vol. 99, p. 398 (1971).
[470] EvidenceBible.com/witnessingtool/missinglinkstillmissing.shtml; accessed January 9, 2024.
[471] Marvin L. Lubenow, *Bones of Contention* (BakerBooks, Grand Rapids, Michigan), 2004, p. 29. Emphasis added.

Here are some of the perplexed statements uttered by evolutionists when observational science doesn't agree with what they "know" about how we evolved:

- "The last common ancestor of chimpanzees and humans remains a holy grail in science."[472]
- "Evidence of humans from this period is sparse and controversial."[473]
- "But with so little evidence to go on, the origin of our genus has remained as mysterious as ever."[474]
- "We thought we had just about nailed human evolution, now everything is up for grabs again."[475]
- "The origin of our own genus remains frustratingly unclear."[476]

As you can see, evolutionists have great faith in deep time, and in the existence of transitional forms of man. And the more evidence we find that militates against evolution, and in favor of creation, the more impressive their staunch faith appears, misguided though it is. So when they, one by one, adopt Creationism, getting to know for themselves the real, live Creator Himself, there will be no stopping them.

One more thing: in the preceding sections, there was a variety of technical names for various hominins: *Homo neanderthalensis, Australopithecines, Pithecanthropus, Homo erectus, Eoanthropus dawsoni, Hesperopithecus haroldcookii, Homo sapiens, Zinjanthropus,* and more. What do these names mean? They are taxonomic descriptors of those fossils thought to be the evolutionary ancestors of modern man, and they are usually in Latin.

You may remember from school that every organism can be classified by its kingdom, phylum, class, order, family, genus, and species, and the names above are specifiers of genus, or genus and species. For example, the taxonomic name for modern man is *Homo sapiens.* What does that mean in the original language? "Homo" is Latin for "man",[477] and "sapiens" means "wise."

---

[472] Viegas, Jennifer. "The Human Family Tree." *Discovery News.* Posted on discovery.com.
[473] Hickman, C. et al. 2011. *Integrated Principles of Zoology,* 15th ed. New York: McGraw-Hill, p. 638.
[474] Wong, K. 2012. "First of Our Kind." *Scientific American.* 306 (4): 30–39.
[475] Yong, E. 2011. "Our Hybrid Origins." *New Scientist.* 211 (7): pp. 34–38.
[476] Wood, B. 2011. "Did early Homo migrate 'out of' or 'in to' Africa?" *Proceedings of the National Academy of Sciences.* 108 (26): p. 10375.
[477] "Homo," the Latin word for "man" (as in "homicide") is not to be confused with "homo," the Greek word which means "same," as in "homogeneous."

So *Homo sapiens* means "wise man." (We named ourselves.) And we think we evolved from apes?

What a perfect confirmation of Paul's statement:

> Romans 1:19–23 (GWORD): What can be known about God is clear to them because he has made it clear to them. [20]**From the creation of the world, God's invisible qualities, his eternal power and divine nature, have been clearly observed in what he made.** As a result, people have no excuse. [21]They knew God but did not praise and thank him for being God. Instead, their thoughts were pointless, and their misguided minds were plunged into darkness. [22]**While claiming to be wise, they became fools.** [23]They exchanged the glory of the immortal God for statues that looked like mortal humans, birds, animals, and snakes.

Touché.

Similarity does not imply relationship—as in "having a common ancestor"—in spite of what evolutionists' taxonomic classification of plants and animals implies. For example, take two people at random: both are women, both are about 5'8" tall, both have long, blonde, straight hair, both are 25 years old, both have slim, athletic body builds, and one was born and raised by Norwegian parents in Norway while the other was born and raised by American parents in Southern California.

These two people may be similar in all external and even internal characteristics, but that does not imply that they are related to each other (unless we are willing to go back to Noah, which most evolutionists would frown upon). Evolution's case is even weaker than this example, because these two women *would* be related if we go back far enough. Why? Because they are the same species: human. The evolutionary claim that *different* species are still related through a common ancestor is simply a matter of evolutionary faith.

As we have already discussed, *not one* beneficial (information-adding) mutation has ever been seen, and *not one* transitional fossil has ever been found, and we will see even more documentation of those two facts in this chapter. But these illusory "relationships"—which would both require countless beneficial mutations and leave countless transitional forms—are what the theory of evolution presumes and implies everywhere: the taxonomic classifications (kingdom, phylum, class, order, family, genus, and species) are based on physical similarities only, and any common ancestors are merely assumed.

Survival of the Fittest

One of the essential doctrines of evolution is "survival of the fittest," which is sometimes also called "natural selection." This was mentioned several times earlier in this book, but not in much depth, so let's actually take a look at this process upon which evolution supposedly depends.

In general, the phrase "natural selection" is not well defined, even among evolutionists. They are still debating what it actually is—is it a law? Is it a principal? Is it a causal agent? Is it a statistical outcome? Numerous evolutionists dismiss natural selection from even being a real entity, because of its inherent contradiction, as is described in the first subhead below. Then in the second subhead below, we will examine how the "filtering" idea of natural selection actually *is* a real thing.

Natural Selection as a Causal Agent

The first usage of "natural selection" I'll discuss is as the phrase, coined by Charles Darwin, to explain how organisms look so incredibly designed, without attributing it to a Designer. *Some* agency had to be the cause of this effect, and Darwin had decided that God couldn't be it.

So, extrapolating from pigeon breeders, whose intentional selection of parents could quickly result in desirable characteristics in the chicks, Darwin imagined that nature itself could select. The problem, self-evident to most people at that time, but largely ignored nowadays, is that nature does not have a mind, or awareness, or will, or intent, and therefore cannot "select" in the way that a human being can. So Darwin's use of "natural selection" was simply an anthropomorphization of nature, like when we say that water "wants" to flow downhill. People have described Darwin's approach as "hiding agency inside an analogy," thus concealing the bait-and-switch.

Let's look a little bit more closely at how the pigeon breeders in Darwin's time bred their pigeons. The breeders had some goal—some attribute of pigeons they considered desirable. As such, these pigeon breeders were using their minds—their mental capacities—to evaluate various attributes, and choose certain ones, and take corresponding action, for the purpose of arriving at the desired goal. So clearly, choosing, or selection, is a process that absolutely depends on a mind to do the considering of options, choosing one, taking actions to effect that choice. Darwin, in his desire to "un-God the universe," promoted the idea of natural selection, as if nature itself was doing the selecting. The obvious problem with us is that nature does not have a mind, and therefore does not have awareness, it cannot evaluate, it cannot choose, it cannot have foresight, it cannot make decisions based on and aimed toward arriving at a desirable goal.

Here is how Jerry Fodor, State of New Jersey Professor of Philosophy, Emeritus, at Rutgers University, put it:

> So what's the moral of all this? Most immediately, it's that the classical Darwinist account of **evolution as primarily driven by natural selection is in trouble on both conceptual and empirical grounds.** Darwin was too much an environmentalist. He seems to have been seduced by an analogy to selective breeding, with natural selection operating in place of the breeder. But this analogy is patently flawed; **selective breeding is performed only by creatures with minds, and natural selection doesn't have one of those.**[478]

The analogy to pigeon breeders Darwin obscured the idea of intelligence being required to select. As people have described since, Darwin "hid agency inside an analogy," as noted above; that is, by creating this analogy to pigeon breeders, Darwin concealed the fact that a mind is necessary to do any kind of selecting.

Intelligent design advocate William Dembski described Darwin's bait-and-switch as follows:

> Before Darwin, the ability to choose was largely confined to designing intelligences, that is, to conscious agents that could reflect deliberatively on the possible consequences of their choices. Darwin's claim to fame was to argue that natural forces, lacking any purposiveness or prevision of future possibilities, likewise have the power to choose via natural selection. **In ascribing the power to choose to unintelligent natural forces, Darwin perpetuated the *greatest intellectual swindle in the history of ideas.* Nature has no power to choose.**[479]

---

[478] Jerry Fodor, "Why Pigs Don't Have Wings," *London Review of Books,* Vol. 29, No. 20, 18 October 2007. [Online at lrb.co.uk/the-paper/v29/n20/jerry-fodor/why-pigs-don-t-have-wings; accessed January 11, 2024.]

[479] Dembski, W. A., *The Design Revolution* (Downers Grove, IL: Intervarsity Press, 2004), p. 263.

And one more, just for good measure. Neil Thomas—who had been an agnostic, but when he pondered the evidence, became a design advocate—said it this way:

> Darwin—against the objections of Wallace and other colleagues who pointed out to him that **there was simply no comparison between what animal breeders did by the use of human ingenuity and how mindless Nature herself acted**—claimed an analogy between the artificial breeding methods of such persons as pigeon-fanciers and the claimed "selection" performed by Nature herself.[480]

In case you didn't notice the title of that article, it is "Natural Selection: A Conceptually Incoherent Term." Touché.

Natural Selection as a Filter

The second usage of the phrase "natural selection" is to designate the process whereby mutations (which always cause damage) tend to reduce the viability of an organism, and the mutated organisms therefore eventually die out. This usage of natural selection is not evolution in any sense, because no new genetic information is created; the best natural selection can do is, as much as possible, to *preserve* information that is already present; hence the synonymous phrase "survival of the fittest." And the preserving of genetic information is itself a losing proposition because of the Second Law of Thermodynamics. Technically, it is "devolution," which leads to extinction.

The phrase "survival of the fittest" is synonymous with this second usage of the phrase "natural selection." These two phrases both mean "Survival of the form that will leave the most copies of itself in successive generations."[481]

It's important to realize that natural selection in the second sense shown above—survival of the fittest—*is* a real thing. But an absolutely critical distinction to understand, as mentioned above, is that natural selection *only preserves—it does not innovate.* In other words, natural selection only weeds out harmful mutations (which, according to observations, is all of them) by preventing those organisms with damaged DNA from thriving and propagating as well as those with undamaged DNA. Natural selection preserves the health-

---

[480] Neil Thomas, "Natural Selection: A Conceptually Incoherent Term" Evolution News & Science Today, December 1, 2021. [Online at EvolutionNews.org/2021/12/natural-selection-a-conceptually-incoherent-term; accessed January 11, 2024.]

[481] en.wikipedia.org/wiki/Survival_of_the_fittest; accessed June 20, 2020.

iest version of DNA because the mutated versions are less fit, and therefore fail to thrive, and so those organisms die out.

Natural selection *cannot*, however, come up with new genetic information for the DNA, any more than randomly scrambling the bytes on your disk drive will yield improved functionality of your computer. Therefore, there is no way that natural selection can be a mechanism that powers Darwinian evolution.

And here is an interesting tidbit: survival of the fittest (also known as natural selection), is a **tautology**. (A tautology is when the same thing is said twice, but in different words, as in, "The guests arrived one after another in succession." That is a tautology because "one after another" means the same thing as "in succession.") Survival of the fittest is a tautology because it predicts that the fittest organisms will have the most offspring, but then it defines "fittest" as "having the most offspring."

So when you get right down to it, survival of the fittest predicts that "the organisms that have the most offspring will have the most offspring." That explains a lot, doesn't it? Not only that, but a corollary truth also emerges: "the organisms that produced the most offspring must have had the characteristics required to produce the most offspring."

The population of any kind of organism is inextricably linked to the number of offspring they have. So what can we learn from the fact that in Manhattan there are roughly 1,600,000 people but 1,230,000,000 ants?[482] That's more than 768 ants for every person. Are ants more "fit" to survive than humans are? As philosopher of science Karl Popper once said: "Darwinism is not really a scientific theory, because natural selection is an all-purpose explanation which can account for anything, and which therefore explains nothing."[483]

Another interesting tidbit: natural selection cannot occur before life begins. In other words, all the fantastically complex processes of assembling exclusively left-handed amino acids of the correct types, assembling them in the correct sequence, to make each of the 2,000 different enzymes that facilitate the construction of tens of thousands of other necessary proteins, all of which require exclusively left-handed amino acids to be assembled in the correct sequence, *must happen simultaneously*, because natural selection can only happen

---

[482] Rob Dunn, "How Many Ants Are There?" [Online at myrmecos.net/2013/08/08/how-many-ants-are-there; accessed November 10, 2020.]

[483] Phillip E. Johnson, *Darwin on Trial* (Regnery Publishing, Washington DC, 1991), p. 22.

CHAPTER 6: MUTATIONS

in the context of life. This beginning of life is a miracle, whether you call it that or not.

All of this has to assemble itself accidentally, and with no mistakes, *before* natural selection can kick in. Just the 2,000 enzymes assembling by chance has been calculated at 1 chance in $(10^{20})^{2,000}$, or $10^{40,000}$, not even counting all the other thousands of required proteins. Since 1 chance in 10^{50} is accepted as being functionally equivalent to "impossible," we can clearly see that life "arising" by itself is impossible.

When astronomer Fred Hoyle calculated the above likelihood, he also commented:

> Any theory with the probability of being correct that is larger than one part in $10^{40,000}$ must be judged superior to random shuffling *[of evolution]*. **The theory that life was assembled by an intelligence has, we believe, a probability vastly higher** than one part in $10^{40,000}$ of being the correct explanation of the many curious facts discussed in preceding chapters. Indeed, **such a theory is so obvious that one wonders why it is not widely accepted as being self-evident. The reasons are psychological rather than scientific.**[484]

And even if life could "arise" by some non-supernatural miracle (saltation?), natural selection, as mentioned, can only preserve *existing* genetic information, it cannot innovate *new* genetic information.

Since this book is discussing whether or not evolution could actually be real, we will discuss survival of the fittest, or natural selection, in the way that evolutionary doctrine defines it, and see where that takes us. So, according to the credo of evolution, the process of survival of the fittest is roughly as follows:

1. An organism's DNA allows it to take in and metabolize nutrients, eliminate waste, respond to stimulus, and reproduce.
2. A random, undirected mutation, usually assumed to be radiation or a DNA-copying "typo," turns out, by sheer luck, to add functional genetic information to the DNA. The resulting new characteristics are beneficial, so the organism is better suited—more "fit"—to survive in its current environment.

---

[484] Fred Hoyle and N. Chandra Wickramasinghe, *Evolution from Space: A Theory of Cosmic Creationism* (New York: Simon and Schuster, 1981), p. 130. Emphasis added.

3. Because the modified organism is more fit, it successfully competes for resources with the old, unimproved version of the organism, eventually driving the old version to extinction, replacing it in its environment.

Of course, the above definition assumes that random-but-beneficial mutations are actually a thing. And remember, no one has ever actually found one.[485] As mentioned earlier, the kind of beneficial mutation we're talking about here is the kind that evolutionary hypothesis depends on: the kind that adds genetic information, as opposed to one that decreases genetic information but serendipitously has some convenient side effect. And even evolutionists acknowledge that beneficial mutations are exceedingly rare: maybe 1 in 10,000 mutations is beneficial.[486] Evolutionists have not even found that many, but their whole theory depends on them happening at least *sometimes*. They're still searching. . .

Comparing this to a comparable event in the computer-science arena, a mutation that adds genetic information is like a complex new subroutine being written by accident, as a result of a data-copying error or a disk-drive head crash,[487] and this new subroutine performs functional and beneficial operations, interfaces properly with already-written code, and does not cause any other aspects of the code to malfunction. Wouldn't that be nice!

But, for the sake of discussion, let's suspend our disbelief and pretend that averaging 9,999 steps backward for every 1 step forward could actually result in progress and improvement. But notice the *enormous* assumption in the simple little recipe above, for survival of the fittest: *we're starting with a viable organism*. Of course, we can't start there, because the obvious, and valid, showstopper question is "How did the first viable organism get there?"

Arrival of the Fittest

Step 1 in the above recipe for survival of the fittest assumed that we had a viable organism to start with. A viable organism, as pointed out, must at least respond to stimulus, metabolize nutrients, eliminate waste, and reproduce. Oh, is *that* all? All of these are *extremely* complex processes that we still don't completely understand.

---

[485] Morris, J. 2008. "The Secrets of Evolution." *Acts & Facts*. 37 (3): 13.
[486] Ernst Mayr, in *Mathematical Challenges to the Neo-Darwinian Interpretation of Evolution*, ed. P. S. Moorhead and M. M. Kaplan (Philadelphia: Wistar Institute Press, 1967), p. 50.
[487] en.wikipedia.org/wiki/Head_crash; accessed September 20, 2020.

Earlier was a short discussion of spontaneous generation, which evolutionists dismiss as impossible, unless they need it to be true. Which they do, because their belief system depends on it. Basically, life arising from non-life, regardless of how long it supposedly takes. But what, exactly, does this entail?

Life "Arising" from Non-Life

When astronomers discover evidence of liquid water on a planet or moon in our solar system, or on an exoplanet (a planet orbiting a star other than our sun), they get all excited and talk about the possibility, the likelihood, or even the inevitability, of life there.

But, as usual, there is more to it than that. While water is certainly *necessary* for life as we know it, it is not *sufficient* for life: other things—*many* other things—are also necessary.

For example, there are many chemicals that are essential for life processes to happen. The trouble is that water *destroys* those very chemicals unless they are inside cell membranes.[488]

Also inside cells are molecular machines. How did these machines come about? Many of these machines manipulate individual atoms of specific types, and when they do, the cell survives; when they fail, the cell dies. No one has yet been able to come up with a scenario in which the laws of physics could, by themselves, assemble machines that avoid decay and diffusion in a given context (like inside a cell). Such decay and diffusion are the normal results of those same laws of physics.

The greater the number of specifications required for life to exist, the less the likelihood that it could arise through natural, undirected means. And we have already discovered *so* many of these requirements, that there is absolutely no way life could have "arisen" just through natural means. The raw materials are there, but they could not have fallen together solely through the operation of natural laws. So, the only other option is that there was some entity—*outside* of the natural laws we know about—Who set life in motion.

Evolutionists may scoff at the idea, but why is it so difficult to believe there may exist beings that they don't know about? Surely they aren't claiming omniscience, as in, "It couldn't exist if I don't know about it." Are they?

---

[488] Thomas, Brian. "Earth Remains the Only Goldilocks Planet" [Online at ICR.org/article/earth-remains-only-goldilocks-planet; accessed June 26, 2020.]

Theoretical physicist and mathematician Freeman Dyson, a member of the U.S. National Academy of Sciences, acknowledges the following:

> **Concerning the origin of life itself,** the watershed between chemistry and biology, the transition between lifeless chemical activity and organized biological metabolism, **there is no direct evidence at all.** The crucial transition from disorder to order left behind no observable traces.[489]

So there is no evidence for the process of evolution, which we've discussed earlier in a variety of contexts. But clearly, we are here. So either we evolved by natural and undirected processes that left no evidence for having happened (which violates all sorts of laws of physics), or we were created (which only violates the preferences of those who don't want God in their lives). The question this book is endeavoring to answer is: which model predicts more accurately what we actually see?

And here, molecular biologist Michael Denton asks a profound and probing question:

> **Is it really credible that random processes could have constructed a reality,** the smallest element of which—a functional protein or gene—is complex beyond our own creative capacities, a reality **which is the very antithesis of chance, which excels in every sense anything produced by the intelligence of man?**[490]

The answer, clearly, is that it is *not* credible to posit that random processes could have created life.

Evolutionist Franklin Harold, Professor Emeritus of Biochemistry at Colorado State University, concurs:

> . . .we must concede that there are presently no detailed Darwinian accounts of the evolution of any biochemical or cellular system, **only a variety of wishful speculations.**[491]

---

[489] Freeman Dyson, *Origins of Life* (New York, NY: Cambridge University Press, 1999), p. 36. Emphasis added.

[490] Michael Denton, *Evolution: A Theory in Crisis* (Bethesda, MD: Adler and Adler, 1985), p. 262. Emphasis added.

[491] Harold, Franklin M. *The Way of the Cell: Molecules, Organisms and the Order of Life*, Oxford University Press, New York, 2001, p. 205.

Another Showstopper: Racemization

Let's talk a bit more about these molecular machines that operate within every living cell. **Amino acids** are the building blocks of proteins: there are about 500 different naturally occurring amino acids, only 20 of which used in living cells. These amino acids are put together in various combinations of 50 to over 30,000, to construct proteins.[492]

You have probably heard the phrase, "It fits like a glove." It is a figure of speech that indicates a perfect fit, an ideal match, a flawless correspondence. But when you buy a pair of gloves, if you put the left glove on your right hand, or vice versa, it is *not* a good fit. Why not? Aren't they the same shape? Yes, but the two gloves are mirror images of each other: you have a left-handed glove and a right-handed glove, and only one of them fits on each hand.

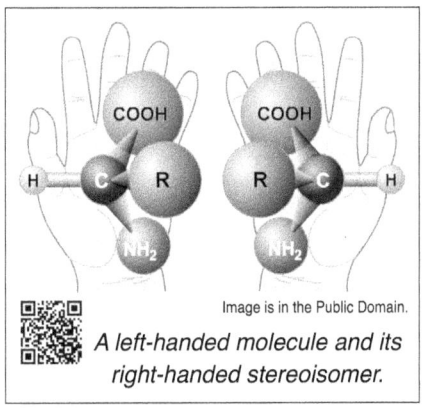

Image is in the Public Domain.
A left-handed molecule and its right-handed stereoisomer.

Molecules are the same way: they exhibit **chirality**, or "handedness" ("chirality" comes from the Greek word for "hand"). In all but the simplest molecules, they come in left-handed versions and right-handed versions, because molecules are three-dimensional. All amino acids (except the simplest one, glycine) have two different **stereoisomers**; in other words, they come in both left-handed and right-handed versions. Each amino acid has an "L" form (from Latin "levulo," which means "left") and a "D" form (from Latin "dexter," which means "right"). For example, the amino acid alanine exists in two forms: L-alanine and D-alanine, indicating the left-handed and right-handed version, respectively.

(By the way, right-handed molecules are called that because polarized light going through a substance made of right-handed molecules is rotated to the right, or clockwise. Similarly, polarized light traveling through a substance composed of left-handed molecules is rotated to the left, or counterclockwise.)

In normal chemical processes, as in creating amino acids in a lab, half of the molecules are left-handed and half are right-handed. This kind of mixture of L-forms and D-forms is called a **racemic mixture**. And, left to themselves,

---

[492] Ken Ham, editor, *The New Answers Book 3,* Answers in Genesis (Master Books, Green Forest, Arkansas, 2010), p. 154.

(that is, not in a living being) random molecular vibrations of the L-forms and D-forms flip them back and forth to the other form, so a quantity of all one kind would eventually become an equal quantity of L-forms and D-forms in a process called **racemization**. That is, even if they all started out the same way—all L or all D—they would eventually end up as a racemic mixture of half and half.

Now here is where this become significant: in living cells, *all* of the amino acids are left-handed. This left-handedness is what gives DNA its helical twist.[493] And these amino acids *must* be all left-handed or the cellular processes fail, and the cell dies. The showstopper question for evolution is: How in the world did they all get left-handed, since random processes (which evolution insists upon) create racemic mixtures?

The same question applies to sugars: All sugars within living cells are right-handed sugars, but random processes can create only racemic mixtures.[494]

A related question is this: The process of creating proteins requires proteins to begin with, so how, using only random processes, did the first protein form? Proteins cannot replicate themselves, so natural selection cannot be the process whereby they "improve."

And here's an intriguing thought: How is DNA replicated during cell division? By the actions of these molecular machines called proteins. But how are proteins constructed? According to instructions in the DNA. Both are dependent upon the existence and correct functioning of the other. *So which one came first?*

And here's another fascinating tidbit. Did you notice a few paragraphs ago, I mentioned that L-forms and D-forms randomly flip back and forth when they are not in a living being? This means that as soon as an organism dies and its life processes stop functioning, the amino acids in its proteins would start flipping back and forth between the two forms in a statistically well-behaved way. How long would it take for the amino acids, initially all the L-form, to become a racemic mixture at 50% L-forms and 50% D-forms?

---

[493] Rosevear, David. "The Myth Of Chemical Evolution" July 1, 1999. [Online at ICR.org/article/myth-chemical-evolution; accessed July 11, 2020.]

[494] Wilder-Smith, A. E. "The Origin of Conceptual thought in Living Systems" *Acts & Facts* February 1, 1993. [Online at ICR.org/article/origin-conceptual-thought-living-systems; accessed July 11, 2020.]

Dr. Larry S. Helmick, a Chemistry Professor at Cedarville College in Ohio, calculates an upper limit of approximately 20 million years.[495] Therefore, any biological matter that has not yet reached 50:50 is younger than 20 million years. Perhaps *much* younger, because heat greatly speeds up the racemization process. But it can't be older than 20 million years.

So imagine evolutionists' surprise when some Miocene sediments (supposedly 30 million years old), some Precambrian sediments (supposedly 1.2 billion years old), and a chert[496] deposit in Africa (supposedly 3 billion years old) all contained *only* left-handed amino acids.[497] So those evolution-based ages are preposterously overblown. But evolutionary faith requires deep time to exist, so it's not surprising that ages are inflated so egregiously.

Where Did the Information Come From?

There was a brief mention in "Chapter 2: The Theory of Evolution" about the information stored in DNA. Let's examine that in a bit more detail.

Imagine a construction site on a bright spring day: piles and piles of lumber, steel girders of various sizes and lengths, hundreds of bags of cement next to a big pile of sand, plus a nearby lake for water to mix with them, stacks and stacks of drywall, a trailer full of electrical supplies, and another trailer filled with plumbing supplies, and so forth. Every raw material needed for the planned building has been delivered.

These raw materials sat there all summer. The scorching sun bore down on them by day, and there was an occasional rainstorm bringing wind and lightning. As the months plodded by, the drywall cracked, the boards warped and started rotting, the pile of sand for the concrete had been partially washed away from the rain, the cement, still in the paper sacks, had solidified into 80-pound paperweights.

In late September, the customer arrives to check out his new building, and is shocked to see that it's not built yet. What went wrong? All the raw materials were there. There had been plenty of sunshine, rain, occasional strong winds, lightning—conditions were *perfect* for a building to evolve. What in the world happened? Why isn't the building there?

---

[495] Helmick, L., "Origins and Maintenance of Optical Activity," *Creation Research Society Quarterly* 12:156–164, December 1975.

[496] Chert is a hard, fine-grained sedimentary rock, of which flint is an example. See "Chert" article in Wikipedia. [Online at en.wikipedia.org/wiki/Chert; accessed May 30, 2022.]

[497] Carl Wieland, "Shaking Hands on a Recent Creation," *Creation* 15(4):24–25, October 1993. [Online at Creation.com/shaking-hands-on-a-recent-creation; accessed May 30, 2022.]

The answer may be obvious to most, but to those who are dedicated to the fable of evolution, maybe not. Evolution assumes that if you have the correct raw materials and plenty of energy available, things will build themselves. After all, *we're* here, aren't we? Doesn't that prove it? In a word, no.

While raw materials and energy are necessary, they are not sufficient. In other words, yes, you need raw material and energy, but they are not the *only* things you need.

For example, suppose you wanted to bake a batch of chocolate-chip cookies. You look in the cupboard, and see that you have flour and chocolate chips, and you conclude that you are good to go. But are you really? Flour and chocolate chips are necessary, aren't they? Yes, they are. But they are not the *only* ingredients that are necessary; you'll also need butter, sugar, eggs, baking soda, and perhaps a few more things. So flour and chocolate chips are *necessary*, but they are not *sufficient*.

In our fictitious construction site, what is missing? *Information.* Undirected energy simply causes disorder, just like the Second Law of Thermodynamics says. The construction site, with all the raw materials and plenty of energy available, still needs the energy to be put to use *in a specified way.* In other words, there needs to be a set of instructions to describe what the building is supposed to look like and how it's supposed to be built. That is, a blueprint. And builders, of course, to take the information in the blueprint and follow those instructions, applying energy in just the right ways to assemble the building.

In the context of biology, the blueprint is the DNA, which stores massive amounts of information on how the organism is built, and the workers are the molecular machines—the proteins—that actually carry out the instructions stored in the blueprint of DNA.

Evolution has been telling us for a long time that all you need are energy and raw materials, and inorganic chemicals will soon spontaneously form a living cell, which will then undergo billions or trillions of beneficial mutations, and become us. According to the story, the Earth was formed 4.5 billion years ago, and life "arose" about 4.3 billion years ago. So, in only 200 million years, inorganic chemicals spontaneously assembled themselves into organisms that could respond to stimulus, metabolize nutrients, eliminate waste, and reproduce.

That's very fast for something as complex as a single-celled organism. Why, at that rate, things as relatively trivial as smartphones and laptops should appear on your front lawn every week or so, having spontaneously assembled themselves out of the raw materials in the soil.

The problem is that *randomness does not generate information.* Clearly, in order to assemble anything, you need raw materials, available energy, *and instructions* so the energy is directed the proper way. Evolutionists are horrified by the prospect of directed energy, because direction implies intent, which requires a Mind. Yes, it does indeed.

Nevertheless, information requires intelligence, because information cannot arise by non-directed processes, as the following scientists confirm:

> **There is no known law of nature,** no known process and no known sequence of events **which can cause information to originate by itself in matter.**[498]

> **Not even one mutation has been observed that adds a little information to the genome.** This surely shows that there are not the millions upon millions of potential mutations the theory [evolution] demands.[499]

> DNA is an information code. . . The overwhelming conclusion is that **information does not and cannot arise spontaneously** by mechanistic processes. **Intelligence is a necessity in the origin of any informational code,** including the genetic code, no matter how much time is given.[500]

The Elephant in the Living Room

This is an excerpt of an interview between writer George v. Caylor, when he interviewed a molecular biologist we will call Sam.[501] Sam apparently did not want his real name to be used, for reasons that will become obvious as you read. The part of the interview quoted below pertains to where the in-

---

[498] W. Gitt, *In the Beginning Was Information* (Green Forest, AR: Master Books, 2006). Emphasis added.

[499] Lee Spetner, *Not by Chance* (New York: Judaica Press, 1997), p. 160. Emphasis added.

[500] L. Lester and R. Bohlin, *The Natural Limits to Biological Change,* (Dallas, TX: Probe Books, 1989), p. 157. Emphasis added.

[501] Walt Brown, *In the Beginning: Compelling Evidence for Creation and the Flood,*, Eighth edition, 2008 (Center for Scientific Creation, Phoenix, Arizona, CreationScience.com), p. 16.

formation in DNA came from: did it evolve randomly, or was it created by an intelligent designer?

George: How did all that information happen to get there?

Sam: Do you mean, did it just happen? Did it evolve?

George: Bingo. Do you believe that the information evolved?

Sam: George, nobody I know in my profession truly believes it evolved. It was engineered by "genius beyond genius," and such information could not have been written any other way. The paper and ink did not write the book. Knowing what we know, it is ridiculous to think otherwise. A bit like Neil Armstrong believing the moon is made of green cheese. He's been there!

George: Have you ever stated that in a public lecture, or in any public writings?

Sam: No. It all just evolved.

George: What? You just told me—?

Sam: Just stop right there. To be a molecular biologist requires one to hold onto two insanities at all times. One, it would be insane to believe in evolution when you can see the truth for yourself. Two, it would be insane to say you don't believe in evolution. All government work, research grants, papers, big college lectures—everything would stop. I'd be out of a job, or relegated to the outer fringes where I couldn't earn a decent living.

George: I hate to say it, Sam, but that sounds intellectually dishonest.

Sam: The work I do in genetic research is honorable. We will find the cures to many of mankind's worst diseases. But in the meantime, we have to live with the "elephant in the living room."

George: What elephant?

Sam: Design. It's like the elephant in the living room. It moves around, takes up an enormous amount of space, loudly trumpets, bumps into us, knocks things over, eats a ton of hay, and smells like an elephant. And yet we have to swear it isn't there.

That's pretty eye-opening, isn't it?

A printed dictionary is *so* much more than a pile of paper with ink on it; the ink has to be placed in precise positions—making letters, words, entries in alphabetic order, cross-references, and so forth—so as to communicate intelligent concepts to the reader. And just like the dictionary, the information inside DNA is *so* much more than the simple, repetitive pattern that self-assembly yields, such as when crystals grow. In DNA is non-repetitive, instruction-laden, coded *information*. It literally is a set of instructions.

We are often impressed with the information-storage capabilities of the latest computer chips, which were purposefully designed by very smart and skilled people, using extremely precise machines, with the intent of creating denser information storage than ever before. As an information-storage device, DNA is 45 trillion times more efficient than silicon memory chips,[502] but DNA fell together all by itself, accidentally, you see.

Anthony Flew, a longtime atheist became a theist when he was confronted with **epigenetic information** in DNA. He states that this information contains:

> . . .genetic instructions [that] are not the kind of information you find in thermodynamics; rather, they constitute semantic information. In other words, **they constitute meaning.**[503]

Evolutionists, because they insist that life arose by chance, by and large believe that life could also arise on other planets, and in a search for intelligent life elsewhere, have spent enormous sums of money on projects like SETI—the Search for Extra-Terrestrial Intelligence. In the SETI program, people use radio telescopes to listen for non-random signals from space, ostensibly from intelligent aliens.

The ironic thing is that evolutionists understand that the presence of non-random, meaningful information coming from space is conclusive proof of intelligence, yet when confronted with non-random, meaningful information in the DNA of every living cell, they claim it arose by accident!

"Junk DNA"

There is much non-gene content in DNA—these **pseudogenes** make up more than 95% of human DNA. This was considered for a long time to be "junk DNA" left over from eons of random mutations that didn't harm the organism, but neither did they contribute anything useful. This "junk DNA"

---

[502] Rosevear, David. "The Myth Of Chemical Evolution" July 1, 1999. [Online at ICR.org/article/myth-chemical-evolution; accessed July 11, 2020.]

[503] Flew, A. and R. A. Varghese. 2007. *There Is a God: How the World's Most Notorious Atheist Changed His Mind.* New York, NY: Harper Collins, p. 129.

didn't code for proteins—that is, they didn't contain instructions for building proteins, which are the molecular machines that do all the work inside cells—so these pseudo-genes were clearly useless.

Evolutionists pounced on this as a chance to mock creationism:

> [Pseudogenes] are genes that once did something useful but have now been sidelined and are never transcribed or translated. . . **What pseudogenes are useful for is embarrassing creationists.** . . [since it] is a remarkable fact that the greater part (95 per cent in the case of humans) of the genome might as well not be there, for all the difference it makes.[504]

Geneticist Susumo Ohno concurred with the consensus about "junk DNA:"

> More than 90% **degeneracy** contained within our genome should be kept in mind when we consider evolutionary changes in genome sizes. What is the reason behind this degeneracy? . . . The earth is strewn with fossil remains of extinct species; **is it a wonder that our genome too is filled with the remains of extinct genes?** . . . Triumphs as well as failures of nature's past experiments appear to be contained in our genome.[505]

When the first drafts of the Human Genome Project were published in 2001, this "junk DNA" was seen all over the genome. People who saw but did not understand this DNA, took it as confirmation of evolution:

> They identified thousands of segments that had the hallmarks of **dead genes.** They found transposable elements by the millions. The Human Genome Project team declared that our DNA consisted of isolated oases of protein-coding genes surrounded by **"vast expanses of unpopulated desert where only noncoding 'junk' DNA can be found."** Junk DNA had started out as a theoretical argument, but now the messiness of our evolution was laid bare for all to see.[506]

---

[504] Dawkins, R. 2009. *The Greatest Show on Earth: The Evidence for Evolution.* New York: Free Press, p. 332–333. Emphasis added.

[505] Ohno, S. 1972. "So much 'junk' DNA in our genome." *Brookhaven Symposia in Biology.* 23: pp. 366–370. [Online at JunkDNA.com/ohno.html; accessed July 10, 2020.] Emphasis added.

[506] Zimmer, C. "Is Most of Our DNA Garbage?" *New York Times.* March 5, 2015. [Online

CHAPTER 6: MUTATIONS

Geneticist Francis Collins was the Director of the Human Genome Project and the Director of the National Institutes of Health. He, too, interpreted "junk DNA" as evidence for evolution:

> Even more compelling evidence for a common ancestor comes from the study of what are known as ancient repetitive elements (AREs). . . . Mammalian genomes are **littered** with such AREs, with roughly 45 percent of the human genome made up of such **genetic flotsam and jetsam.** . . . **Of course, some might argue that these are actually functional elements placed there by a Creator for a good reason, and our discounting them as 'junk DNA' just betrays our current level of ignorance.**[507]

Note the descriptor applied to the repetitive elements: they are called "ancient" repetitive elements. The very name is based on the non-negotiable presumption that deep time is a real thing and that evolution is the only possible process by which life came to be.

Also note the underlying, but almost certainly unintentional, mindset here. In a nutshell, the above statements reveal a mindset of "We don't know how junk DNA is used, so therefore it is useless." If you think about it, that attitude presumes omniscience: "If I don't know what it does, it doesn't do anything." That sounds like arrogance on steroids.

But even as evolutionists far and wide were gloating over the purposelessness of "junk DNA" being proof for evolution, there were already published articles documenting its purpose. Collins' final sentence in the above quote turned out to be more prophetic than he intended: the evolutionists' discounting of junk DNA actually *did* betray their current level of ignorance.

The uselessness of "junk DNA" had been the prevailing opinion for over 30 years, until bioinformaticians (scientists who study the information required for life) discovered that virtually *all* of DNA is used.[508] What used to be considered "junk DNA" (simply because we didn't yet know what it did) now is known to contain very useful information: epigenetic factors that are

---

at NYTimes.com/2015/03/08/magazine/is-most-of-our-dna-garbage.html?_r=4; accessed July 10, 2020.] Emphasis added.

[507] Collins, F. S. 2006. *The Language of God: A Scientist Presents Evidence for Belief.* New York: Free Press, pp. 135–137.

[508] Brian Thomas, "Epigenetics: More Information than Evolution Can Handle" January 30, 2009. [Online at ICR.org/article/epigenetics-more-information-than-evolution-can-ha; accessed June 21, 2020.]

313

required to access, interpret, and process genetic information elsewhere in the DNA.

Carl Zimmer, a reporter for the *New York Times*, wrote in 2015 about Frances Collins' change of opinion:

> In January [2015], Francis Collins, the director of the National Institutes of Health, made a comment that revealed just how far the consensus has moved. At a health care conference in San Francisco, an audience member asked him about junk DNA. "**We don't use that term anymore,**" Collins replied. "**It was pretty much a case of hubris to imagine that we could dispense with any part of the genome—as if we knew enough to say it wasn't functional.**" Most of the DNA that scientists once thought was just taking up space in the genome, Collins said, "**turns out to be doing stuff.**" [509]

So there are genes that code for proteins—that is, they contain the instructions for creating proteins that carry on the countless tasks necessary for the organism to survive. But there are also numerous other functions that are necessary for life processes to continue: functions involved in information transmission, code translation, error correction, and data duplication. Instructions for these functions are stored in what used to be called "junk DNA," and without these functions, the information in the protein-coding genes would be useless.

So now the question arises: which evolved first? The protein-coding genes? The transmission mechanism? The translation mechanism? The error-correction mechanism? The duplication mechanism? Since a cell cannot survive without all five kinds of processes being present and functioning properly, *they all must have evolved at the same time,* or the cell would die, and any further evolution would stop. So what is the likelihood that all these functions would evolve instantaneously, successfully, compatibly, and simultaneously? That's right: zero.

Joseph Kuhn, a surgeon at Baylor University Medical Center, in a presentation before the Texas State Board of Education, pointed out that cells can stay alive only because of the information that exists in DNA—genes, pseudo-

---

[509] Zimmer, C. "Is Most of Our DNA Garbage?" New York Times. March 5, 2015. [Online at NYTimes.com/2015/03/08/magazine/is-most-of-our-dna-garbage.html?_r=4; accessed July 10, 2020.] Emphasis added.

genes, what used to be called "junk DNA"—all of which contains very non-natural information. DNA has almost none of the randomness that random processes produce. And, of course, evolutionists insist on random (non-directed, non-intentional) processes. As mentioned earlier in this chapter, when natural, random processes *do* act on these DNA molecules, they lose vital information, and the organism loses robustness, gets sick, or dies, because mutations are never beneficial.[510]

Kuhn goes on to say that the information in DNA is an actual language system. There are symbols, specific meanings for each symbol, grammatical rules that determine how those symbols should be interpreted, message sources, message destinations, and purposeful effects coming from the information that was communicated. These kinds of processes *always* come from an intelligent entity, not from natural causes.

In his paper, Kuhn continues:

> Based on an awareness of the inexplicable coded information in DNA, the inconceivable self-formation of DNA, and the inability to account for the billions of specifically organized nucleotides in every single cell, **it is reasonable to conclude that there are severe weaknesses in the theory of gradual improvement through natural selection (Darwinism) to explain the chemical origin of life.** Furthermore, **Darwinian evolution and natural selection could not have been causes of the origin of life,** because they require replication to operate, and there was no replication prior to the origin of life.

Richard Dawkins, one of the most passionate enthusiasts for evolution, has even admitted:

> . . .the most profound unsolved problem in biology is the origin of life itself.[511]

School students in Texas—as well as everywhere else in the world—should have access to such information about evolution's deal-breaking shortcomings. But after the Texas Board of Education's decision to include such information, 90% of the revised textbooks still failed to adequately include descriptions of

---

[510] Kuhn, J. A. 2012. "Dissecting Darwinism." *Baylor University Medical Center Proceedings.* 25 (1): 41–47. [Online at ncbi.nlm.nih.gov/pmc/articles/PMC3246854; accessed June 25, 2020.]

[511] Dawkins, R. 2009. "Evolution: The next 200 years." *New Scientist.* 2693: p. 41.

evolution's weaknesses. For all the weaknesses of the theory itself, its proponents are doggedly determined to keep the belief system alive.[512]

Irreducible Complexity

Irreducible complexity is yet another reason that Darwinian evolution is unable to result in the life we see all around us. Biochemist Michael Behe, in his book *Darwin's Black Box,* described the all-or-nothing systems in living organisms as "irreducibly complex."[513]

In any system, there is a level of complexity at which everything present is necessary for the continued functioning of the system. In other words, at this point of irreducible complexity, if *any* one part fails or is removed, the entire system fails.

For example, consider a simple mousetrap, as shown here. It has nine parts: a base, a trigger, a restraining bar, a hammer, a spring, three narrow staples, and one wide staple (for the trigger). A moment's reflection confirms that if *any one* of these pieces is removed, the entire mechanism would fail to function. (True, the artwork on the base is not strictly necessary, but since it is part of the base, which *is* necessary, let's ignore the artwork.)

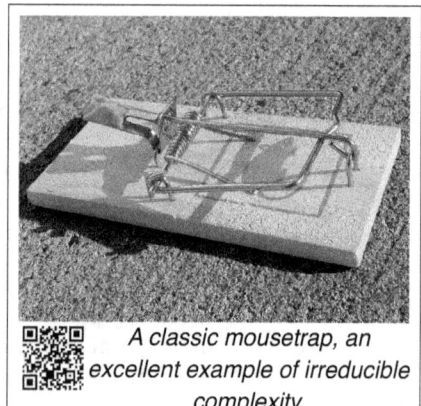

A classic mousetrap, an excellent example of irreducible complexity.

And keep in mind that, in discussing the above nine components whose *presence* is necessary for the correct functioning of the mousetrap, we haven't even mentioned the *size* of the components, the *shape* of the components, the *composition* of the components, or the relative *locations* of the components. All of these have to be correct also, for the mousetrap to function. For example, if the spring were too small, or the restraining arm was bent wrong, or if the hammer was made from cotton, or the staples were in the wrong places, the whole thing would fail.

---

[512] "An Evaluation of Supplementary Biology and Evolution Curricular Materials Submitted for Adoption by the Texas State Board of Education." Discovery Institute. [Online at Discovery.org/a/16971; accessed June 25, 2020.]

[513] Behe, M. 1996. *Darwin's Black Box: The Biochemical Challenge to Evolution.* New York: Free Press, p. 42.

CHAPTER 6: MUTATIONS

Every living organism also has a level of irreducible complexity. If a person loses a finger, although it is certainly disfiguring, it is not life-threatening. But if a person loses functionality of the nerve pathway between the brain and the heart, the heart stops beating and the person dies. So this particular nerve pathway is one part of an irreducibly complex system that keeps people alive.

I've seen multiple cases where evolutionists attempt to downplay or invalidate the idea of irreducible complexity of some design by describing a simpler, less complex design. That doesn't invalidate the principle, because they're basically missing the point; they're making an error in logic by refuting a statement that creationists didn't make.

Specifically, the evolutionists are not addressing the same thing creationists are. Creationists who object to evolution on the basis of irreducible complexity are not saying a simpler design could not be conceived (which is what the evolutionists refute, even though creationists don't claim that). The creationists are saying that for *any* given design, there is some level of irreducible complexity, below which the whole design fails. In the context of biology, this means—*at least*—that the organism dies. But infinitely more likely, the organism would not have even gotten started.

Even if an evolutionist presents a less complex design, there is a level of irreducible complexity with that design too, below which the whole mechanism fails. It very well may be a simpler design, but it too has a point of irreducible complexity. That is what creationists are stating.

This pertains directly to the alleged process of evolution because evolution is supposed to add tiny features little bit by little bit, over many generations. But the point is that, in *so* many systems we see in the natural world, none of the pieces would be of any benefit until they are *all* in place. Only then would the mechanism/process/organism function properly.

So what are some examples of irreducibly complex systems in the body of a human or animal? Here are some: The neurological system, vision, hearing, balance, smell, taste, touch, the skin, the endocrine system, the respiratory system, the gastrointestinal system, the reproductive system, the circulatory system, the excretory system, the musculoskeletal system, blood clotting, the fight-or-flight response, reflexes, the immune system, and speech, to name a few.[514] *All* of these systems comprise scores, hundreds, or even thousands of separate components, the absence of any one of which would make the whole

---

[514] Geoffrey Simmons, M.D., *What Darwin Didn't Know*, 2004 (Harvest House Publishers, Eugene, Oregon).

system fail. There is simply no way for irreducibly complex systems like the above to develop bit by bit.

If evolution actually did happen, adding genetic material enough to create components of any multi-part mechanism would be pointless until all the parts were there. Suppose an organism has some biological process that needs ten components to work. Whichever component was first to evolve would be useless without the other nine, so natural selection would not tend to propagate that mutated version, because of the "baggage" of that one useless component.

But suppose that one mutation stayed around, through a stroke of amazing luck, and then three or four others of the needed components fortuitously happened to mutate into existence. The organism still wouldn't have all ten components required for the mechanism, so there's just that much more genetic baggage, hindering the organism's fitness to survive. Because *all* the pieces would be considered detrimental baggage until the final component happened along, evolution would not favor organisms that needed life processes of any more than one component.

The trouble is, all life processes are enormously complex—*many* more than one component. Which means they would never get started at all by natural means.

It's interesting to note that Charles Darwin, in his *Origin of Species,* made this statement:

> If it could be demonstrated that any complex organ existed, which could not possibly have been formed by numerous, successive, slight modifications, **my theory would absolutely break down.**[515]

The presence of even *one* irreducibly complex system in a living organism would confirm evolution's impossibility. But because we see *numerous* irreducibly complex systems all over the biosphere, we can unequivocally conclude that *Darwin's theory **has** absolutely broken down:* decisively, undeniably, and permanently.

Joseph Kuhn, mentioned above, wrote a paper entitled "Dissecting Darwinism." In it, he described several showstopper problems with Darwinian

---

[515] Charles Darwin, *On the Origin of Species by Means of Natural Selection,* 1859, Chapter VI: "Difficulties on Theory," p. 80.

CHAPTER 6: MUTATIONS

evolution.[516] His analysis of the information contained in DNA showed that it could not possibly have arisen through natural, random means, so the only possible conclusion is that the information must have come from outside nature.[517]

To the evolutionists who roll their eyes at such a faith-based concept, this takes much less faith than to believe such intricate, interdependent, and yes, irreducibly complex systems could arise through random inputs of energy.

In his paper, Kuhn describes the inability of evolution to account for the irreducible complexity that shows up in all living things. Referencing the work of Geoffrey Simmons, another medical doctor, he points out that there are at least seventeen different all-or-nothing systems in the human body.[518] There are *seventeen* different systems in the human body that are irreducibly complex, so if even a single component of any of these systems were removed or failed to function, the whole system would fail.

Which means, of course, that all of these components would be just so much unhelpful genetic baggage until all the pieces are in place and functioning properly. And *that* means, of course, that all the pieces of these interdependent mechanisms would all have had to evolve instantaneously and simultaneously. But wait: doesn't that sound like. . .?

Kuhn noted in his paper that "virtually every aspect of human physiology has regulatory elements, feedback loops, and developmental components that require thousands of interacting genes leading to specified protein expression." Stating it another way, "the human body represents an irreducibly complex system on a cellular and an organ/system basis."[519]

Evolution can't even satisfactorily explain how a *single* irreducibly complex mechanism could evolve through random, undirected mutations. How in the world could it explain the interdependent network of distinct, irreducibly complex systems found in every human's body? So could evolution happen?

---

[516] Kuhn, J. A. 2012. "Dissecting Darwinism." *Baylor University Medical Center Proceedings.* 25 (1): 41–47. [Online at ncbi.nlm.nih.gov/pmc/articles/PMC3246854; accessed June 25, 2020.]

[517] Thomas, B. "Baylor Surgeon 'Dissects' Darwinism." *ICR News.* February 3, 2012. [Online at ICR.org/article/6607; accessed June 25, 2020.]

[518] Simmons, G. and W. Dembski. 2004. *What Darwin Didn't Know: A Doctor Dissects the Theory of Evolution.* Eugene, OR: Harvest House Publishers.

[519] Kuhn, J. A. 2012. "Dissecting Darwinism." *Baylor University Medical Center Proceedings.* 25 (1): 41–47. [Online at ncbi.nlm.nih.gov/pmc/articles/PMC3246854; accessed June 25, 2020.]

Kuhn states that such a process "would require far more than could be expected from random mutation and natural selection."

A few paragraphs ago, vision was mentioned as one of the systems in the body that is irreducibly complex: it has so many interdependent components, the lack of any one of which would cause the whole visual system to fail, that it can clearly be categorized as irreducibly complex.

Here's what Charles Darwin said about the eye—the most evident component in the visual system:

> **To suppose that the eye,** with all its inimitable contrivances for adjusting the focus to different distances, for admitting different amounts of light, and for the correction of spherical and chromatic aberration, **could have been formed by natural selection, seems, I freely confess, absurd in the highest possible degree.**[520]

But then Darwin goes on to say that he believes it anyway, for the following reasons. If *removing* components, one at a time, from the visual system could in every case leave a system that would still be useful to the organism, even at a lower level, surely they could be *added* one at a time by natural selection, to cause the eye to evolve by natural selection.

Do you see the two showstopper problems with Darwin's reasoning? First, components *can't* be removed from the visual system and leave a functioning system, even at a less effective level. Vision has been shown to be an irreducibly complex system, so if any one of its functions is removed from the genetic code, the whole system would fail. Now granted, genetics was unknown in Darwin's day, but it seems quite a leap of faith to imagine that visual components could be repeatedly removed and have an eye that still functions at any level.

Second, even assuming that Darwin's first assumption was true, concluding that, since functionality could be *removed* easily, it could be *added* easily, goes against the Second Law of Thermodynamics, which people have been dealing with every day for thousands of years. It's like saying that since a stray spark could *burn down* a house, it's just as possible for a stray spark to *build* a house. Or, since a disk crash could *destroy* 10,000 lines of functional computer code, another disk crash could *write* 10,000 lines of functional computer code.

---

[520] Charles Darwin, *On the Origin of Species by Means of Natural Selection, or the Preservation of Favoured Races in the Struggle for Life*, 1859, Chapter VI: "Difficulties on Theory", p. 80.

Or, since a rock can roll downhill, it could just as easily roll uphill. Or, since a river can flow downstream, it could just as easily flow upstream.

The ridiculousness of the Second Law of Thermodynamics being bidirectional is the whole basis for the absurdity in this nursery rhyme:

> There was a man of our town,
>   and he was wondrous wise.
> He jumped into a bramble bush
>   and scratched out both his eyes.
> And when he saw his eyes were out,
>   with all his might and main,
> He jumped into another bush
>   and scratched them in again.

Could random lacerations inflicted by sharp objects actually restore functionality to damaged eyes? Of course not. The silliness of the idea that randomness could result in improvement is intuitively recognized even by small children. It's mind-boggling that someone could actually believe that the Second Law of Thermodynamics is bidirectional, that order and complexity could be created as easily as they are destroyed.

So, just for fun, let's take a look at some actual examples of irreducible complexity in living organisms. The next three sections describe processes involved in vision, blood clotting, and reproduction. Please don't be intimidated by the big words (there will not be a quiz), but just ponder the incredible intricacy of the processes, and the number of steps involved, *all* of which need to happen, and happen correctly, for the respective processes to work.

The "Light-Sensitive Spot"

A few paragraphs ago, Charles Darwin was quoted as saying that assuming the eye could evolve by numerous, small, random changes was "absurd in the highest possible degree." But then he said he believed it anyway. Many evolutionists have glibly talked about how simple and natural it would be for an eye to evolve, and many of the fanciful accounts contain a reference to the process starting with "a simple, light-sensitive spot."

A "simple" light-sensitive spot? Let's actually think about that. What is required for living tissue to be light-sensitive? Below is a description of the biochemical steps involved in the light-sensing process in a photoreceptive cell like the retina of the eye.[521,522] Note that this description does *not* include the lens and its controlling muscles, the iris and its controlling muscles, the eye-aiming muscles, the blood supply, and so forth; this is just the "simple" part:

1. A particle of light—a photon—strikes a small organic molecule of **11-cis-retinal**, which is changed, in only a few picoseconds, into **trans-retinal**. (A picosecond is one trillionth of a second, the amount of time light takes to go less than $1/3$ of a millimeter.)
2. The change in the previous step causes the protein **rhodopsin**, tightly bound to the trans-retinal, to change shape, which changes how it behaves; the rhodopsin becomes **metarhodopsin II**.
3. Because of a "stickiness" in the metarhodopsin II, it attaches itself to another protein called **transducin**.
4. When the metarhodopsin II binds to the transducin, the reaction dislodges a small molecule called **guanosine diphosphate** (**GDP**), to which the transducin had previously been bound.
5. In the location where the GDP disconnected, a very different molecule called **guanosine-5′-triphosphate** (**GTP**) connects.
6. The GTP-transducin-metarhodopsin II complex now binds with another protein called **phosphodiesterase**, on the inner membrane of the cell.
7. When the above binding takes place, the phosphodiesterase gains the ability to "cut" molecules of **cyclic guanosine 3′-5′ monophosphate** (**cGMP**), which is related to both GDP and GTP.
8. Initially, there is a large number of cGMP molecules inside the cell, but the phosphodiesterase, since it is was bound to the GTP-transducin-metarhodopsin II complex, cuts them apart, reducing the number of them in the cell.
9. There is a different protein, called an **ion channel**, in the cell membrane, which brings sodium ions inside the cell. Another protein sends them out, and these two proteins, working together, maintain the optimum concentration of sodium ions within the cell.
10. The cleavage of cGMP by the phosphodiesterase shuts down the ion channel, so the number of sodium ions inside the cell is reduced.

---

[521] Michael Behe, *Darwin's Black Box* (New York, Touchstone Books, SimonAndSchuster.com, 1998), p. 22.

[522] "Rod Cell: Photoreception" [Online at en.wikipedia.org/wiki/Rod_cell#Photoreception; accessed November 26, 2021.]

11. The unbalanced concentration of sodium ions inside the cell, as compared to the concentration outside the cell, causes an electrical current to be generated, which, when propagated along the optic nerve to the brain, is interpreted by the brain and it "perceives" the light.
12. Then there is another process, almost as complex, that allows the system to be restored to its resting state, so it can detect another photon, and the process can start all over again.

Now seriously: could anyone really claim that the above process could "fall together" accidentally? Read the process again, and notice how many precisely functioning components there are! If any of those protein molecules were absent, or had the wrong number of amino acids, or the wrong sequence of amino acids, or any of the amino acids were right-handed instead of left-handed, or any of them were in the wrong configuration, or the wrong concentration, *the whole process would fail,* and even something as "simple" (as evolutionists describe it) as a light-sensitive spot would fail to work.

The above process is mentioned in the irreducible complexity section because any subset of the components above would be useless for light sensitivity until they were *all* there. So why would any one of these components, so precisely designed to eventually be interdependent with the others, assemble by accident, and then stay around for who knows how long, surviving natural selection (which weeds out useless genetic information), just waiting until all the other components accidentally assembled as well? They wouldn't; they would have to evolve all at the same time for the process of light-sensitivity to work at all. And yes, that sounds an awful lot like creation.

So the process of light sensitivity is yet *another* fatal flaw to the theory of evolution.

The Clot Thickens. . .

Another example of an irreducibly complex mechanism is that of blood clotting. This is such a common thing, we often don't even think about what a marvelous and intricate process it is. We get a paper cut, or even cut our finger with a knife while chopping onions, we simply put pressure on it, maybe put on a band-aid, and soon it stops bleeding and the repair is underway. But what a marvelous repair process it is!

Once you think about it, it doesn't take much contemplation to realize that if blood clotting went wrong in one direction, you could bleed to death from a paper cut, and if it went wrong in the other direction, your whole blood supply could clot from the smallest injury. And if a clot formed in the wrong place, it could cause a heart attack or a stroke. What governs this process? What initiates it? What determines where it happens? How does it know when to stop? Let's look.[523]

1. Not counting red blood cells, about 2–3% of the protein that is in your blood is called **fibrinogen,** which is the substance from which blood clots are made. Fibrinogen is a composite of six different protein chains: a pair each of three different kinds.

2. When you are injured, another protein, called **thrombin,** cuts off pieces of two of the three pairs of protein chains in fibrinogen, converting it to **fibrin.**

3. The areas of the fibrin where the cut-off pieces used to be are "sticky" (but only to other fibrin molecules), and this allows large numbers of fibrin molecules to stick together in strings, forming the beginning of a blood clot.

4. Because of the shape of fibrin molecules, the long strings overlap one another in a specific way, forming a meshwork patch that can entrap blood cells, instead of forming an irregular blob, which would require much more material to work with, and also likely plug up more blood vessels than it should.

A couple steps ago, thrombin was mentioned as the protein that starts coagulation by converting fibrinogen to fibrin. You may have thought, very astutely, "What keeps the thrombin from cutting the fibrin before it's supposed to?" This is an excellent question, because if it did so too early, your entire blood system would clot, and you would die. So let's expand Step 2, above, into a more complete description, appropriately called "the blood-clotting cascade," because of its intrinsic domino effect:

A. To avoid thrombin making fibrin too early, thrombin exists in the bloodstream in a dormant form called **prothrombin.** If this were not the case, your entire blood supply would clot, as mentioned above. But if the prothrombin were never activated, you could bleed to death from even a small wound.

B. A protein called **Stuart factor** cleaves prothrombin, turning it into thrombin. However, if this happened too early, again, the entire blood

---

[523] Michael Behe, *Darwin's Black Box* (New York, Touchstone Books, SimonAndSchuster.com, 1998), p. 72.

CHAPTER 6: MUTATIONS

supply would clot. So, Stuart factor also initially exists in a dormant form.

C. However, even activated Stuart factor can't convert prothrombin into thrombin at a fast enough rate to prevent you from bleeding out, so another protein, **accelerin,** is needed as a catalyst.

D. But accelerin can't do its thing too early either, so, to avoid *all* the blood clotting, accelerin also comes in an inactive form, dubbed **proaccelerin.**

E. So proaccelerin needs to be converted into accelerin, so it, working with the activated Stuart factor, can convert the prothrombin to thrombin fast enough to form a clot before you bleed to death.

F. So what converts proaccelerin into accelerin? Thrombin. But wait a minute: thrombin is way down the line in this control cascade! That's like you arranging the circumstances of your own father's childhood! But that is indeed the way it works: Because Stuart factor cleaves prothrombin very slowly, there is always a trace of thrombin in the bloodstream. Therefore, blood clotting is an **auto-catalytic** process: proteins in the process stimulate the production of more of the same kinds of proteins.

G. But actually, prothrombin can't be converted to thrombin, even if activated Stuart factor and accelerin are present, until ten specific amino acid residues—namely **glutamate (Glu) residues**—are converted into **Y-carboxyglutamate (Gla) residues**. The required modification can be illustrated by a half of a pair of pliers (which can't grasp anything) being properly attached to the other half, enabling the assembly to do its intended job of grasping. In this case, the Gla residues bind to calcium, allowing the prothrombin to stick to cell surfaces. Only then can it be cleaved by activated Stuart factor and accelerin to produce thrombin.

H. But how is Glu converted to Gla? By the presence of **Vitamin K. . .**

Okay, you get the idea. And the above is not even the whole process. But as you can see, the above process of blood clotting absolutely could *not* have arisen bit by bit in an evolutionary way because, as in all irreducibly complex systems, *none* of the components are of any use until *all* of them are in place. And even if it could, it would be pointless without the other supporting irreducibly complex systems of a vascular system, lungs to oxygenate the blood, a heart to circulate it, sensory nerves to detect when the lungs and heart should operate more (or less) energetically, a brain to process said information, motor nerves to communicate the commands to the various destinations, and so forth.

Biochemistry professor Russell Doolittle, of the University of California (San Diego) Center for Molecular Genetics, attempted in his Harvard Ph.D. thesis to figure out how the blood-clotting process could have evolved bit by bit. Then, still unable to answer the question more than three decades later, in an article[524] directed at the world's foremost authorities on blood clotting, Doolittle asks the very probing question:

> How in the world did this complex and delicately balanced process evolve? . . . The paradox was, if each protein depended on activation by another, how could the system ever have arisen? Of what use would any part of the scheme be without the whole ensemble?[525]

Clearly, such intricate and elaborate systems, with all their interdependent components, could not have evolved by random processes; such an occurrence is absolutely, utterly, incontrovertibly impossible. The likelihood is *exactly* 0%.

Reproduction

As another example of irreducible complexity, let's consider the process of reproduction. So it doesn't get overly complex, let's consider the reproductive process of a single-celled **prokaryote**—a cell that doesn't have a true nucleus or other organelles; a typical bacterium, in this example.[526]

Let's take a look at the major aspects of this process:

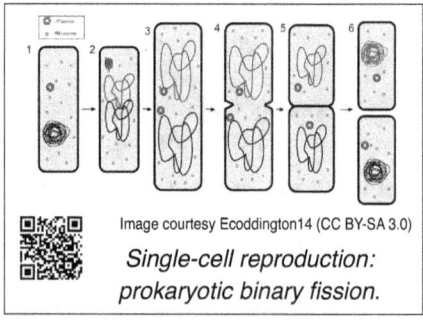

Image courtesy Ecoddington14 (CC BY-SA 3.0)
Single-cell reproduction: prokaryotic binary fission.

1. Let's start with a live, viable bacterial cell. This is an enormous "given," but let's just assume, like Darwinian evolution does, that *somehow* the cell came into existence, that *somehow* the staggeringly

---

[524] Russell Doolittle "The Evolution of Vertebrate Blood Coagulation: A Case of Yin and Yang," *Thrombosis and Haemostasis, 70* (1993), pp. 24–28.

[525] Behe, Michael J., *Darwin's Black Box: The Biochemical Challenge to Evolution*, [Simon and Schuster, Free Press, 2006], p. 115. Kindle Edition.

[526] Even the name "prokaryote" assumes evolution, because "pro" means "before," and this follows the evolutionary tenet of faith that simpler things existed *before* they evolved into the more complex things that came later—because things spontaneously get more complex as time goes on, you see.

complex information in the DNA wrote itself, and that *somehow* the proteins required to do all the normal cellular tasks, including processing the information in the DNA, accidentally appeared. Let's just assume all that. This cell contains information-packed DNA, **ribosomes** (molecular machines that assemble proteins according to instructions in the DNA), and **plasmids** (extrachromosomal DNA that contains instructions on how to cope with certain environmental conditions).

2. The bacterium ascertains that it is "safe" to reproduce—if it is dehydrated or is malnourished, it likely wouldn't even start the process because of the energy that would be consumed on a process that is likely to fail anyway.
3. If conditions are satisfactory, a "decision" is made to start the reproductive process.
4. The tightly coiled DNA strand uncoils, so splitting is possible.
5. The plasmids, using instructions intrinsic to them, duplicate.
6. Ribosomes, using instructions in the DNA, synthesize the proteins necessary to split the double helix of DNA.
7. Those proteins separate the two "uprights" of the twisted ladder of DNA, leaving half-rungs exposed. These half-rungs specify the components that need to be added to make the DNA complete again. The half-rungs are matched to their respective components, completing the DNA duplication.
8. The cell increases in length, preparing for splitting in half.
9. One strand of the newly duplicated DNA moves toward one end of the bacterium, while the other strand moves toward the other end.
10. Similar to the DNA, half the plasmids move to one end of the bacterium, and the other half to the other end.
11. Similar to the DNA and plasmids, half the ribosomes move to one end of the bacterium, and the other half to the other end.
12. A band of the cell wall in the middle of the bacterium begins to constrict and pinch, initiating the physical separation process.
13. The constricted area in the middle closes completely, pinching the cell into two daughter cells.
14. The still-unraveled DNA in both daughter cells now coils up into its "resting" condition.

So *all* of the above steps have to happen correctly, and in the proper sequence, for the bacterium to reproduce. In fact, it's more complicated than the above, because each step above is actually a whole orchestration of smaller subprocesses. And of course, there is again the question of where the information in the DNA came from—the information on how to create the proteins to do all the necessary tasks. The above is a perfect example of irreducible complexity: if any step in the above recipe fails, reproduction doesn't happen and the organism dies.

But here's the point: The above process absolutely cannot evolve. Why? Because biological evolution supposedly happens through numerous tiny changes accumulating over many generations. *But you won't have any generations at all until reproduction works.* Which means that *all* of the steps above would have to "fall into place" at the same time—within the lifetime of that one single bacterium—or else reproduction couldn't happen, and the species would go extinct after the original, first organism.

"All the necessary pieces falling into place at the same time" sounds suspiciously like creation, does it not?

And keep in mind that the above example is *asexual* reproduction of a single-celled bacterium. Sexual reproduction, where genetic information from both the male and female is required, is *much* more complex—just like the organisms themselves—be they humans, birds, reptiles, fish, mammals, amphibians, or whatever. Just think about it: for sexual reproduction to evolve, all of the "machinery" mentioned above would have to evolve *instantly, twice, simultaneously,* and *compatibly,* for the process to work. Just how much suspension of disbelief is required to believe in evolution?

Quite a bit, actually. Read these comments from various scientists about the likelihood of sexual reproduction evolving:

> This book is written from a conviction that **the prevalence of sexual reproduction** in higher plants and animals **is inconsistent with current evolutionary theory.**[527]

> Indeed, **the persistence of sex is one of the fundamental mysteries** in evolutionary biology today.[528]

---

[527] George C. Williams, *Sex and Evolution* (Princeton, New Jersey: Princeton University Press, 1975), p.v. Emphasis added.

[528] Gina Maranto and Shannon Brownlee, "Why Sex?" *Discover,* February 1984, p. 24. Emphasis added.

> **Sex is something of an embarrassment to evolutionary biologists.** Textbooks understandably skirt the issue, keeping it a closely guarded secret.[529]
>
> **The evolution of sex is one of the major unsolved problems of biology.** Even those with enough hubris to publish on the topic often freely admit that they have a little idea of how sex originated or is maintained. **It is enough to give heart to creationists.**[530]

So if unicellular reproduction can't evolve—and we saw above that it can't, unless all the components evolve instantly and simultaneously—and if sexual reproduction can't evolve—and we saw above that it can't, unless all the components evolve instantly, twice, compatibly, *and* simultaneously—how did we get here? The only other option: We were created.

And, by the way, this is the additional monkey wrench foreshadowed earlier in the chapter, in the section talking about our 200-part organism evolving via random mutations. We allowed for 200 generations of the organism to take place every second, but *there would be no generations at all without reproduction.* And reproduction is *so* much more complex than a paltry 200 parts. This means that our little organism would not have gotten started at all, so yet again, evolution couldn't possibly have given rise to life—there is absolutely no way for life to "arise" from naturalistic random processes.

Kinesin: Life's Delivery Service

In the previous section we discussed the reproduction of prokaryotes—cells without nuclei. In this section, we'll consider **eukaryotes**—cells *with* nuclei.

One of the fascinating things about eukaryotic cells is that they use a **motor protein** called **kinesin**. A motor protein, as its name implies, is a protein that exhibits movement, and its name is related to **kinetic** (pertaining to motion), **kinesiology** (the study of the mechanics of bodily movement), **kinematics** (the study of the mechanics of the motion of objects), and so forth.

So what does kinesin do? It is a delivery service that takes packages from one location within the eukaryotic cell to another location within the cell. As modern research has shown, living cells are breathtakingly complex, far from

---

[529] Cathleen McAuliffe, "Why We Have Sex," *Omni,* December 1983, p. 18. Emphasis added.
[530] Michael Rose, "Slap and Tickle in the Primeval Soup," *New Scientist,* Volume 112, October 30, 1986, p. 55. Emphasis added.

the "blob of protoplasm" they were considered not too long ago. And in these cells, there are numerous processes that must constantly take place for the cell to survive, and many of these processes are accomplished through the actions of kinesin.

Structurally, kinesin is a complex protein, intricately folded in just the right way so as to do its job. Functionally, it could be thought of as roughly "person shaped"—that is, on one end it has two appendages (its "arms") with which it grasps the package it is taking somewhere, and on the other end it has two more appendages (its "legs") with which it walks along, during the process of delivering its cargo to its destination.

How does the kinesin know where to go? The answer to that has several components, each of which, by itself, is a deathblow to random evolution as the source of this amazing substance. The first component is the "road" along which kinesin travels: these are called **microtubules**, composed of long chains of another protein called **tubulin**. These microtubules are not so much "pipes" through which substances flow, as much as hollow "cables" along which the kinesin proteins walk, hauling their cargo.

Here is a typical use of kinesin, both on a cellular level, and illustrated (in italics) by comparable tasks done by human being in the course of a day:[531]

1. Some part of a cell, often a protein, is damaged, usually directly or indirectly because of the Second Law of Thermodynamics, and a replacement protein is needed to repair the damage. *Joe Regularperson is working away in his office, in some building of his high-tech campus, and his machine breaks.*
2. A "caretaker" part of the cell close to the damage notices, identifies the part that was damaged, sends a message to the ribosome. *Joe notices which part broke, and emails the manufacturing floor, requesting a replacement part.*
3. Included in the request for the replacement part is the location to which the replacement part is to be shipped. *Joe includes his email signature file so the factory knows which building and which office to send the replacement part to.*
4. The ribosome can manufacture the part, but it doesn't have the instructions for making it; that information is in the DNA in the cell's nucleus, so the ribosome requests the specs from the nucleus. *The factory floor has manufacturing capability, but it doesn't keep the database of blueprints*

---

[531] Calvin Smith, "Incredible Kinesin!" June 26, 2012. [Online at Creation.com/incredible-kinesin; accessed January 24, 2022.]

CHAPTER 6: MUTATIONS

that describe how to manufacture replacement parts, so it requests a database lookup.

5. Proteins in the nucleus take the request, and the genetic instructions are recognized, and the requested information is located within the vast repository of information in the DNA strand. *The manufacturing floor sends a part number to the computer center, which brings up the correct blueprint from the database.*

6. Accessing the DNA, proteins make a copy of the requested assembly instructions, in the form of a strand of RNA, and send the copy back to the ribosome. *The blueprint is photocopied (they wouldn't dream of sending the original) and sent to the manufacturing floor.*

7. Using the genetic information in the RNA, the ribosome creates the desired protein. *Using the blueprint, the manufacturing floor assembles the replacement part.*

8. The Golgi apparatus puts the new protein into a vesicle. *The manufacturing floor then packages the replacement part for delivery.*

9. The Golgi apparatus also chemically imprints the destination location on the surface of the vesicle. *Using Joe's building and office number, received in Step 3, the manufacturing floor addresses the package so the courier knows where to deliver it.*

10. The ribosome summons a kinesin molecule to deliver the vesicle to its destination. *The factory contacts the shipping department to send a courier, and the courier soon arrives.*

11. The kinesin protein bonds to the vesicle with two of its appendages, interprets the destination address, determines the optimal route along the microtubules, and walks the vesicle to its destination. *The courier comes to the manufacturing floor, picks up the package, reads the destination address, figures out how to get there, and, keeping to the sidewalks and off the grass, walks the package to Joe's office.*

12. The replacement protein is extracted from the vesicle and installed in place of the damaged protein. *Joe unwraps package and installs the replacement part in his machine so it works again.*

This whole process contains numerous deathblows to the notion that such an intricate and irreducibly complex system could "fall" into place accidentally, but let's take a closer look at just the delivery system: the courier named kinesin. The kinesin actually *walks* along the microtubule, in the standard bipedal manner: putting one foot in front of the other.

331

In the illustration at right, notice how the two "legs" of the kinesin protein alternate which one is forward, effecting the walking motion. The large circle at the top is the vesicle—the "wrapper" containing the recently manufactured goods being delivered. Although the precise mechanism is not yet known, microbiologists have observed that one molecule of ATP (adenosine triphosphate) is consumed for each step the kinesin takes.[532]

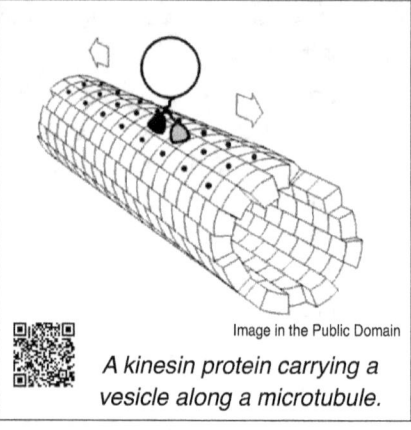

Image in the Public Domain

A kinesin protein carrying a vesicle along a microtubule.

A kinesin molecule is about 70 nanometers long (1nm is one billionth of a meter), and it takes steps about 8nm long. How fast is this walking process? Quite fast, actually. A kinesin molecule can take about 100 steps per second; a human being walking at that rate would be traveling more than 200 miles per hour! These steps are pretty tiny, though: a kinesin molecule takes 125,000 steps to go 1 millimeter![533]

What happens when a kinesin molecule is done delivering cargo to the destination? Several things can happen. First, it can go into a "power saving" mode, where it is on standby, and pauses its consumption of ATP fuel pellets while waiting for the next task.[534] Or, it can be ferried back to the nucleus to get assigned to another delivery job. Or, it can be disassembled and built into something else that is needed.[535]

Just considering the cargo-delivering capabilities of kinesin: note how many things had to "evolve" simultaneously so the process could work. One ATP had the right amount of energy to move the feet without dislodging them; the kinesin had to have appendages of two types; all appendages had to be able to hang on and let go when required; the kinesin had to be able to "read" the address label on the package; the microtubules had to be constructed such that kinesin could travel along them; and *so* much more. With-

---

[532] "Kinesin" [Online at en.wikipedia.org/wiki/Kinesin; accessed January 24, 2022.]

[533] "The Kinesin Linear Motor" [Online at Creation.com/media-center/youtube/the-kinesin-linear-motor; accessed January 24, 2022.]

[534] Kaan, H.Y.K, et al., "The Structure of the Kinesin-1 Motor-Tail Complex Reveals the Mechanism of Autoinhibition," *Science* 333(6044):883–885, 2011.

[535] Gutierrez-Medina, B., et al., "Kinesin: an ATPase that steps along microtubules," Stanford.edu, accessed January 2012.

out kinesin, numerous essential processes within the cell would fail, and the cell would die. We sure are lucky all those components randomly evolved within the lifetime of a single cell, huh?

Chapter 6 Summary

Here are the most important points in this chapter:

- **Spontaneous generation is a "disproven" and "obsolete" doctrine** ("doctrine" is Wikipedia's word), but *slow* spontaneous generation is **essential for biological evolution.**
- Evolutionism prefers the word "abiogenesis" to indicate life from non-life, since "spontaneous generation" to indicate life from non-life is a disproven doctrine.
- There are two different things called "beneficial mutations:"
  - A mutation that loses genetic information, but has serendipitously beneficial outcome.
  - A mutation that actually *adds* genetic information, and causes an increase in overall "fitness" to survive.

  It is this second type that evolution requires to happen, *trillions* of times, **but which has never actually been seen.**
- **Even "neutral" mutations are actually detrimental,** because they degrade DNA.
- **Harmful mutations *far* outnumber beneficial mutations.** (Remember, no one has actually seen a beneficial mutation.)
- **Evolutionists give lip service to beneficial mutations, but don't really believe it.** If they did, they would wear radioactive jewelry, because just *think* of all the wonderful mutations that would result!
- **The error-correction algorithms** that operate during the DNA replication process **couldn't have evolved,** because their very purpose is to prevent mutation.
- In human beings, approximately **63 new mutations are added to each generation** of offspring. This ever-increasing burden of damaged DNA is called **genetic load.**
- With the genetic load that would have built up in the 2.4 million years mankind has supposedly been evolving, and the observed rate of more than 60 mutations per generation, the resulting **7.2 million mutations would have caused humanity's extinction long ago.**

- **Evolutionists acknowledge that only about 1 in 10,000 mutations is beneficial.** They haven't actually seen one, but they've just *gotta* be there somewhere. And evolution still allegedly made enough progress to turn inorganic molecules into humans.
- What used to be called "junk DNA" is actually **essential for life.**
- Any mechanism has some point of **irreducible complexity:** that point at which, if any one piece fails or is removed, the entire mechanism fails. *Numerous* examples exist in every living thing. For example:
    - The **"light-sensitive spots"** that evolutionists glibly say could transform themselves into eyes **could never have evolved,** because the process of light sensitivity is enormously complex and all components must be in place before any of them can contribute any benefit—that is, be sensitive to light.
    - The process of **blood clotting** is so intricate, and with so many steps, feedback loops, and control mechanisms that it **could never have fallen together accidentally.** The tiniest alteration to this process would result either in bleeding to death from a miniscule cut, or the entire blood supply clotting.
    - **Asexual reproduction could never have evolved,** because evolution assumes small changes over many generations, and you wouldn't have *any* generations until reproduction worked. And reproduction of any kind is an enormously intricate, irreducibly complex process.
    - The cellular "package-delivery" system of **kinesin could never have evolved, because of so many intricately-designed interdependent parts:** kinesin's "arms" and "legs," its ability to deliver to a particular address, its ability to go into standby mode when not needed, its walking motion along special-purpose microtubules, each step powered by an individual molecule of ATP, and so forth. Again, none of these would be useful until all of them were present.
- **How could sexual reproduction have evolved?** We already saw that *asexual* reproduction is an irreducibly complex process, and therefore could not have evolved, so how could something even *more* complex have evolved? If the incomprehensably complex process of asexual reproduction could not have evolved even once, how could it evolve twice, instantaneously, simultaneously, and compatibly? It couldn't. Period.

Chapter 7:

The Fossil Record

The fossil record, you would expect, should be the evolutionists' best friend. After all, if all life on Earth evolved from a single-celled organism somewhere in the dim, dark past, the fossil record should be filled with transitional, half-this-and-half-that fossils. After all that time, the transitional forms should outnumber the stable forms many to one. (Or, more accurately, there would be no such thing as a "stable form" if mutations are our friends as much is evolution claims they are.)

But, alas, that is not what we see. There is an absolute famine of fossils showing a transitional form of anything changing into anything else. This was predicted by the Creation model, but it must be explained by the evolution model: "Why, if evolution is true, do we not see fossils of transitional forms?" They should be everywhere.

This paucity of fossil evidence is acknowledged by the evolutionists themselves. For example:

> Since 1859 one of the most vexing properties of the fossil record has been its obvious imperfection... **The inability of the fossil record to produce the 'missing links' has been taken as solid evidence for disbelieving the theory.**[536]

---

[536] A. J. Boucot, *Evolution and Extinction Rate Controls* (Amsterdam, Elsevier Scientific Publishing Co., 1975), p. 196.

The author is correct: the absence of fossil evidence *has* been taken as solid evidence for disbelieving evolution. And very justifiably so.

Stephen Jay Gould is a paleontologist from Harvard, as well as a philosopher of science. Here is his observation:

> All paleontologists know that **the fossil record contains precious little in the way of intermediate forms**; transitions between major groups are characteristically abrupt.[537]

By "precious little," Gould means "nothing," and by "characteristically," he means "always." But it's understandable why he wouldn't use such strong words: it makes it even more obvious that evolution is a bankrupt theory.

If the Creation model is accurate, the fossil record should show all its fossils as fully formed organisms: each type is not there, and then suddenly it is, fully developed, with no transitional forms. And this is exactly what the fossil record shows.

If the Darwinian evolution model is accurate, the fossil record should show innumerable fossils that are partly one thing and partly another. It should contain fossils of Organism A, some that are 90% Organism A and 10% Organism B, some that are 80% A and 20% B, some that are 70% A and 30% B, and so forth. Only with a lot smaller granularity than 10% jumps. But they're not there. *Anywhere.*

The Geologic Column

The geologic column is the series of rock strata (layers), often containing characteristic types of fossils, that we find all over the world. These are called **systems.** Also, depending on your paradigm of history, the geologic column may also be thought of as a way to assign various geological strata (rock layers) to certain timeframes—deep ones, of course. In evolutionary thinking, the timeframes deal with millions and billions of years, and the various sizes of timeframes, large to small, are called Eons, then Eras, then Periods, then Epochs, and then Ages.

An important distinction here is that the rock layers are real: the strata do exist, and they often contain characteristic fossils. But the timeframes associated with those layers (those systems of strata) are just speculations and guess-

---

[537] Stephen Jay Gould, "The Return of Hopeful Monsters," *Natural History,* Vol. LXXXVI, June–July 1977, p. 24.

CHAPTER 7: THE FOSSIL RECORD

work based on the presumption of deep time necessitated by a philosophical worldview that is desperate to exclude God at all costs.

So, for example, the Jurassic *system* is a real thing: that rock layer exists, and certain types of dinosaur fossils are often found in it. However, the Jurassic *period* is a timeframe that is a figment of the imagination based on presumptions that are themselves supported primarily by a philosophical paradigm (naturalism, which states there is no God) that evolutionists assume to be unquestionable.

As you research geologic timeframes in greater detail, you will find that some of the starting dates and durations are different, depending the sources you are referencing. Continuing research finds more and more observations and evidences that don't match evolutionary thinking, so take the timeframe in the illustration with a grain of salt. Or *lots* of grains.

In the "Geochronology" entry of the *Encyclopedia Britannica*, it says this:

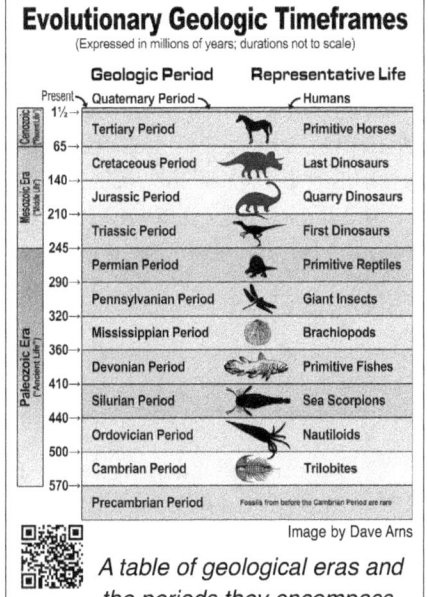

A table of geological eras and the periods they encompass, according to evolutionary thought, along with a life form that represents that timeframe.

> Dating, in geology, determining a chronology or calendar of events in the history of Earth, **using to a large degree the evidence of organic evolution in the sedimentary rocks** accumulated through geologic time in marine and continental environments.[538]

In other words, "We know how old the rocks are because of fossils we find in them." Then a little later, in the same article, it states:

> Just as the use of the fossil record has allowed a precise definition of geologic processes in approximately the past 600 million years, absolute ages allow correlations back to Earth's

---

[538] Britannica.com/science/dating-geochronology; accessed June 18, 2020.

oldest known rocks formed more than 4 billion years ago. In fact, **even in younger rocks, absolute dating is the only way that the fossil record can be calibrated.**

In other words, "We know how old the fossils are because of the rocks we find them in." What's wrong with this picture?

In the field of logic, there is a fallacy of thought that is called "begging the question" or sometimes "assuming the conclusion." This type of logical fallacy assumes the truthfulness of the conclusion and uses that assumption in its own proof. In other words, you can prove the proposition is true if you first assume it's true.

It's an example of circular reasoning: the conclusion precedes and is used in the proof, which then "proves" the conclusion. This is what the *Britannica* article does above: the rocks determine the age of the fossils, but the fossils determine the age of the rocks. Which simply means that both conclusions are unreliable. I first saw this error in a hardcover version of the *Encyclopedia Britannica* in the mid-'70s in the Colorado State University library. I was astonished back then as well, to see two things "proving" the validity of each other, while simultaneously *assuming* and *requiring* the validity of each other.

But I am certainly not the first to discover this circular—and therefore completely untrustworthy—reasoning:

> **It cannot be denied that** from a strictly philosophical standpoint **geologists are here arguing in a circle.** The succession of organisms has been determined by a study of their remains embedded in the rocks, and the relative ages of the rocks are determined by the remains of organisms that they contain.[539]

> **A circular argument arises:** interpret the fossil record in the terms of a particular theory of evolution, inspect the interpretation, and note that it confirms the theory. **Well, it would, wouldn't it?** . . .the fossils do not form the kind of pattern that would be predicted using a simple NeoDarwinian model.[540]

---

[539] R. H. Rastall, "Geology," *Encyclopaedia Britannica,* Vol. 10, 1954, p. 168.

[540] Thomas S. Kemp, "A Fresh Look at the Fossil Record," *New Scientist,* Vol. 108, December 5, 1985, p. 66.

CHAPTER 7: THE FOSSIL RECORD

Fast Geological Processes

The phrase "geological process" is rather broad, so the topics in this section will not be simply different flavors of the same basic thing. But what they do have in common is that they are all geological processes, they all influence the geologic column, and they all show a much younger Earth than evolutionists want to believe.

Polystrate Fossils

Speaking of geologic "columns," have you ever heard of a **polystrate fossil**? They are sometimes also called "polystratic" or "multistratic" fossils, but what does that mean?

Image courtesy Ian Juby (Creative Commons CC0 1.0 Universal Public Domain Dedication)

A cast of the famous Tennessee polystrate fossil trunk, on exhibit outside of Creation Evidence Museum.

The "poly" (or the "multi") part means "many," and the "strate" (or "stratic") part pertains to rock "strata," or layers. So a polystrate fossil is a single fossil that crosses many rock layers.

Just the fact that such fossils exist is extremely significant, because here is yet another discovery that single-handedly collapses the deep-time dogma of evolution. According to long-age enthusiasts (evolutionists), water-borne sediment is laid down and then, after hundreds of thousands or maybe millions of years, the sediment hardens into sandstone or some other sedimentary rock.

339

OH, EVOLVE! (GOOD LUCK WITH THAT...)

The problem, of course, is those pesky polystrate fossils. If a tree died, it wouldn't wait hundreds of thousands of years or more to decompose, while waiting for many layers of sediment to build up around it. No, most trees completely decompose in a century or less, and some tougher strains take up to about 125 years.[541] Which means that the tree would be totally decayed away and gone eons before its own layer of sediment hardened into rock, let alone future layers.

Polystrate fossil tree trunk in the Fossil Cliffs of Joggins, Nova Scotia. Note the hammer at the base of the trunk for a size reference.

For example, in the image above, from Joggins, Nova Scotia, those layers supposedly (using evolutionary deep time) took ten million years to lay down. Those were *very* hardy trees, wouldn't you say?

Trees aren't the only polystrate fossilized items that strengthen even further the already conclusive failure of evolution. There are polystrate examples of:

- Fossilized leaves standing on edge in layers of diatoms,
- A fossilized whale buried in layers of diatomaceous earth,
- Jellyfish fossilized across seven layers of sediment allegedly deposited over a million-year period (they must have been *really* tough jellyfish),
- Numerous tadpoles in layers of diatoms,
- Aquatic mesosaur reptile in "annual" **varves** (thin layers of sediment supposedly showing winter/summer deposition of a single year),
- A fossilized pod of whales buried across four strata,
- Delicate spines radiating through layers of solid rock,
- Nautiloids in the Redwall Limestone which allegedly formed at 4,000 years per inch,
- A fossilized school of extinct trout-perch species,
- Three-dimensional (not crushed) trilobites in limestone,
- Communities of organisms found in limestone,
- Dinosaur footprints that deform multiple rock layers,

---

[541] NorthernWoodlands.org/knots_and_bolts/tree-falls-in-a-forest; accessed July 1, 2020.

. . .and more. For photos, descriptions, and documentation, see the excellent website kgov.com/list-of-the-kinds-of-polystrate-fossils.

The upshot of the existence of polystrate fossils is that the deposition of sedimentary layers didn't take *nearly* as long as evolutionists maintain. The existence of individual fossils that cross dozens of layers of sedimentary rock could only be caused by very rapid layering, caused by, say, perhaps a worldwide flood.

Folded Rock Strata

According to evolutionary time expectations, the layers of sand and silt we see today were laid down and solidified into sedimentary rock during periods of time millions of years in length. But if that is true, how do evolutionists explain folded rock strata?

Image Courtesy Sergey Kotov and Creative Commons ShareAlike 3.0 Unported License

Folded rock strata in the Pay-Khoy Ridge in Russia

Consider the images at right. All of them exhibit *folding*, not *breaking*, of sedimentary rock layers. If the evolutionary deep-time assumption is correct, and it takes millions of years for layers of soft sediment to solidify into rock, these rock layers would have to have been solidified long before the layers were bent. But if the rock layers were hard—as they would have been in deep time—the forces that bent these rocks would have shattered them into rubble.

Image courtesy Ali Ogden (CC BY-SA 2.0)

Folded rock strata on Gulann nan Osna in Scotland.

How did these layers get bent so far, and with such small radii of curvature (sharp bends), without shattering? Clearly, they must have been still soft when they were bent. Then, after the bending of the still-pliable layers was done, the solidification process hardened them into the rock layers we see here. This tells us that the geologic processes that formed the layers, as well as the processes that caused the bending, didn't take nearly as long as the assumptions of evolution would dictate.

Image courtesy Kerk Phillips. Public domain.
Folded rock strata in Provo Canyon, Utah.

Because the rock layers bent without breaking, the bending had to occur less than a few thousand years after being laid down.[542] Perhaps *much* less.

Image courtesy Jonathan Wilkins (CC BY-SA 2.0)
Folded rock strata in the Precambrian Amlwch Beds in Wales.

---

[542] Austin, S. A. and J. D. Morris, "Tight folds and clastic dikes as evidence for rapid deposition and deformation of two very thick stratigraphic sequences," *Proceedings of the First International Conference on Creationism*, vol. II, Creation Science Fellowship (1986), Pittsburgh, PA, pp. 3–15. [Online at static.ICR.org/i/pdf/technical/Tight-Fold-and-Clastic-Dikes-Rapid-Deposition-Deformation.pdf; accessed June 30, 2020.]

Bioturbation

That sounds like twenty-dollar word, doesn't it? What in the world is bioturbation? The first part, "bio," means "life," the second part, "turb," means "stirred up" (as in the word "turbulence"), and the third part, "-ation," is a suffix that means "the process of." So bioturbation is the process of live organisms stirring up sediment, either land-based or on the ocean floor.

Sedimentary layers clearly seen in the Badlands of South Dakota.

Suppose a flood happens, and washes much sediment along with it. Sooner or later, the water loses its momentum, and the sediment settles out, forming one or more distinct layers. As more and more layers are deposited, the lower layers are compressed by the weight of the overlying sediment, and the water is squeezed out and, if there is a cementing agent present, the sediment turns into sedimentary rock, like we see all over the world.

So **bioturbation** is the process by which life stirs things up, and specifically, it refers to a freshly deposited layer of sand, dirt, silt, and so forth, brought in by the action of water. Surface layers of sediment, whether on land or underwater, teem with both plant life and animal life. Plant roots push through the sediment, and animals—moles, worms, clams, etc.—dig through it. This churns, mixes, and turns over the sediment, so any remnant of sedimentary layers is soon obliterated.

Sedimentary layers in the Antelope Canyon erosion channel in Arizona.

Sir Charles Lyell said "The present is the key to the past," and in so doing, he gave a strong impetus to the evolutionary infatuation with uniformitarianism. Lyell, a friend and colleague of Charles Darwin, published in 1830 his classic book entitled *Principles of Geology*, and it was in that book that he promoted uniformitarianism. So if it is true that these layers took so long to form and petrify, what kept animals and plants (and weather, for that matter) away from the fragile sedimentary layers until they solidified, perhaps millions of years later?

No. The plants and animals that move sediment around as they grow and live would completely erase any layered appearance *long* before the layers had a chance to solidify, if deep-time uniformitarianism were a real thing. But since we see sedimentary layers all over the world, that is a very strong indicator that deep time is an imaginative fabrication invented by those who are highly motivated to keep God out of their lives.

Original-Tissue Fossils

One of the things that has astonished paleobiologists is the discovery of soft tissue and undamaged strands of DNA in organisms that are assumed to be hundreds of thousands of years old, or even millions of years old.

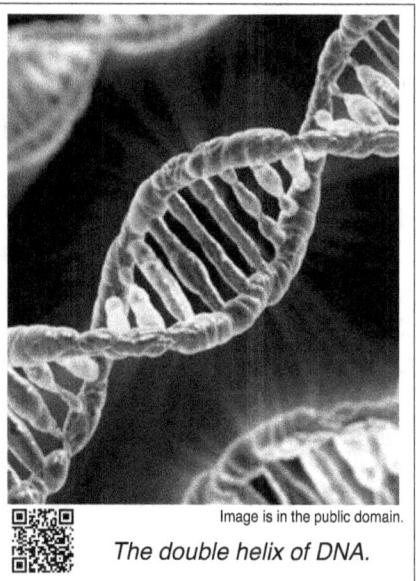

Image is in the public domain.
The double helix of DNA.

The problem is that DNA experts are adamant in claiming that DNA cannot exist in natural environments for longer than 10,000 years. So how can we be finding strands of DNA, *still intact*, in fossils claimed to be much older than 10,000 years? Fossils like the bones of Neanderthals and even dinosaurs, as well as insects preserved in amber.[543] Bacteria that evolutionists claim to be 250 million years

---

[543] Cherfas, J., "Ancient DNA: still busy after death," *Science* 253:1354–1356 (20 September 1991).

Cano, R. J., H. N. Poinar, N. J. Pieniazek, A. Acra, and G. O. Poinar, Jr. "Amplification and sequencing of DNA from a 120–135-million-year-old weevil," *Nature* 363:536–8 (10 June 1993).

Krings, M., A. Stone, R. W. Schmitz, H. Krainitzki, M. Stoneking, and S. Pääbo, "Neandertal DNA sequences and the origin of modern humans," *Cell* 90:19–30 (Jul 11, 1997).

Lindahl, T. "Unlocking nature's ancient secrets," *Nature* 413:358–359 (27 September 2001).

old have been revived with apparently no damage to their DNA.[544] Soft tissue and even blood cells from a *T. rex* fossil have flabbergasted paleontologists.[545]

Something is clearly wrong here. Either the deep-time age guesses are wrong, or the DNA cannot still exist. Yet we have found such DNA many times.

A team of researchers analyzed 158 leg bones of extinct moa birds that used to live on the South Island of New Zealand. Such a large sample size— *much* larger than paleoanthropologists get to work with—makes for unprecedented accuracy of the results.[546]

Allentoft and his team discovered that it takes only about 10,000 years for DNA to disintegrate so badly that DNA sequencers can't process it anymore. They also found that DNA degrades on a logarithmic scale, similar to the radioisotopes discussed earlier in the "Carbon-14" section. As a result, it makes sense to talk about the half-life of DNA. And what is the half-life of DNA? 541 years.[547]

Previous attempts to determine DNA's rate of decay have resulted in inconsistent results, in all likelihood because inconsistencies in location, the presence of water, other chemicals close by, and so forth. Much better results were obtained by Allentoft and his team because the bones were of the same kind, had similar burial conditions, and because they had such a large sample size.

This presents evolutionists with a difficult problem. This research was well done, using peer-reviewable methods, and a large sample size, while minimizing "contaminating" factors. The problem is that it yielded a result that conflicts with evolutionists' highest goal: preserving deep time, at all costs. It

---

[544] Vreeland, R. H., W. D. Rosenzweig, and D. W. Powers, "Isolation of a 250 million-year-old halotolerant bacterium from a primary salt crystal," *Nature* 407:897–900 (19 October 2000).

[545] Schweitzer, M., J. L. Wittmeyer, J. R. Horner, and J. K. Toporski, "Soft-Tissue vessels and cellular preservation in *Tyrannosaurus rex*," *Science* 207:1952–1955 (25 March 2005).

[546] Allentoft, M. E. et al. "The half-life of DNA in bone: measuring decay kinetics in 158 dated fossils." *Proceedings of the Royal Society B.* October 10, 2012. [Online at pubmed.ncbi.nlm.nih.gov/23055061; accessed June 30, 2020.]

[547] Specifically, this is their determined half-life for a 242-base-pair segment of mitochondrial DNA called the **control region**. The researchers calibrated this result using time in years from carbon dating the fossil bones. Although carbon dating is unreliable in older samples, it often provides reasonable age information for objects within the relatively recent time range of these moa bones. See Aardsma, G. A. 1989. "Myths Regarding Radiocarbon Dating." *Acts & Facts.* 18(3). [Online at ICR.org/article/myths-regarding-radiocarbon-dating/; accessed June 30, 2020.]

is their highest goal because if they don't preserve deep time, they'll be forced to acknowledge God in the equation, and that is unacceptable.

Surtsey, Iceland

One of the reasons that evolutionists dismiss the idea of a worldwide flood as preposterous superstition, is that they claim the ecosystem could not possibly have recovered from such an event that quickly. According to biblical chronology, Noah's flood happened around 4500 years ago. Is it possible for the world ecosystem to recover to its current state in only 4500 years?

Image courtesy Wikimedia Commons. Public Domain.

The 1963 volcano, off the coast of Iceland, that created the island of Surtsey.

The island of Surtsey, off the coast of Iceland, helps answer that question. Surtsey is an island that appeared when a volcanic eruption broke the surface of the ocean in 1963. While the vulcanism that created the island is also fascinating, the question at hand is: Has anything started to grow on this brand-new volcanic island?

Actually, yes. And quite a bit. Here is a quote from the Surtsey page of a UNESCO website, dated October 30, 2011:

> Surtsey, a volcanic island approximately 32 km from the south coast of Iceland, is a new island formed by volcanic eruptions that took place from 1963 to 1967. It is all the more outstanding for having been protected since its birth, providing the world with a pristine natural laboratory. Free from human interference, Surtsey has been producing unique long-term information on the colonisation process of new land by plant and animal life. Since they began studying the island in 1964, scientists have observed the arrival of seeds carried by ocean currents, the appearance of moulds, bacteria and fungi, followed in 1965 by the first vascular plant, of which there were 10 species by the end of the first decade. By 2004, they numbered 60 together with 75 bryophytes, 71 lichens and 24 fungi. Eighty-nine species of birds have been recorded on Surtsey, 57 of which breed else-

where in Iceland. The 141 ha island is also home to 335 species of invertebrates.[548]

(Note that the area of the island was expressed above in **hectares,** abbreviated "ha." This metric unit of area is equal to 10,000 m², or 2.471 acres. So 141 hectares is a bit more than 348 acres, or a bit more than half a square mile.)

Did you notice how many kinds of plants and animals were already on Surtsey in 2011, and even in 2004? The speed at which this new land, so recently nothing but red-hot magma, has developed an ecosystem is startling to many secular biologists. A writer in the May/June 2008 issue of *Science Illustrated* commented: "Surtsey always provides surprises. We discover about 20 new species each year."[549] That's not just 20 new organisms; that's 20 new *species* of organisms. Every year!

Image courtesy Michael F. Schönitzer (Creative Commons Attribution-Share Alike 2.0 Generic)

Surtsey Island as it appeared in 2011. Note the tiny research station at the top of the hill.

In the same article, it was noted that loose volcanic ash had already turned into **palagonite tuff,** a hard, glassy mineral. This was astonishing to geologists, because the process had been expected to take thousands of years. *But it only took a few decades.* How many other processes are a lot faster than we've been told?

So if all the above happened in only about fifty years, would the Earth's ecosystem have had any problem getting to where it is today, with 4500 years of growth? Apparently not.

---

[548] unesco-gforpcrossing.blogspot.com/2011/10/iceland-surtsey.html; accessed July 2, 2020.
[549] "An Island Laboratory." 2008. *Science Illustrated.* May/June, 42–47.

Mount St. Helens, United States

Since 1830, the idea of uniformitarianism—"The present is the key to the past," so passionately promoted by James Hutton and Sir Charles Lyell—has held sway in the field of geology. In geology, things just *don't* happen quickly. Everybody knows that.

But then on May 18, 1980, Mount St. Helens, in Washington state, erupted. At 8:32AM local time, a magnitude 5.1 earthquake occurred. Mount St. Helens shook, and one side of the mountain—about half a cubic mile of rock—slid away. Volcanic pressure was released, and superheated water instantaneously flashed into steam, releasing an explosion equivalent to a 20-megaton nuclear bomb. Within six minutes, 150 square miles of timber was flattened like toothpicks. And that was just the steam.

Image courtesy Austin Post and USGS. Public Domain.
Mount St. Helens erupting on May 18, 1980.

With the side of the mountain gone, and the steam explosion finished, the volcano itself erupted. North of the volcano, one-eighth of a cubic mile of rock slid into Spirit Lake, causing a gigantic water wave that sped across the lake and roared up the mountain on the opposite side, stripping trees from the mountainside as high as 850 feet above the normal water level.

In all, thirteen hundred feet of the peak of the volcano collapsed or blew outwards. The resulting avalanches of debris filled 24 square miles of the valley below. 250 square miles

Image courtesy Larry Moore and Creative Commons ShareAlike 3.0 Unported License
Spirit Lake, with a mat of logs and tree trunks stripped off the mountainside by the water wave powered by an eruption-caused landslide. This photo was taken five years after the eruption.

of nearby land—mostly recreation land, timber land, and private land—were damaged by St. Helens' lateral blast. An estimated 200 million cubic yards of material was deposited directly by **lahars** (volcanic mudflows) into nearby river channels, and the total amount of energy released that day—May 18, 1980—was equivalent to 400 million tons of TNT. That's about 20,000 Hiroshima-sized atomic bombs.

I was living in Fort Collins, Colorado, during the Mount St. Helens eruption. I remember it well: ash from the volcano in Washington state blew all the way to Colorado, causing sunrises and sunsets to be very orange, and leaving a fine layer of dust on everything. As a joke, I drew a smiley-face with my finger in the dust on the hood of a friend's car, not realizing how sharp and jagged those ash particles were. Without realizing it, I had *scratched* a smiley-face in the paint of his hood. I won't do that again. . .

On May 18, 1980, geologists and other scientists were given an enormous laboratory experiment sitting right in front of them. And over the following days, months, and years, many of them were stunned to realize that geological processes, far

Mount St. Helens, as shot from the International Space Station in 2017. Note the Log Mat, still covering a large part of Spirit Lake.

from requiring millions of years, could happen abruptly. Catastrophic occurrences in geology—or **catastrophism**, which creationists had been teaching all along—was a thing! A *real* thing! Geological processes, once thought to require long ages, happened within minutes, days, or weeks, right before their eyes.

Fast Deposition

In classical (uniformitarian) geology, laminated strata, or layers, were thought to take a great deal of time to form, with one layer accumulating per year, or maybe two thin ones, at best, and in other cases, thousands or millions of years. But with Mount St. Helens, up to 400 feet of strata have built up since 1980! The material for these strata were supplied by ash falling from the air, **pyroclastic** flows (rock blown out of a volcano), landslides, and stream water. Mount St. Helens trounced uniformitarianism by depositing a 25-foot-thick series of thin layers in just a few hours![550]

Further research has shown that such rapid deposition is the *usual* way for layers to form: not the exception, but the norm. Evolutionary thinking in geology, archaeology, paleontology, and other sciences have used the idea of very slow deposition of sediment settling out of unmoving water as "proof" of evolution. Many decades' worth of school students have been indoctrinated with the long-age mantra, but now that argument has fallen by the wayside as well. Empirical evidence since Mount St. Helens has revealed that laminated clays need energetic settings and moving water, but not long periods of time.[551]

Rapid Erosion

Another things that we've been told takes "millions and millions of years" is erosion. Rock is pretty hard stuff, so water digging out a canyon surely takes a really long time, right?

Or not. The North Fork of the Toutle River was blocked by almost a cubic mile of rock, mud, and debris from the 1980 eruption of Mount St. Helens, covering 23 square miles of the North Fork of the Toutle River Valley, up to 300 feet thick. The downstream flow of the Toutle River was completely cut off. But of course, the river kept flowing from upstream, so a large lake built up behind the new volcanically-caused dam. The multi-layered sediments from the eruption solidified into solid rock within two years.[552]

A little less than two years after the 1980 eruption, as the water level in the new lake approached the top of the debris dam, another minor eruption of Mount St. Helens melted a large quantity of snow and ice, causing a massive mudflow to rush into the growing lake and causing it to overtop (flow over) the dam from 1980's main eruption. The water and mud quickly eroded the dam, widening the gap, releasing more water, until a thunderous torrent was flowing downstream. In less than *nine hours*, the new lake was mostly emptied,

---

[550] Austin, S. A. 1986. "Mt. St. Helens and Catastrophism." *Acts & Facts.* 15 (7). [Online at ICR.org/article/mt-st-helens-catastrophism; accessed July 2, 2020.]

[551] Schieber, J., J. Southard, and K. Thaisen. 2007. "Accretion of Mudstone Beds from Migrating Floccule Ripples." *Science.* 318 (5857): 1760–1763. [Online at Science.ScienceMag.org/content/318/5857/1760.full; accessed July 2, 2020.]

[552] Brian Thomas, June 30, 2015, Acts & Facts [Online at ICR.org/article/monuments-catastrophe-from-mount-st; accessed September 22, 2020.]

its roaring water having carved out a new canyon—140 feet deep, 1,000 feet wide, and 2,000 feet long.[553,554] *In nine hours!*

Like the real Grand Canyon, this "Little Grand Canyon" shows hundreds of sedimentary layers in the walls of the newly-cut canyon. Where did they come from? They came from the volcanic action that had taken place in 1980—less than two years before. These two events show how fast water can dig a canyon.[555]

If water can dig such a canyon up to 140 feet deep in a single day, could it be that the real Grand Canyon may have taken less than the "millions and millions of years" we've been told all our lives? That is an excellent question, and it will be covered later in this chapter.

So how fast does erosion happen in "normal" cases, where there's not a flood going on? That's another excellent question, and it's another serious problem for deep time. With the observed non-exceptional erosion rates, all the continents should have been eroded into the sea, many times over, by now.[556] In fact, as mentioned earlier in the section entitled "That Pesky Erosion," one recent study confirmed that the average erosion rate of rock that is exposed to the elements is about 40 feet every million years.[557] Doing the math, most continents would be completely eroded in less than 50 million years, yet they're still here. Something—or perhaps I should say, yet *another* something—about deep time doesn't make sense.

Restoration of Flora and Fauna

How is the area around Mount St. Helens recovering? Have any plants or animals started to come back? Yes, and with a vengeance.

Right after the eruption, it was indeed bleak:

> When Mount St. Helens erupted in 1980, it destroyed every living thing around it. Gas, ash and rock, heated to over

---

[553] Eric Hovind [Online at CreationToday.org/mount-st-helens—a-model-of-the-grand-canyon; accessed September 22, 2020.]

[554] Steven Austin, July 1, 2008, Acts & Facts [Online at ICR.org/article/red-rock-pass-spillway-bonneville-flood; accessed September 22, 2020.]

[555] Austin, S. A. 1986. "Mt. St. Helens and Catastrophism." *Acts & Facts.* 15 (7). ICR.org/article/mt-st-helens-catastrophism; accessed July 3, 2020.]

[556] Blatt, H., G. Middleton, and R. Murray. 1980. *Origin of Sedimentary Rocks,* 2nd ed. Englewood Cliffs, NJ: Prentice-Hall, Inc.

[557] Portenga, E. W. and R. R. Bierman. 2011. "Understanding Earth's eroding surface with ^{10}Be." *GSA Today.* 21 (8): 4–10. [Online at GeoSociety.org/gsatoday/archive/21/8/article/i1052-5173-21-8-4.htm; accessed July 3, 2020.]

1000 degrees Fahrenheit, sterilized a 60-kilometer square area, leaving a gray lunar-looking landscape devoid of plants and animals. Within a year, the first plant life had started to return, just as ecologists predicted it would.[558]

So the above was "within a year." Soon after, other plants recovered quickly—even in lahars (mudflows)—and recovery speed was influenced by what other plants were closeby:

> ...there are striking differences—the forest-surrounded lahar has recovered much faster and has pines and firs atop it, while the more isolated lahar is still mostly covered by grasses, early-stage colonizers.[559]

Today, much of the area around Mount St. Helens has recovered nicely. Much of the old debris is still there, but much new vegetation has covered the once-seared landscape. After 20 years, most of the animals had returned, and today, more than 40 years after the eruption, many places have lush trees again.

Another name for uniformitarianism is **actualism,** as if "the way things *actually* happened was slowly and steadily over the eons." But now geology is faced with undeniable evidence that catastrophism *actually* happens too, and is a major contributor to Earth's current geologic state.

---

[558] "Mount St. Helens Recovery Slowed By Caterpillar." University of Maryland, College Park. *ScienceDaily*. [Online at ScienceDaily.com/releases/2005/11/051116085634.htm; accessed July 3, 2020.]

[559] Thompson, A. "Mount St. Helens Still Recovering 30 Years Later." *Live Science*. May 17, 2010. [Online at LiveScience.com/6450-mount-st-helens-recovering-30-years.html; accessed July 3, 2020.]

Arches National Park

In eastern Utah is the Arches National Park, which offers some of the most beautiful and impressive natural sandstone arch formations in the world—over 2000 of them. If you are a frequent visitor to this park, you may hear from time to time about a rock arch that has collapsed, just from normal erosion and weathering. But that brings up a good point: how often do they collapse? At what rate are these arches collapsing?

Image courtesy Creative Commons CC BY-SA 3.0

Delicate Arch at sunset, in Arches National Park in Utah.

A fascinating article[560] looks into that very question. Between 1970 and 2015 (when the article was written), 43 arches collapsed in the park. That's 43 arches in 45 years—almost one every year! At that rate, they would all be gone in around 2000 years. But the official story from the evolution-informed park service is that the features are five million years old.

Do you see the problem? If the arches are collapsing at a rate of about one per year, how are *any* of them still here? Remember, evolutionary thinking is very passionate about uniformitarianism, where processes continue at a pretty constant rate over enormous periods of time, so evolutionists would not be in favor of only a few arches falling down in the first 4,998,000 years of their supposed lifespans, and then suddenly accelerating to the observed rate of almost one per year just in the last century. But if, as uniformitarianism demands, they had been falling down all along at the currently-observed rate of about one per year, they would all have been gone eons ago! However, more than 2000 of them are still here, falling down at a rate of about one per year. This is another observation that evolution can't explain, another place where the worldview of evolution disagrees with repeatable observation. Which one are we going to trust?

Other Fast Processes

Most of us who have been educated in public schools were indoctrinated with the idea of geological processes taking "millions and millions of years"—

---

[560] David Catchpoole, "A Dangerous View," *Creation* 37(2):12–15, May 2015. [Online at Creation.com/a-dangerous-view; accessed May 30, 2022.]

it is definitely one of the most sacred articles of faith in all of evolutionism. But as we'll see again in this section, we have been sold a bill of goods.

Rock Formation

How long does it take for rocks to form out of mud and sand? According to evolution, "it takes millions and millions of years!" But does it really?

- Near the South Jetty at Westport, Washington, a clock, embedded in rock, was found in 1975.[561] How likely is it that the clock was "millions of years" old?
- In 1998, in Ventura Country, California, a marine spark plug (estimated to be from the 1950s) was found encased in rock.[562] How likely is it that the spark plug was "millions of years" old?
- In 1852, a wooden ship named *Isabella Watson* sank near Victoria, Australia. After being underwater for less than 150 years, its bells were discovered firmly cemented in a rock matrix.[563] How likely is it that this ship's bells were "millions of years" old?
- The famous Petrifying Well in Knaresborough, 13 miles west of York, England, has been attracting tourists since the 1600s. Its claim to fame? The conditions are such that teddy bears dangled in the waterfall there become encased in stone in *three to five*

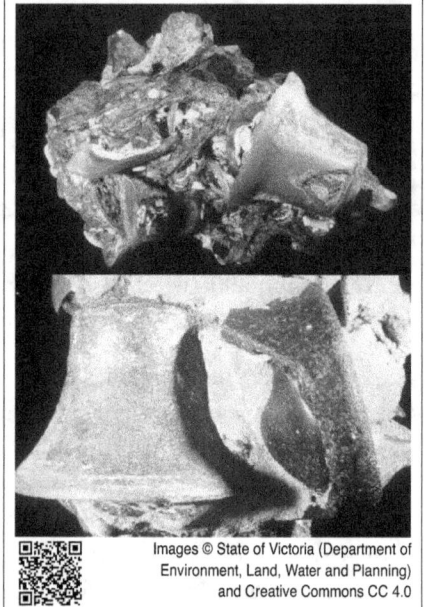

Images © State of Victoria (Department of Environment, Land, Water and Planning) and Creative Commons CC 4.0

Isabella Watson's bells, encased in rock, after less than 150 years underwater.

---

[561] "The Clock in a Rock," *Creation*, 19, no 3 (June 1997). [Online at AnswersInGenesis.org/geology/catastrophism/the-clock-in-the-rock; accessed August 4, 2020.]

[562] "Sparking Interest in Rapid Rocks," *Creation*, 21, no 4 (September 1999). [Online at AnswersInGenesis.org/geology/geologic-time-scale/sparking-interest-in-rapid-rocks; accessed August 4, 2020.]

[563] "Bell-ieve It: Rapid Rock Formation Rings True," *Creation*, vol. 20, no. 2, March 1998: p. 6. [Online at AnswersInGenesis.org/geology/catastrophism/bell-ieve-it-rapid-rock-formation-rings-true; accessed August 4, 2020.]

months![564] How likely is it that the teddy bears that people hang there are "millions of years" old?

Clearly, evolutionists' religious conviction that it takes "millions and millions of years" for rock to form is not supported by observations. It does *not* take nearly that much time; it just requires the proper conditions.

Coal and Oil

How long does it take for coal to form out of biomatter? According to evolution, "it takes millions and millions of years!" But does it really?

Image by Sarah Stierch, Creative Commons CC BY-SA 4.0

Coal.

- The Argonne National Laboratory has shown that, far from the "millions of years" we've all heard it takes for coal to form, they've done it in less than a year. They used wood, water, and acidic clay, heated to about 300°F for about 1–9 months and it formed black coal.[565] Note that none of these conditions is uncommon in nature.

- Peat has also been turned into coal, using peat from Indonesia, the Everglades in Florida, Okefenokee Swamp in Georgia, and other places.[566]

---

[564] Monty White, "The Amazing Stone Bears of Yorkshire," *Answers in Genesis,* June 1, 2002. [Online at AnswersInGenesis.org/articles/cm/v24/n3/stone-bears; accessed August 4, 2020.]

[565] R. Hayatsu, R. L. McBeth, R. G. Scott, R. E. Botto, R. E. Winans, *Organic Geochemistry,* vol. 6 (1984), p. 463–471.

[566] W. H. Orem, S. G. Neuzil, H. E. Lerch, and C. B. Cecil, "Experimental Early-stage Coalification of a Peat Sample and a Peatified Wood Sample from Indonesia," *Organic Geochemistry* 24(2):111–125, 1996;

A. D. Cohen and A. M. Bailey, "Petrographic Changes Induced by Artificial Coalification of Peat: Comparison of Two Planar Facies (Rhizophora and Cladium) from the Everglades-Mangrove Complex of Florida and a Domed Facies (Cyrilla) from the Okefenokee Swamp of Georgia," *International Journal of Coal Geology* 34:163–194, 1997;

S. Yao, C. Xue, W. Hu, J. Cao, C. Zhang, "A Comparative Study of Experimental Maturation of Peat, Brown Coal and Subbituminous Coal: Implications for Coalification," *International Journal of Coal Geology* 66:108–118, 2006.

And how long does it take for oil to form out of biomatter? Like the coal above, it takes (according to evolution) "millions and millions of years!" But. . .

- Oil can be created in much less than millions of years; the entire process takes only two hours:

    > Turkey and pig slaughterhouse wastes are daily trucked into the world's first biorefinery, a thermal conversion processing plant in Carthage, Missouri. On peak production days, 500 barrels of high-quality fuel oil **better than crude oil** are made from 270 tons of turkey guts and 20 tons of pig fat.[567]

- Oil can also be made from algae in less than an hour:

    > Researchers at the Pacific Northwest National Laboratory (PNNL) in Washington State have pioneered a new technology that makes diesel fuel from algae—and their cutting-edge machine produces the fuel in just minutes. . . Simply heat pea-green algal soup to 662°F (350°C) at 3,000 psi for almost 60 minutes.[568]

Crude oil.

Note that such pressures and temperatures are not at all uncommon beneath the surface of the earth.

There are other examples as well. The point is that it simply *doesn't* take millions of years to produce coal or oil—it takes less than a year, or even less than an hour; this has been shown repeatedly. The hard part is proving that it can happen over millions of years.

---

[567] Dr. Andrew Snelling, "The Origin of Oil," Answers 2, no. 1, December 27, 2006, p. 74–77. [Online at AnswersInGenesis.org/geology/the-origin-of-oil; accessed August 4, 2020.] Emphasis added.

[568] Brian Thomas, "One-Hour Oil Production," ICR, January 13, 2014. [Online at ICR.org/article/7874; accessed August 4, 2020.]

Cave Formations

If you've ever taken a tour through a beautiful cave, the tour guide probably told you not to touch the **speleothems** (cave formations), although he may not have used that term. The reasoning usually given is that it takes "millions of years" for them to form. Here's another case of an evolutionary belief system that is demonstrably wrong.

Image courtesy Dave Bunnell/Under Earth Images and Creative Commons CC BY-SA 2.5

The six main types of speleothems.

- In the early 1950s, a worker left a lemonade bottle in the Jenolan Caves in New South Wales, Australia. By the 1980s, the bottle was coated in a 3mm layer of calcite—about one millimeter per decade—and continues to grow.[569]
- Dr. Emil Silvestru, a cave geologist, comments:
  > In the Cripple Creek Gold Mine in Colorado, stalagmites and stalactites over a meter long have grown in less than 100 years![570]
- Jerry Trout is a cave specialist with the United States Forest Service, and he says that he has watched, and has photographic evidence, that a stalactite can grow several inches in a matter of days.[571]
- Bodie Hodge, while touring the Cosmic Caverns in northwest Arkansas, stepped around a large stalagmite—about three feet tall—right in the middle of the walking path. The tour guide said that it had formed the previous year (2011) because of unusually heavy rain.[572]

---

[569] Editors, "Bottle Stalagmite," *Answers in Genesis,* March 1, 1995, originally published in Creation, vol. 17, no. 2, March 1995:6. [Online at AnswersInGenesis.org/geology/caves/bottle-stalagmite; accessed August 4, 2020.]

[570] Emil Silvestru, *The Cave Book* (Green Forest, AR: Master Books, 2008), p. 46.

[571] Marilyn Taylor, "Descent," *Arizona Highways,* January 1993, p. 11.

[572] Ham, Ken and Hodge, Bodie: *A Flood of Evidence: 40 Reasons Noah and the Flood Still Matter* (Master Books, Green Forest, Arkansas, 2016), p. 126.

- In a small English village named Salterforth is a quaint old establishment called the Anchor Inn, which contains the Canalside Pub and Eatery. Originally built in 1655, this structure was rebuilt in 1795, when the nearby Leeds and Liverpool Canal was constructed. Since then, in the cellar of the pub, hundreds of "straws"—thin, hollow stalactites—have formed from mineral-laden water dripping from the 12-foot ceiling.

Limestone stalactites, in the cellar of the Canalside pub in the Anchor Inn.

Notice the brick wall behind the stalactites: imagine trying to build that wall (or any of the other ones) *after* the stalactites were formed, without breaking any of them! This shows that the stalactites formed after the walls were constructed. But the fact that the stalactites exist at all *in a man-made structure* demonstrates they don't take as long as we've been told. Many of these have reached the floor already, spanning the entire 12-foot distance from ceiling to floor in the roughly 225 years since the pub was rebuilt! That averages out to almost $2/3$ of an inch *per year!*[573] Unless, of course, we believe the deep-time mantra, in which case the pub must be several million years old. . .

Short, stubby stalagmites that formed underneath the stalactite straws growing from the cellar ceiling in the Anchor Inn's Canalside Pub.

---

[573] Gavin Cox, "Defying Deep-Time Dogma: Stunning Stalactites in a Pub Cellar," *Creation* 42(4), pp. 12–14, October 2020. [Online at Creation.com/stalactites-in-a-pub-cellar; accessed November 21, 2021.]

And there are numerous other examples of stalagmites, stalactites, and other speleothems forming in only a few decades.[574,575,576,577] Millions of years to form speleothems? Nah. . .

Gemstones

How long does it take to form diamonds or other precious stones? Millions of years? Evolutionists seem to think so, but let's look into it.

Image by Mauro Cateb (Creative Commons Attribution-Share Alike 2.0 Generic)

Assorted gemstones.

- Diamonds are constantly being made in the laboratories of Apollo Diamond Incorporated in Massachusetts, using chemical vapor deposition.[578] This process only takes a few days.
- Gemesis, in Sarasota, Florida, takes a different approach: high temperature and pressure, so this method is more similar to the conditions in which natural diamonds form underground. In this process, graphite—a form of carbon—breaks down into atoms and travels through a metal solvent to bond to a tiny diamond seed, crystallizing layer by layer. Three or four days later, the resulting stone is cut and polished in the usual manner.[579] Like the previous process, the resulting stones are scientifically, chemically, and aesthetically identical to naturally-formed diamonds.

---

[574] Stephen Meyers and Robert Doolan, "Rapid Stalactites," *Creation* 9(4):6–8, September 1987. [Online at Creation.com/rapid-stalactites; accessed June 2, 2022.]

[575] Don Batten, "'Instant' stalagmites!" *Creation* 19(4):37, September 1997. [Online at Creation.com/instant-stalagmites; accessed June 2, 2022.]

[576] Michael Oard, "Rapid Cave Formation by Sulfuric Acid Dissolution," Journal of Creation 12(3):279–280, December 1998. [Online at Creation.com/rapid-cave-formation-by-sulfuric-acid-dissolution; accessed June 2, 2022.]

[577] Dave Woetzel, M.S., and Brian Thomas, Ph.D., "Carlsbad Caverns National Park: Fast Formations," October 29, 2021. [Online at ICR.org/article/carlsbad-caverns-national-park-fast-formations; accessed June 2, 2022.]

[578] Greg Hunter and Andrew Paparella, "Lab-Made Diamonds Just Like Natural Ones," September 9, 2015, ABC News Internet Ventures, produced for *Good Morning America*. [Online at ABCNews.go.com/GMA/story?id=124787; accessed August 4, 2020.]

[579] Greg Hunter and Andrew Paparella, "Lab-Made Diamonds Just Like Natural Ones," September 9, 2015, ABC News Internet Ventures, produced for *Good Morning America*. [Online at ABCNews.go.com/GMA/story?id=124787; accessed August 4, 2020.]

- You can even turn the cremated remains of your loved ones into diamonds. LifeGem (LifeGem.com) and Algordanza (MyMemorialDiamond.com) both will do this service for you, and it likely will take less than a few million years.
- Opals supposedly take millions of years to form, but again, it's not true. Andrew Snelling remarks:

    > A committed Christian, Len [Cram] has discovered the secret that has enabled him to actually "grow" opals in glass jars stored in his wooden shed laboratory, and the process takes only a matter of weeks![580]

- Other gemstones—rubies, garnet, topaz, emeralds, etc.—are made all over the world. Rubies have been grown since the late 1800s, and in the 1930s, Germany was using a process called the "flux-grown method" to grow emeralds.[581]
- Volcanoes often quickly and naturally produce gemstones. For example, when Mount St. Helens erupted in 1980 and 1982, numerous precious stones were produced, which are on sale in the Gift Shop. On the Mount St. Helens Gift Shop's website is this statement:

    > Volcanoes are an incubator for many of the world's treasures. Other gems commonly associated with volcanic origins include emerald, diamond, garnet, peridot, and topaz.[582]

Again, the point is that creating gemstones, like the other processes mentioned above, *does not take that long,* despite the apparent juggernaut of deep time that evolution peddles so strongly.

---

[580] Dr. Andrew Snelling, "Creating Opals," *Creation Ex Nihilo* 17, no. 1, Dec. 1994: p. 14–17. [Online at AnswersInGenesis.org/geology/rocks-and-minerals/creating-opals; accessed August 5, 2020.]

[581] Editors, "Synthetic and Artificial Gemstone Growth Methods: In-Depth Treatment Information," Jewelry Television online, July 2012. [Online at JTV.com/library/article/gemstone-natural-artificial; accessed August 5, 2020.]

[582] Editors, Mount St. Helens Gift Shop Website. [Online at Mt-St-Helens.com/obsidianite.html; accessed August 5, 2020.]

Petrification

When you hear of a petrified forest, or petrified wood, does the phrase "millions and millions of years" also come to mind? If so, the deep-time marketing department has done its job very effectively. But maybe we should check, just to make sure such things really *do* take millions and millions of years. Because it's always possible that they don't.

Image by Andrew Kearns
(Creative Commons CC BY-SA 2.0)

Logs in the Petrified Forest National Park.

- In Spray Mine, in Australia, a miner's lost hat completely turned to stone in only 50 years, and is now in a mining museum in Tasmania. In New Zealand, a bowler hat, a ham, and a bag of flour buried in ash from the Mt. Tawawera volcanic eruption in 1886, turned to stone in only 20 years! [583]

- In the Minnesota Museum of Mining in Chisholm, Minnesota, one of the displays is of a bundle of candles used by miners to see in the dark mines in years gone by. The fascinating thing is that these candles are completely petrified! The candles, left behind when the mine shut down in the early 1940s, were surrounded by mineralized water, which infused the candles, turning them to stone in only a few decades. [584] I'm pretty sure that the candles were not left in the mine "millions of years ago," aren't you?

- In 1986, a man named Hamilton Hicks was awarded a patent from the US Patent Office for creating a process that petrifies wood in a way similar to naturally occurring petrified wood. Using minerals and an acidic solution, both of which are often found near volcanoes, wood can be petrified. [585] Note: it doesn't take millions of years.

- Researchers at the Department of Energy are using a petrification process using temperatures and environmental conditions comparable to those near a volcanic eruption, to create petrified wood in a few

---

[583] John Mackay, "Fossil Bolts and Fossil Hats," *Creation Ex Nihilo,* vol. 8, Nov., p. 10. [Online at CreationMoments.com/sermons/the-fossilized-hats; accessed August 4, 2020.]

[584] Jonathan O'Brien, "Candles Turned to Stone," *Creation* 42(2), p. 56, April 2020. [Online at Creation.com/candles-turned-to-stone; accessed November 22, 2021.]

[585] Hamilton Hicks, Sodium silicate composition, United States Patent Number 4,612,050, September 16, 1986. [Online at google.com/patents/US4612050; accessed August 5, 2020.]

days.⁵⁸⁶ In the article, they say that nature takes "millions of years" to accomplish this, but that has not actually been observed, so it's just an evolutionary religious belief. We *do* know, however, that it can happen in a few days; that *has* been observed.

- Petrification of wood has been observed in the alkaline springs of Yellowstone National Park, at rates of up to 4mm per year.⁵⁸⁷
- The chapel of Santa Maria of Health (Santa Maria de Salute) in Venice, Italy, was built in 1630 to celebrate the end of the Black Plague. Venice is built on water-saturated sand and clay, so the massive stone-block chapel was constructed on 180,000 wooden pilings as reinforcements for the foundations. Still solid after all these years, how have the wooden pilings lasted this long? It's simple: they petrified. The chapel now rests on stone pilings.⁵⁸⁸

Petrification of animal matter, more commonly known as fossilization, is also claimed to take millions and millions of years. But if you think about it, if the process were really that slow, the animal's carcass—even the bones—would have completely rotted away eons before it could petrify. Fossils require fast burial, like a flood, to keep oxygen out and prevent decay.

Animal fossils, even of soft tissue, can be formed within a few weeks, as the Ediacara formation in Australia shows us: *millions* of jellyfish and other marine life, fossilized in great detail, in a deposit 300 miles in length. Since jellyfish decay into shapeless blobs within hours of being stranded on land, we know that this burial was *very* fast.

Reporters for the *New York Times*' Science Watch report:

> Scientists have for the first time produced fossils of soft animal tissues in a laboratory. In the process they discovered that **most of the phosphate required for the fossilization of small animal carcasses comes from within the animal itself.**⁵⁸⁹

---

⁵⁸⁶ Editors, "Petrified Wood in Days," Phys.org, January 25, 2005. [Online at Phys.org/news/2005-01-petrified-wood-days.html; accessed August 5, 2020.]

⁵⁸⁷ Andrew Snelling, "'Instant' Petrified Wood," *Creation*, vol. 17, no. 4, September 1995, p. 38–40, September 1, 1995. [Online at AnswersInGenesis.org/fossils/how-are-fossils-formed/instant-petrified-wood; accessed August 5, 2020.]

⁵⁸⁸ Segment on 'Burke's Backyard,' Channel 9 TV, Sydney, June 1995. Quoted in Andrew Snelling, "'Instant' Petrified Wood," *Creation*, vol. 17, no. 4, September 1995, p. 38–40, September 1, 1995. [Online at AnswersInGenesis.org/fossils/how-are-fossils-formed/instant-petrified-wood; accessed August 5, 2020.]

⁵⁸⁹ "Man-Made Fossils," *New York Times*, Science Watch, March 9, 1993. [Online at NYTimes.com/1993/03/09/science/science-watch-man-made-fossils.html; accessed August 5, 2020.]

Fossils have even been made *in a single day* by simulating conditions that would be expected in a rapid sediment-deposition setting (say, for example, a worldwide flood).[590]

In fact, induced fossilization is becoming *so* fast and *so* easy nowadays, that another problem is arising—faked fossils.[591] Read what this paleontologist says:

> Red-faced and downhearted, paleontologists are growing convinced that they have been snookered by a bit of fossil fakery from China. **The "feathered dinosaur"** specimen that they recently unveiled to much fanfare apparently **combines the tail of a dinosaur with the body of a bird**, they say. "It's the craziest thing I've ever been involved with in my career," says paleontologist Philip J. Currie of the Royal Tyrrell Museum of Paleontology in Drumheller, Alberta.[592]

And actually, it's worse than just "a bit" of fossil fakery from China; it's *much* worse:

> Another much more serious problem, however, is posed by forged, faked and manipulated specimens—such as *National Geographic*'s Archaeoraptor—which are becoming increasingly common. . . . The problem of faked fossils in China is serious and growing. . . . An investigative report published in *Science* in 2010 revealed that as many as **80 percent of marine reptile fossils on display in Chinese museums had been altered or manipulated**.[593]

Petrification and fossilization: two more "millions of years" theories bite the dust. . .

Emphasis added.

[590] Ron Neller, "Fossils in a Day?" *Creation* 41(3):46–47, July 2019. [Online at Creation.com/fossils-in-a-day; accessed June 10, 2022.]

[591] Randy J. Guliuzza, "Major Evolutionary Blunders: The Imaginary Archaeoraptor" August 31, 2016. [Online at ICR.org/article/major-evolutionary-blunders-imaginary; accessed June 3, 2022.]

[592] Monastersky, R. 2000. "All Mixed Up Over Birds and Dinosaurs." *Science News*, 157 (3): 38. [Online at ScienceNews.org/article/all-mixed-over-birds-and-dinosaurs; accessed June 3, 2022.]

[593] John Pickert, "How Fake Fossils Pervert Paleontology," *Scientific American,* November 15, 2014. [Online at ScientificAmerican.com/article/how-fake-fossils-pervert-paleontology-excerpt; accessed August 5, 2020.] Emphasis added.

Glaciers

Glaciers, too, are considered by secular scientists to be very slow-forming objects. Wikimedia states that "Glaciers are a valuable resource for tracking climate change over long periods of time because they can be hundreds of thousands of years old."[594]

How long does it take glaciers to form? National Geographic says that they formed "over many centuries," and that most of them are left over from the Ice Age, which "ended more than 10,000 years ago."[595] The Encyclopedia Britannica says that in cold climates, the process of changing snow to glacial ice "takes several thousand years and burial to depths of about 150 metres."[596]

But why do we think that? Is this another "fact" that is based only on the assumed long-age time spans that are required by the presupposition of evolution? It looks that way, based on the discovery of the "Lost Squadron" of World War II.

On July 15, 1942, a squadron of six P-38 "Lightning" fighter planes and two B-17 "Flying Fortress" bombers took off from a Army air base in Greenland, heading for Britain. They planned on refueling in Iceland, but a severe blizzard made landing there impossible. They turned around to return to their Greenland base before their fuel ran out, but that base, too, was closed because of the blizzard. They had no choice but to crash-land on the snow and ice of Greenland's eastern coast, most of them doing belly landings. Fortunately, all the crew members survived and were rescued by dogsled nine days later, but the planes were abandoned.[597]

---

[594] "Glacier" [Online at en.wikipedia.org/wiki/Glacier#Climate_change; accessed April 30, 2022.]

[595] "Glacier" [Online at NationalGeographic.org/encyclopedia/glacier; accessed April 30, 2022.]

[596] "Formation and Characteristics of Glacier Ice" [Online at Britannica.com/science/glacier/Heat-or-energy-balance; accessed April 30, 2022.]

[597] Carl Wieland, "The lost squadron: Deeply buried missing planes challenge 'slow and gradual' preconceptions" *Creation* 19(3):10–14, June 1997. [Online at Creation.com/the-lost-squadron; accessed May 1, 2022.]

Fast-forward to 1988. Using ice-penetrating radar, a salvage expedition located the planes beneath the ice. What was astonishing to the members of the expedition is that the planes were 75 meters down—under 250 feet of ice! Why were they so astonished? Because "everybody knows" that ice takes thousands of years to build up to those depths. But here, in only 46 years, 250 feet of ice had built up! More than five feet per year!

Some people, in an effort to rescue evolution from yet more contrary evidence, suggested that the planes sank through the ice, and that the ice didn't really build up that fast. But airplanes, in order to maximize stability during flight, have their centers of mass (primarily engines) toward the front. This means that if the planes had actually sunk through the ice, the noses would have sunk faster, being heavier. Thus, they would have all been pitched forward, nose down, when they were uncovered. They weren't—they were all still as horizontal as they day they belly-landed.

So it looks like observations and measurements of glacial ice, like the other fast processes already mentioned, are loudly objecting to the assumptions of the very slow buildup that evolution desperately needs.

Coral Reefs

Coral reefs are another item in the deep-time "millions and millions of years" category. At least, according to evolution. Let's check into that. . .

Ohio's Cedarville University geologist and professor Dr. John Whitmore writes:

Image by Chris Menzel from Pixabay

A coral reef, with some of its attendant marine life.

> Corals which build coral reefs have been reported to grow as much as 99 to 432mm per year.[598] Large coral accumulations have been found on sunken World War II ships only after several decades.[599] *Acropora* colonies have reached 60–80cm in diameter in just 4.5 years in some ex-

---

[598] A. A. Roth, *Origins* (Hagerstown, MD: Review and Herald Publishing Association, 1998), p. 237.

[599] S. A. Earle, "Life springs from death in Truk Lagoon," *National Geographic* 149(5):578–603, 1976. [Online at NationalGeographicBackIssues.com/national-geographic-may-1976; accessed August 5, 2020.]

perimental rehabilitation studies.[600] At the highest known growth rates, the Eniwetok Atoll (the thickest known reef at 1400m) would have taken about 3,240 years to rise from the ocean floor.[601]

So all of these geological processes, which we've been told take "millions and millions of years," don't. Good to know.

The Grand Canyon

The Grand Canyon is a canyon in Arizona that was carved out by the Colorado River. It is about 275 miles long, 18 miles wide in some places, and up to a mile deep. It is a beautiful canyon, as you can see in this photo.

Image courtesy Pixabay.com.

View of the Grand Canyon

The Grand Canyon is often held up as a poster child of evolution, because of the enormous amounts of time supposedly required to dig out such a geological feature. The enormous amounts of time are absolutely essential to a belief in the theory of evolution, because if creatures appeared too suddenly, or if processes happened too quickly, that would be too obvious of a pointer toward a Creator. The thought goes, "With enough time, anything is possible." But as we saw in the previous chapter, even the entire alleged age of the planet is utterly insufficient to evolve even the simplest of organisms, let alone the incredible complexity we observe everywhere in the world.

The belief that the Grand Canyon supports evolution's required enormous timeframe seems to be ubiquitous among proponents of evolution. For example:

---

[600] H. E. Fox, "Rapid Coral Growth on Reef Rehabilitation Treatments in Komodo National Park, Indonesia," *Coral Reefs* 24:263, 2005. [Online at ResearchGate.net/publication/248159460_Rapid_coral_growth_on_reef_rehabilitation_treatments_in_Komodo_National_Park_Indonesia; accessed August 5, 2020.]

[601] John Whitmore, "Aren't Millions of Years Required for Geological Processes?" in *The New Answers Book 2,* Ken Ham, gen. ed. (Green Forest, AR: Master Books, 2008), p. 240–241.

- "Nearly two billion years of Earth's geological history have been exposed as the Colorado River and its tributaries cut their channels through layer after layer of rock while the Colorado Plateau was uplifted."[602]
- ". . .several recent studies support the hypothesis that the Colorado River established its course through the area about 5 to 6 million years ago."[603]
- "A new theory published in 2011 contends that this ancient northeast river system carved the Grand Canyon as early as 70 million years ago."[604]
- "Some scientists believe that the Grand Canyon is 70 million years old. Others contend that the natural wonder is only between five and six million years old. Both are right."[605]
- "Writing in the journal Science, they [Rebecca Flowers of the University of Colorado and Brian Wernicke of the California Institute of Technology] claimed that an ancient river flowing west to east carved the western part of the Grand Canyon almost to modern depths some 70 million years ago, during the Late Cretaceous period."[606]

. . .and on and on. Whether the answer is only 5 or 6 million years, 70 million years, or two billion years, there seems to be a consensus that it had to have taken a Really Long Time™. After all, that's an awful lot of rock that

---

[602] Wikipedia article "Grand Canyon" (accessed December 8, 2019), quoting "Geologic Formations of the Grand Canyon," Archived April 3, 2007, at the Wayback Machine National Park Service, Retrieved November 17, 2009.

[603] Wikipedia article "Grand Canyon" (accessed December 8, 2019), quoting:
- Karlstrom, Karl E.; Lee, John P.; Kelley, Shari A.; Crow, Ryan S.; et al. (2014). "Formation of the Grand Canyon 5 to 6 million years ago through integration of older palaeocanyons". *Nature Geoscience*. 7 (3): 239–244;
- Darling, Andrew; Whipple, Kelin (2015). "Geomorphic constraints on the age of the western Grand Canyon". *Geosphere*. 11 (4): 958–976; and Spencer, J. E.; Patchett, P. J.; Pearthree, P. A.;
- House, P. K.; Sarna-Wojcicki, A. M.; Wan, E.; Roskowski, J. A.; Faulds, J. E. (2013). "Review and analysis of the age and origin of the Pliocene Bouse Formation, lower Colorado River Valley, southwestern USA". *Geosphere*. 9 (3): 444–459.

[604] Article "How Old Is the Grand Canyon?" at Geology.com/articles/age-of-the-grand-canyon.shtml, accessed December 8, 2019.

[605] Article "The Real Age of the Grand Canyon" at ExploreTheCanyon.com/the-real-age-of-the-grand-canyon sponsored by the Grand Canyon Visitor Center. Accessed December 8, 2019.

[606] Article "How Old is the Grand Canyon? That Depends, Scientists Say" at History.com/news/how-old-is-the-grand-canyon-that-depends-scientists-say.

had to be carried away by one little river. How many millions of years must it have taken?

There is a rather simple way to answer this question, and all we need are three quantities:

- The volume of the Grand Canyon,
- The flow rate of the Colorado River, and
- The percentage of sediment of the Colorado River.

Let's discuss these three quantities below, and then do some math on them.

The Volume of the Grand Canyon

So what is the volume of the Grand Canyon? In other words, how much rock had to be removed to leave a canyon the size of the Grand Canyon?

Because of the irregular shape and numerous tributaries of the Grand Canyon, there's not an easy formula into which you can plug in some values. However, by estimating the volumes of many small sections of the canyon, and adding them together, we can come up with a reasonable ballpark figure.

Here are some estimates of the volume of the Grand Canyon:

- This estimate, from the National Park Service, is 5.45 trillion cubic yards.[607] Since there are 1760 yards in a mile, we can calculate the number of cubic miles:

$$V \approx 5.45 \times 10^{12} \, yd^3$$
$$\approx 5.45 \times 10^{12} \div 1760^3 \, mi^3$$
$$\approx 1000 \, mi^3$$

So according to this estimate, the volume of the Grand Canyon is about 1000 cubic miles.

- Another estimate, from Yahoo!Answers,[608] calculates the volume by multiplying its length, average depth, and average width:

$$V \approx \text{Length (277 mi, or 1,462,560 ft)} \times$$
$$\text{Average depth (5000 ft)} \times$$
$$\text{Average width (10 mi, or 52,800 ft)}$$
$$\approx 386 \times 10^{12} \, ft^3$$

---

[607] nps.gov/grca/learn/management/statistics.htm#geography.
[608] Article "Volume of the Grand Canyon?" at answers.yahoo.com/question/index?qid=20130605073302AASGcuy.

$\approx 386 \times 10^{12} \div 5280^3 \text{ mi}^3$

$\approx 2623 \text{ mi}^3$

So according to this estimate, the volume is around 386 trillion cubic feet, or 2623 cubic miles.

Of course, other estimates exist, but the majority of them place the volume at somewhere between roughly 1000 and 2600 cubic miles.

Colorado River Flow Rate

The average flow rate of the Colorado River—the river that cut the Grand Canyon—is:

- Wikipedia:[609] 22,500 ft³/sec, which is about 16.3 million acre-feet per year. (An **acre-foot** is the amount of water that would cover one acre of area—or $1/640$ of a square mile—to a depth of one foot; this comes out to 43,560 cubic feet.)
- Encyclopedia Britannica:[610] 17 million acre-feet per year.
- United States Bureau of Reclamation:[611] 16.4 million acre-feet per year.

So the consensus seems to be that the Colorado River flow rate is 16–17 million acre-feet per year.

Colorado River Sediment Content

The final thing we need for our calculations is the percentage of the Colorado River sediment content.

In any river, the flowing water carries sediment—dirt, mud, sand, and so forth—along with it. In a very clean, pure river, the sediment content would be close to zero, and the more sediment that is being carried along by the water, the higher the percentage of sediment there would be.

So what is the sediment rate of the Colorado River? Most studies of sediment concentration are very detailed, noticing how the sediment concentra-

---

[609] Wikipedia article "Colorado River" at en.wikipedia.org/wiki/Colorado_River; quoting from Nowak, Kenneth C. (April 2, 2012). "Stochastic Streamflow Simulation at Interdecadal Time Scales and Implications to Water Resources Management in the Colorado River Basin" (PDF). *Center for Advanced Decision Support for Water and Environmental Systems.* University of Colorado, p. 114. Accessed December 8, 2019.

[610] Article "Colorado River" at Britannica.com/place/Colorado-River-United-States-Mexico

[611] Executive Summary: *Colorado River Basin Water Supply and Demand Study* at usbr.gov/watersmart/bsp/docs/finalreport/ColoradoRiver/CRBS_Executive_Summary_FINAL.pdf.

tions vary by month, in any number of tributaries of the Colorado River, in various locations, and at various depths, and so forth. The one source[612] I found that had aggregated the partial answers into the answer to the single question we have here, states the sediment percentage of the Colorado River averages out to be 31%.

In other words for every gallon of river water, only 69% of that gallon is actually water; the other 31% of that gallon is sediment.

Doing the Math

With our raw numbers above, we are now able to do the calculations. This will happen in two stages: the first stage is to convert all quantities into similar units (so we're not comparing apples and oranges), the second stage is to answer the original question of "How long would it take the Colorado River to dig out the Grand Canyon?"

The volume of the Grand Canyon was estimated to be between 1000 and roughly 2600 cubic miles. To avoid the appearance of slanting the data to make a point, let's assume the largest of the estimates for the volume: 2623 cubic miles.

The flow rate of the Colorado River was between 16 and 17 million acre-feet per year. Again, to avoid the appearance of slanting the calculations to make a point, I will assume the smallest estimated annual flow rate of 16 million acre-feet. As you remember, an acre-foot of water is the amount of water required to cover one acre to a depth of one foot, so a cubic mile would be 640 × 5280 acre-feet.

So to calculate the flow rate F of the Colorado River, expressed in cubic miles:

$$F \approx 16 \times 10^6 \text{ acre-feet/yr}$$
$$\approx 16 \times 10^6 \div (640 \times 5280) \text{ mi}^3/\text{yr}$$
$$\approx 4.73 \text{ mi}^3/\text{yr of river water}$$
$$\approx 1.47 \text{ mi}^3/\text{yr of sediment}$$

So the Colorado River, at current flow rates and sediment percentages, removes almost 1½ cubic miles of sediment per year. So how many years would it take to remove 2600 cubic miles of sediment? If it removed exactly

---

[612] Lucas Siegfried, "Sediment Supply and Flow in the Colorado River Basin" at watershed.uc-davis.edu/education/classes/files/content/page/7%20Siegfried_Colorado%20River%20Sediment%20Supply%20Transport.pdf, page 4, accessed December 10, 2019.

one cubic mile of sediment per year, it would take 2600 years. Since it removes *more* than one cubic mile of sediment per year, the number of years to dig out the Grand Canyon would be less than 2600:

$T \approx 2600 \text{ mi}^3 / (1.47 \text{ mi}^3/\text{yr})$
$ \approx 1771 \text{ yr}$

What? It would take less than *2000 years* to dig out the Grand Canyon? Yes. Do the math yourself to confirm. That is certainly a far cry from the millions and millions of years that evolution keeps telling us. Note that the starting values (canyon volume, river flow rate, and sediment percentage) were taken from sources that were not trying to support the creationist model in any way, but they gave numbers that we can calculate with.

And even if the Grand Canyon were twice as big, and the Colorado River flow rate was half as much, we'd still arrive at an answer in the ballpark of 7,000 years. This is much closer to the Creation model's stated 6,000-year age of the Earth than the evolution model's 4½-billion-year age or its 5-to-70-million-year task of digging out the Grand Canyon. And that is without even considering catastrophic events such as floods (or the Flood), which would shorten the time even further! And actually, it is likely that a catastrophistic event caused a large part of the rock removal of the Grand Canyon, and other canyons as well.[613,614]

Given the Colorado River's current flow rate and sediment content, plus the uniformitarianism that evolution holds so dear, there's no need for it to take the 5 million years evolutionists claim. It would take 5 million years to dig out if the Colorado River flow rate were reduced by 99.96%, but then, why would it flow so slowly for most of those 5 million years, and only come up to its current flow rate when humans started watching it?

Yet another case of evolution's numbers just not adding up. . .

Other Fast Canyons

Reading the previous sections, we realize that it doesn't really need to take "millions and millions of years" for a canyon to form. Well, at least for the Grand Canyon to form. But that's only one canyon. Are there others that were dug out in less than "millions and millions of years?"

---

[613] Steven Austin, July 1, 2008, Acts & Facts, ICR.org/article/red-rock-pass-spillway-bonneville-flood; accessed September 22, 2020.

[614] Emil Silvestru, September 5, 2007, Creation.com/wild-wild-floods; accessed September 22, 2020.

Glad you asked.

Canyon Lake Gorge

In the Texas hill country lies Canyon Lake, fed by the Guadalupe River. The lake was not exceptionally noteworthy until 2002, when record-breaking rainfall filled it beyond its capacity and water poured through the spillway. The usual flow of 350 cubic feet of water per second suddenly became 70,000 cubic feet per second.

Image courtesy Larry D. Moore and Creative Commons CC BY-SA 4.0

Canyon Lake Gorge in Texas.

The water rushed down the riverbed and carved a large steep-walled canyon in the solid limestone bedrock. In only a few days, the water gouged out a canyon more than a mile long and up to *50 feet deep* (averaging 20 feet deep), and displacing numerous rocks three feet in diameter or more.[615,616] Evolutionists across the world are being confronted with the clearly observable fact that it *doesn't* take the "millions and millions of years" they've claimed for decades, and even centuries.

Field geologist Sanjeev Gupta, of Imperial College London, in an interview by *Science News*, stated:

> Geology is typically about events that happened long ago and very slowly. This is the *[sic]* one of the first studies to study the effects of a single canyon-cutting event.[617]

Gupta was right about it being a "canyon-cutting event"—it was a single event that gouged out a significant canyon in short order.

---

[615] Brian Thomas, "Texas Canyons Highlight Geologic Evidence for Catastrophe" July 8, 2010. [Online at ICR.org/article/texas-canyons-geologic-catastrophe; accessed August 9, 2020.]

[616] Ham, Ken and Hodge, Bodie: *A Flood of Evidence: 40 Reasons Noah and the Flood Still Matter* (Master Books, Green Forest, Arkansas, 2016), p. 115.

[617] Perkins, S. 2010. "Even a newborn canyon is big in Texas." *ScienceNews*. 178(2): 15. [Online at ScienceNews.org/article/even-newborn-canyon-big-texas; accessed August 9, 2020.]

Geologists Michael Lamb and Mark Fonstad examined the brand-new Canyon Lake Gorge, and published a paper in *Nature Geoscience*. Their observations:

> . . .show that the [Canyon Lake] flood moved metre-sized boulders, excavated [about **seven meters**] **of limestone** and transformed a soil-mantled valley into a bedrock canyon **in just [about three] days.**[618]

Geomorphologist Alan D. Howard, at the University of Virginia in Charlottesville, has studied landforms on Mars that look like they were created by running water. Considering the Canyon Lake Gorge, he said:

> It doesn't take millions of years to create an impressive channel. **Flowing liquid can do a lot of work in a short period of time.**[619]

It certainly can.

Mount St. Helens' Little Grand Canyon

This was discussed earlier, in the "Mount St. Helens" section. On March 19, 1982, a minor eruption melted snow and ice and unleashed a mudflow, that flowed into a lake caused by water backing up behind a debris dam deposited by the big eruption in 1980. The dam breach resulted in a canyon being dug out at a remarkable speed: the canyon was 1,000 feet wide, 2,000 feet long, up to 140 feet deep, and *it was formed in only nine hours* of violent catastrophistic water flow! It looks curiously like a $1/40$-scale model of the real Grand Canyon, replete with numerous sedimentary layers visible in the new canyon's walls.

This was described in more detail in the section "Rapid Erosion" above; refer back to that if you wish to review.

---

[618] Lamb, M. P. and M. A. Fonstad. 2010. "Rapid formation of a modern bedrock canyon by a single flood event." *Nature Geoscience*. 3 (7): pp. 477–481. [Online at Nature.com/articles/ngeo894; accessed August 9, 2020.] Emphasis added.

[619] Perkins, S. 2010. "Even a newborn canyon is big in Texas." *ScienceNews*. 178(2): 15. [Online at ScienceNews.org/article/even-newborn-canyon-big-texas; accessed August 9, 2020.] Emphasis added.

Burlingame Canyon

Near Walla Walla, Washington, is Burlingame Canyon. Caused by heavy rains and plant-choked canals, an irrigation ditch no deeper than ten feet and no wider than six feet, was dramatically enlarged in 1926.

Because of the factors mentioned above, catastrophism happened, and serious erosion began. The ditch became a gully, which became a gulch, which became a canyon. In just *six days*, the irrigation ditch became a canyon 1500 feet long, up to 120 feet deep, and 120 feet wide. Around 5 million cubic feet of sediment were removed in those six days.[620]

Image courtesy Creative Commons CC BY-SA 3.0
Touchet layers of the kind in Burlingame Canyon, in the Walla Walla valley.

Providence Canyon

Providence Canyon in Georgia is not the result of a catastrophistic event like a flood, but it still formed in a short amount of time, geologically speaking.

This canyon, in southwest Georgia, near the town of Lumpkin, is a deep chasm with nine tributaries. The canyon is up to 160 feet deep, 600 feet wide, and 1300 feet long. It was caused by erosion due to poor farming practices in the 1800s. In less than a century, even without catastrophism, this geological feature was formed.[621]

Image courtesy Robbie Honerkamp and Creative Commons CC BY-SA 3.0
Providence Canyon in Georgia.

Measurements in 1984 and 1994 confirm that the canyon is still growing, mostly in width.

---

[620] John Morris, "A Canyon in Six Days!" September 1, 2002. [Online at AnswersInGenesis.org/geology/natural-features/a-canyon-in-six-days; accessed August 9, 2020.]

[621] Rebecca Gibson, "Canyon Creation" September 1, 2000. [Online at AnswersInGenesis.org/geology/natural-features/canyon-creation; accessed August 9, 2020.]

The Cambrian Explosion

As mentioned earlier, one of the most vexing problems for the evolutionist is that the fossil record does not at all support the idea of one kind of organism gradually changing, over innumerable generations, into some other kind of organism.

As shown in the diagram of geological timeframes in the section "The Geologic Column," the lowest ("oldest") period of complex fossils is the Cambrian. Before this—in the Precambrian period—were just single-celled organisms, or occasionally, colonies of them, but nothing "complex." (As if a single-celled organism that can respond to stimuli, metabolize nutrients, eliminate waste, and reproduce is "simple.")

And then comes the Cambrian period. And suddenly, abruptly, fossils abound. Fossils of clams, snails, trilobites, sponges, brachiopods, worms, jellyfish, sea urchins, sea cucumbers, swimming crustaceans, sea lilies, and other complex invertebrates. First, there's nothing, and then *bam,* all these organisms, fully formed. Almost as if they were created. The arrival of all these types of organisms is so sudden that the appearance of all the life forms that became these fossils has been named "the Cambrian Explosion."

And again, proponents of the creation model predicted that this would be the case, but proponents of the evolutionary model now need to explain why, if their theory is correct, does the fossil record show otherwise?

The abruptness of the appearance of all these fossils is indeed exasperating for those who trust in evolution. As one prominent evolutionist states:

> Establishing the presence of biological activity during the very early Precambrian clearly poses difficult problems. . . **Skepticism** about this sort of evidence of early Precambrian life **is appropriate.**[622]

Skepticism is indeed appropriate, because the fossil record, which should be *loaded* with transitional forms in the Precambrian rocks, if Darwinian evolution actually happened, is devoid of such transitional forms.

---

[622] A. E. J. Engel, et al, *Science* 161:1008 (1968). Emphasis added.

Daniel Axelrod, on the Geology staff of the University of California, Los Angeles, wrote:

> One of the major unsolved problems of geology and evolution is the occurrence of diversified, multicellular marine vertebrates in Lower Cambrian rocks on all the continents and **their absence** in rocks of greater age... However, when we turn to examine the Precambrian rocks for the forerunners of these early Cambrian fossils, **they are nowhere to be found.** Many thick (over 5,000 feet) sections of sedimentary rock are now known to lie in unbroken succession below strata containing the earliest Cambrian fossils. These sediments apparently were suitable for the preservation of fossils because they are often identical with overlying rocks which are fossiliferous, yet **no fossils are found in them.**[623]

Even Richard Dawkins, the British biologist and staunch evolutionist, has stated:

> ...the Cambrian strata of rocks, vintage about 600 million years [evolutionists are now dating the beginning of the Cambrian at about 530 million years], are the oldest in which we find most of the major invertebrate groups. And we find many of them already **in an advanced stage of evolution, from the very first time they appear.** It is as though **they were just planted there, without any evolutionary history.** Needless to say, this appearance of sudden planting has delighted creationists.[624]

Note that Dawkins is loath to imply these things that were in "advanced stages" from "the first time they appear" were actually created: he doesn't call it a "sudden planting," but an "*appearance* of sudden planting." As in, "It wasn't actually created; it just *looks* that way." And *of course* this delights the creationists; this is exactly what the creation model predicts, and the complete opposite of what the evolutionary model predicts. Again, notice the undertone: even though Dawkins acknowledges that all these fossils, from the very first time they appear, are in "an advanced stage" of evolution, yet "without any evolutionary history," he cannot bring himself to acknowledge the possi-

---

[623] Daniel Axelrod, *Science* 128:7 (July 4, 1958). Emphasis added.
[624] Richard Dawkins, *The Blind Watchmaker* (New York: W. W. Norton, 1987), p. 229.

bility of a Creator. He simply says that they "evolved" virtually instantaneously. Sounds like creation to me. . .

Author Douglas Futuyma is an ardent anti-creationist, yet in his book on evolutionary biology, he states:

> It is considered likely that all the animal phyla became distinct before or during the Cambrian, for **they all appear fully formed, without intermediates** connecting one form to another.[625]

That's amazing. They weren't created—they evolved—but there are *no intermediate forms* (which is essential to evolutionism). They just "all appear fully formed." So why is he still an evolutionist?

There are numerous other quotes that could be here, of staunch evolutionists acknowledging that the animals in the Cambrian Explosion appeared abruptly, and without transitional forms. But still they hang onto a theory that requires animals to appear gradually, generating countless transitional forms; still, they hang onto this theory for which they can find no supporting evidence.

Punctuated Equilibrium

The Cambrian Explosion is so contrary to evolution that something had to be done. All these plant and animal forms appearing in the fossil record "fully formed, without intermediates" was *so* suggestive of the Creation model, that it was very disturbing to those who had already decided that Creation was "unthinkable."

Yes, something had to be done. Since it was not permissible to even consider the Creation model as the source of all creation, an evolutionary explanation had to be invented. But how do they deal with all these fully formed creatures appearing so suddenly in the Cambrian Explosion? We need another rescue device to keep evolution afloat!

Of course: let's just say the organisms evolved suddenly! True, we'll have to fudge a little on the uniformitarianism thing, because something clearly non-uniformitarian happened in the Cambrian Explosion. But even though it all happened in a sudden burst, we'll still say it's evolution, because we all know that Creation couldn't be real.

---

[625] Douglas Futuyma, *Evolutionary Biology,* second edition. (Sunderland, Massachusetts: Sinauer Associates, Inc., 1986), p. 325. Emphasis added.

And so, "Punctuated Equilibrium" was born. A new flavor of evolution in which things could evolve suddenly and abruptly, adding huge amounts of new genetic information that comes from—well, somewhere. Sort of a "catastrophist uniformitarianism," if you will. A scenario in which organisms change slowly unless they don't. In Punctuated Equilibrium (also known as "Punk Eek"), large populations of plants and animals stay stable for long periods of time, until some environmental catastrophe happens and then, inbreeding within small "founder" populations happens, which stimulates abrupt evolution. In other words, the reason we don't see evidence for evolution in the present is that it happens too slowly, and the reason we don't see evidence of evolution in the past is because it happened too quickly.

What kind of geological catastrophes would cause these evolutionary "punctuations?" Floods would do it, and that would also explain all the millions of fossils we see in the sedimentary rock layers. Not a global flood, mind you: that is getting way too close to the Biblical story, and we can't condone that. These were local floods that just happened to occur all over the world, and they caused bursts of speedy evolution.[626]

One naturalistic theory that attempts to explain the gaps in the fossil record is called **saltation**. Saltation (from the Latin word *saltus*, "to leap") claims that huge increases (leaps) in complexity, with its requisite huge increases in genetic information, can occur in a single generation. As we've seen, such gaps in the fossil record would be nonexistent if Darwinian evolution had actually happened.

On the one hand, evolutionists dismiss saltation from consideration because of the impossibility of the requisite genetic information appearing out of nowhere. This dismissal is entirely justified. On the other hand, Punctuated Equilibrium basically says the same thing: it requires saltation to be real, although it's usually called something different (like, say, "Punctuated Equilibrium") so as to appear less impossible.

The rescue device called Punctuated Equilibrium is gaining in popularity; it's an acceptable compromise between evolution and evidence. The astounding thing about punctuated equilibrium is that it is not based on evidence, but on the *lack* of evidence! (But actually, most rescue devices are.)

The thinking goes like this: Since the fossil record reveals a very incriminating absence of transitional forms, evolution had to happen fast. The Sec-

---

[626] Morris, Henry. "Revolutionary Evolutionism" November 1, 1979. *Acts & Facts*. [Online at ICR.org/article/revolutionary-evolutionism; accessed July 11, 2020.]

ond Law of Thermodynamics says that all systems, left to themselves, go from order to disorder, so the greater organization required by punctuated equilibrium must, through some unknown process, arise out of the devastation of a more quickly disintegrating system! Where, oh where, have all the scientists gone?

In the next several sections, which cover the supposed evolution from one kind of animal to another, you'll notice a common thread: the transitions are all highly imaginative. In other words, based on the presumption that evolution is the only possible way animals could have gotten here, the vast in-between areas are just "filled in" with what the evolutionists wanted to put in there. These flights of fancy are not founded on any evidence, but the transitions "must have" happened, because evolution is the only possible way animals could have gotten here. Isn't that convenient, that the assumption of factuality is required in order to come to a conclusion of factuality?

Stephen J. Gould, the evolutionary theorist, stated that paleontologists use:

> . . .a set of procedures for making strong **inferences** about phyletic *[evolutionary]* history from data of an imperfect record that cannot, in any case, 'see' past causes directly, but can only **draw conclusions** from preserved results of these causes.[627]

He goes on to say that paleontologists explain what surely must have happened in the past through the use of:

> . . .large-scale results by **extrapolation** from short-term processes. . . *[and]* **extrapolation** to longer times and effects of evolutionary changes actually observed in historic times (usually by analogy to domestication and horticulture).

Did you catch that? Paleontologists use inferences, draw conclusions, and employ extrapolation. Based on what? Based on "evolutionary changes actually observed in historic times." But, as we've shown in numerous ways earlier in this book, evolutionary changes have *not* been observed. Even Richard Dawkins, passionate advocate of evolution, when he stated that evolution *has* been observed, had to add the caveat, "It's just that it hasn't been observed

---

[627] Gould, S.J. 2002. *The Structure of Evolutionary Theory.* Cambridge, MA: Harvard University Press, p. 59.

while it's happening."[628] In other words, evolution has *not* been observed. So, when evolution is inferred, concluded, and extrapolated, based on missing evidence and lack of observation, it is reduced to merely "making stuff up." It is fabricated, imaginary, and pretend.

As we saw earlier in this book, evolutionist Stephen J. Gould admits that "Our ways of learning about the world are strongly influenced by the **social preconceptions and biased modes of thinking.**"[629] You'll see this in *every one* of the following sections that talk in detail about how things evolved, but in which, for some strange reason, we have no evidence of said evolution.

From Invertebrates to Fish

After the Cambrian Explosion, animals with spinal chords appeared; millions of fossils attest to that fact. Did they evolve? If so, there should be numerous intermediate forms. But of all the fossils we've found, *not one* intermediate form is among them.

Image courtesy Pixabay
Did invertebrates evolve into fish?

Here's a quote on how **chordate** (having a spinal cord) fish evolved, from the *Life Nature Library:*

> How this earliest chordate stock evolved, which stages of development it went through to eventually give rise to truly fishlike creatures, **we do not know.** Between the Cambrian when it probably originated, and the Ordovician when the first fossils of animals with really fishlike characteristics appeared, **there is a gap of perhaps 100 million years which we will probably never be able to fill.**[630]

That's pretty amazing: 100 million years of evolution, and 100 million years of missing fossil evidence. Are we seeing a pattern here?

---

[628] Bill Moyers and David Brancaccio interviewing Richard Dawkins on NOW, the weekly newsmagazine from PBS. [Online at PBS.org/now/transcript/transcript349_full.html#dawkins; accessed July 24, 2020.]

[629] Gould, Stephen J., *Natural History* 103(2):14, 1994.

[630] F. D. Ommanny, *The Fishes,* Life Nature Library (NewYork: Time-Life, Inc., 1964), p. 60. Emphasis added.

Errol White, an evolutionist and an expert on fishes said, in his presidential address on lungfishes to the Linnean Society of London:

> But whatever ideas authorities may have on the subject, **the lungfishes, like every other major group of fish** is that I know, have **their origins firmly based in nothing**. . . I have often thought **how little I should like to have to prove organic evolution in a court of law**. . . **We still do not know the mechanics of evolution** in spite of the over-confident claims in some quarters, nor are we likely to make further progress on this by the classical methods of paleontology or biology; and we shall certainly not advance matters by jumping up and down shrilling "Darwin is God and I, So-and-so, am his prophet"—the recent researches of workers like Dean and Henshelwood (1964) already suggest the possibility of **incipient cracks in the seemingly monolithic walls of the Neo-Darwinian Jericho.**[631]

This is impressive—a creationist would be hard-pressed to point out evolution's inabilities more eloquently.

And notice that the above quote refers to research done in 1964—more than *half a century ago*—and people were already publishing articles questioning evolution's validity. And in the past half a century and more, hundreds of other problems have been discovered with the theory of evolution. Why don't we hear about these problems in the mainstream media? And in our school systems? Clearly, because there is an anti-God sentiment in much of modern society: we don't *want* there to be a God, so we will not acknowledge any evidence that points to Him.

---

[631] Errol White, *Proceedings of Linnean Society of London* 177:8 (1966). Emphasis added.

From Fish to Amphibians

So far, we've gotten from no life to single-celled organisms in the Precambrian period, through complex invertebrates in the Cambrian period, to fish in Devonian period, by way of the Ordovician and Silurian periods. And as far as transitional forms go, evolution is batting zero.

Undaunted, however, evolutionists confidently claim that fish evolved into reptiles by way of amphibians. Robert Caroll assumes that primitive rhipidistian fish evolved into amphibians, yet he admits:

Did fish evolve into amphibians?

> **We have no intermediate fossils** between rhipidistian fish and early amphibians...[632]

A "rhipidistian" fish is a fish similar to the Coelacanth shown here. It was considered a very "primitive" fish that had gone extinct 65–70 million years ago, until live ones were caught off the coast of Madagascar in 1938.

The Coelacanth, believed to be extinct for 70 million years. This specimen was caught near the Comoros Islands in 1974.

Malcolm Browne wrote an article in the *New York Times*, commenting on a paper by a couple of other paleontologists, discussing their opinions on the ancestor of amphibians. He states:

> **No one can be certain** which group or groups of fishes was the first to make the transition to land, or what their evolutionary pathways may have been... The transition from water to land occurred long ago, and various family trees suggested by the fossil record are so tangled that scientists ac-

---

[632] R. L. Caroll, *Vertebrate Paleontology and Evolution* (New York: W. H. Freeman and Company, 1988), p. 138. Emphasis added.

knowledged **they may never be able to sort them out definitively.**[633]

Why may scientists "never be able to sort them out definitively?" Simply because there are no transitional forms. If there *were* transitional forms, simply noticing what was being transitioned *from,* and what was being transitioned *to,* would clearly show the family tree. But as it is, paleontologists merely put forth their preferred opinions and pet theories—there's nothing else to go on.

Speaking of **Lissamphibia** (which includes frogs, toads, salamanders, newts, and caecilians), paleontologist Alfred S. Romer states:

> Between them [the Lissamphibia] and the Paleozoic group is **a broad evolutionary gap not bridged by fossil materials.**[634]

Concerning **Lepospondyls** (amphibians with spool-shaped vertebrae), evolutionist Robert Caroll concurs:

> We have not found any Lepospondyl fossils later than the lower Permian. **There is a surprising gap in the record of small amphibians until the appearance of frogs and salamanders** with the Jurassic.[635]

Hmm. *Still* no evidence that something evolved into something else. . . And notice he said there is a "surprising" gap between fish and frogs/salamanders. Why is it a surprising gap? Because evolution is presumed to be unquestionable by those who labor under that belief system, so it is genuinely unexpected to find evidence that contradicts it. Even though it is *everywhere.*

---

[633] Malcolm W. Browne, "Biologists Debate Man's Fishy Ancestors," *New York Times,* March 16, 1993, page C-1.

[634] H. S. Romer, *Vertebrate Paleontology,* Third Edition (Chicago: Chicago University Press, 1966), p. 98. Emphasis added.

[635] R. L. Caroll, *Vertebrate Paleontology and Evolution* (New York: W. H. Freeman and Company, 1988), p. 180. Emphasis added.

From Amphibians to Reptiles

So, no fossil evidence so far, for anything evolving into anything else. But what about from amphibians into reptiles? Surely there's hope, isn't there?

Quoting evolutionist Robert Caroll again, as he comments on **amniotes** (an amniote is an animal such as a reptile or bird, that lays **amniotic eggs**, which are eggs that do not need to be in water, like amphibian eggs do):

Did amphibians evolve into reptiles?

> The early amniotes are sufficiently distinct from all Paleozoic amphibians that **their specific ancestry has not been established.**[636]

The above quote shows that there are no amphibian-to-reptile transitional forms, because if there were any, the line of ancestry would be clear.

Walking Reptiles to Flying Reptiles

The transition from walking reptiles to flying reptiles supposedly took about 150 million years.[637]

Let's think about this proposed 150-million-year process of walking dinosaurs evolving into flying dinosaurs. According to evolutionists, many tiny changes over the ages would gradually cause the finger bones to elongate and stretch until eventually, the organism could fly.

Artist's conception of the pterodactyl, a flying dinosaur.

But during the middle 95% or so of that transition process, there would have been ever-growing arms and fingers that were not yet flight-capable. Such

---

[636] Ibid., p. 198. Emphasis added.

[637] Duane Gish, *Evolution: The Fossils Still Say No!* (Institute for Creation Research: El Cajon, California), 1995, p. 103.

long, lanky extremities would have been a great hindrance to walking and grasping, and because they weren't yet flight-ready, not only would these hypothetical animals have a difficult time catching their own food, but they would have been highly vulnerable to predators, with all this "baggage" of partially formed but not-yet-functional wings. In other words, such mutations would have made bad legs long before they made good wings. Clearly, such mutationally crippled organisms, no longer able to run from predators or catch their food nearly as well as before, would have gone extinct many times over, had this evolutionary process actually started to happen.

This is another case of natural selection resisting the growth of wings, because until *all* the changes were in place, and flight was actually possible, the incomplete states would have made the reptile much less effective in getting through each day unscathed—much less "fit to survive." Like the reproduction discussion earlier, *all* the components had to be in place at the same time for *any* of them to be of benefit.

There's another problem, too. Remember that evolution, by its very definition, requires *undirected* mutations—any overall directional tendency (except for the ever-increasing decay caused by the Second Law of Thermodynamics) would imply a lack of randomness, which implies intention in whomever, or whatever, is causing the changes. And, of course, intention implies a Mind, which evolutionists vigorously resist.

In a hypothetical walking reptile that is "evolving" into a flying reptile, let's say that only a hundred mutations are required (in actuality, this number is very conservative). After the first successful mutation that slightly elongated the finger bones, the next mutation, being random, has just as much chance of *shortening* the previously lengthened finger bones, as it does of lengthening them again. Which means that in order for forelegs to evolve into wings, a hundred mutations *in the same direction* would need to occur. But randomness will not allow that (at least, within the supposed 4.3-billion-year age of the earth), and non-random directionality implies a governing Mind, which is disallowed by evolutionists.

Assuming that it took 150 million years for land-based dinosaurs to evolve into flying dinosaurs (the current evolution-based guess), and assuming that each generation took 50 years (although it was likely much less than that) that means that there were *at least three million generations* in which to evolve into flying dinosaurs. What might that imply? Obviously: lots and lots of intermediate forms in the fossil record.

> ...all the Triassic pterosaurs were **highly specialized** for flight and were very similar to later rhamphorhyncoids in most features. They provide **little evidence of their specific ancestry and no evidence of earlier stages in the origin of flight.**[638]

Isn't that interesting? "No evidence of the earlier stages in the origin of flight." In other words, there are *no intermediate forms* of non-flying dinosaurs (for example, velociraptors) partially evolved into **rhamphorhyncoids**; that is, into flying dinosaurs (for example, pterodactyls).

But perhaps there are no intermediate forms because we haven't yet found very many fossils of flying reptiles yet. Could that be? Well, no:

> Pterodactyloids were already numerous and diverse in the Upper Jurassic, as we can see from fossils from Solnhofen (Wellenhofer, 1970). These remains include some of the smallest pterosaurs and the ancestors of the largest flying reptiles. **No forms are known that are intermediate** between rhamphorhyncoids and pterodactyloids. . . The fossil record of pterosaurs extends for approximately 150 million years, and **nearly 90 species have been reported from every continent** except for Antarctica.[639]

Nearly 90 species from almost every continent—no, the absence of evidence is not for lack of fossils; it's simply for lack of evolution. Every one of these species appears fully formed in the fossil record. Just like the creation model predicts.

---

[638] R. L. Caroll, *Vertebrate Paleontology and Evolution* (New York: W. H. Freeman and Company, 1988), p. 336. Emphasis added.
[639] Ibid.

From Reptiles to Birds

Another common thought among evolutionists, and therefore among the general public, is that reptiles evolved into birds.

The idea of reptiles evolving into birds was first speculated by Thomas Huxley in the mid 1800s. Since then, the idea has been more accepted or less accepted, as evolutionary fads wax and wane through the years. But now it seems to be in fashion again.

Image courtesy Pixabay
Did reptiles evolve into birds?

As of this writing, in fact, Wikipedia bluntly states that birds *are* dinosaurs:

> **Birds are** a group of feathered theropod **dinosaurs,** and constitute **the only living dinosaurs.** Likewise, **birds are considered reptiles** in the modern cladistic sense of the term, and their closest living relatives are the crocodilians.[640]

Birds *are* dinosaurs? Birds *are* reptiles? What would be required for an actual reptile to evolve into a bird? Let's think about that.

- **Blood Temperature:** First, there's the problem of changing from **ectothermic**, or cold-blooded (like reptiles are) to **endothermic**, or warm-blooded (like birds are). This is not just a difference in blood temperature, but an ability in birds (or any other warm-blooded animal) to *maintain* its internal body temperature at a specific value. This ability to maintain a constant internal temperature requires an entirely new interdependent set of physical attributes (for example, a way to perceive the body's internal temperature, a way to generate more heat on cold days, a way to get rid of excess heat on hot days, and so forth), plus genetic information concerning how to use those attributes. This is another case of irreducible complexity: none of these features would be of any benefit until *all* were in place, requiring the whole system to "evolve" all at once. To get around this, some evolutionists have spec-

---
[640] en.wikipedia.org/wiki/Bird; accessed August 16, 2020.

ulated a rescue device—that maybe dinosaurs were warm-blooded—but like most rescue devices, this is not supported by evidence.[641,642]

- **Hip Bones:** One way that dinosaurs are classified is by the shape of their hip bones. In **saurischians,** or "lizard-hipped dinosaurs," the pubic bone is directed toward the front, similar to how it is in modern reptiles and mammals. In **ornithischians,** or "bird-hipped dinosaurs," the pubic bone is directed toward the rear, similar to how it is in birds.

Image courtesy Christian-wittmann-1964 and Creative Commons CC BY-SA 4.0
Brachiosaurus: 32–64 tons

The problem is that bird-hipped dinosaurs (including the enormous *Brachiosaurus* and *Diplodocus* shown here) are even less birdlike than the lizard-hipped bipedal **theropod** ("beast-footed") dinosaurs.[643]

- **Fingers:** Most vertebrates develop embryologically with a hand design based on five fingers. But if three-fingered theropods evolved into three-fingered birds, you'd expect the same three fingers to be used, right? But theropods retain fingers 1, 2, and 3 (finger 1 being the thumb), but birds retain fingers 2, 3, and 4.[644]

Image courtesy Christian-wittmann-1964 and Creative Commons CC BY-SA 4.0
Diplodocus: 12–16 tons

---

[641] A. Feduccia, "Dinosaurs as reptiles," *Evolution* 27:166–169,1973; A. Feduccia, *The Origin and Evolution of Birds,* 2nd Ed., Yale University Press, New Haven, Connecticut, 1999.

[642] N. R. Geist and T. D. Jones, "Juvenile skeletal structure and the reproduction habits of dinosaurs," *Science* 272:712–714, 1996.

[643] Ken Ham and Bodie Hodge, *The New Answers Book 1,* Answers in Genesis (Master Books, Green Forest, Arkansas, 2008), p. 299.

[644] A. Feduccia, T. Lingham-Soliar, and J. R. Hinchliffe, "Do feathered dinosaurs exist? Testing the hypothesis on neontological and paleontological evidence," *Journal of Morphology* 266:125–166, 2005.

- **Lungs:** Reptilian lungs breathe in and out, but birds have one-way ventilation through their lungs.[645] Needless to say, *numerous* simultaneous changes had to be made in order to make such respiration possible. Sure is "lucky" that all the required mutations happened at the same time, wouldn't you say? But perhaps the reptiles-turning-birds simply held their breath for a few million years until all the necessary respiratory changes evolved into being. . .
- No animals except birds have anything even remotely resembling feathers, yet by and large, evolutionists think reptiles evolved into birds. So somewhere along the line, we could expect to see scales that are "half-evolved" into feathers. But no. In spite of many reports of feathered dinosaurs from the Liaoning province in China, further examination shows the claims to be false. For example, "hair-like" structures that were announced to be "protofeathers" turned out to be subcutaneous collagen fibers of connective tissue. Evolutionist Feduccia bemoans the research practices he sees:

    . . .the major and most worrying problem of the feathered dinosaur hypothesis is that the integumental structures have been homologized with avian feathers **on the basis of anatomically and paleontologically unsound and misleading information.**[646]

    Translated, this means, "the supposed similarities between these fibers and feathers are based on unreliable and deceptive information."
- **Archaeopteryx Role Revised:** The *Archaeopteryx* has long been the poster boy of transitional fossils, but now has been shown to be a true bird, teeth notwithstanding. However, according to the evolutionary timeframe, the *Archaeopteryx* was around earlier than the reptiles now claimed to have evolved into birds. So here is another case of evolutionary descendants being older than their ancestors. *Archaeopteryx* and other real birds—with real feathers—have been found in Upper Jurassic rocks secularly dated as being 147 million years old, while *Velociraptor* and *Deinonychus*, the alleged "ancestors" of birds, are found in Cretaceous rocks secularly dated as being only 110 million years

---

[645] J. A. Ruben, T. D. Jones, N. R. Geist, and W. J. Hillenius, "Lung structure and ventilation in theropod dinosaurs and early birds," *Science* 278:1267–1270, 1997.

[646] A. Feduccia, T. Lingham-Soliar, and J. R. Hinchliffe, "Do feathered dinosaurs exist? Testing the hypothesis on neontological and paleontological evidence," *Journal of Morphology* 266:125–166, 2005. Emphasis added.

old.⁶⁴⁷ Curious, isn't it, how the *Archaeopteryx* is almost 40 million years older than its ancestors?

In the above list, and in all the other imagined cases of something evolving into something else, there are many more problems that were omitted for the sake of brevity—problems that would prevent evolution from ever having happened.

From Reptiles to Mammals

So far, we have the single celled Precambrian fossils, then complex invertebrates, then fish, amphibians, reptiles, and birds. The fossils we find of all of these types of organisms are numerous and widespread. However, the fossils of any transitional forms have yet to be found.

Did reptiles evolve into mammals?

And now the evolutionists attempt to address the imaginary transition from reptiles into mammals. And although the fossil record is rich with examples of both categories, the fossil record of the expected evolutionary transition is similarly vacuous.

For example, paleontologist Edwin H. Colbert, a protégé of Henry Fairfield Osborn (mentioned earlier), noted:

> **It is not easy to determine the precise line of mammalian ancestors** among the theriodont reptiles. Some theriodonts were far advanced toward the mammals in certain characters, but comparatively primitive in others; and among all the theriodonts the mixtures of advanced and conservative characters are so varied that **it is not possible to point to any one particular group and define it as progressing** most positively in the direction of mammals.⁶⁴⁸

---

⁶⁴⁷ Clarey, Tim: *Dinosaurs: Marvels of God's Design* (Master Books, Green Forest, Arkansas, 2015), p. 125.

⁶⁴⁸ E. H. Colbert and M. Morales, *Evolution of the Vertebrates,* Third Edition (New York: John Wiley and Sons, 1980), p. 246. Emphasis added.

Theriodonts ("the ones with beast teeth")[649] are so named because they have mammalian-looking teeth, and are supposedly the ancestors of mammals. But notice what Colbert says: "it is *not possible* to point to any particular group and define it as progressing" from reptilian to mammalian. Why not? Because there are absolutely no fossils of transitional forms.

From Walking Mammals to Swimming Mammals

According to the theory of evolution, at some point, land-based, walking mammals transitioned into water-based, swimming mammals. After all, since they're both mammals, that should have been pretty easy, right?

Image courtesy Pixabay
Did walking mammals evolve into swimming mammals?

Let's think about that. Here are the words of creationist Douglas Dewar, as he points out the ludicrousness of such a multi-faceted transition happening all by itself:

> Let us notice what would be involved in the conversion of a land quadruped into, first a seal-like creature and then into a whale. The land animal would, while on land, have to cease using its hind legs for locomotion and to keep them permanently stretched out backwards on either side of the tail and to drag itself about by using its forelegs. During its excursions in the water, it must have retained the hind legs in their rigid position and swim by moving them and the tail from side to side. As a result of this act of self-denial we must assume that the hind legs eventually became pinned to the tail by the growth of membrane. Thus the hind part of the body would have become like that of a seal. Having reached this stage, the creature, in anticipation of a time when it will give birth to its young under water, gradually develops apparatus by means of which the milk is forced into the mouth of the young one, and meanwhile a cap has to be formed round the nipple into which the snout of the young one fits

---

[649] en.wikipedia.org/wiki/Theriodontia; accessed June 25, 2020.

tightly, the epiglottis and laryngeal cartilage become prolonged downwards so as tightly to embrace this tube, in order that the adult will be able to breath while taking water into the mouth and the young while taking in milk. These changes must be effected completely before the calf can be born under water. Be it noted that there is no stage intermediate between being born and suckled under water and being born and suckled in the air. At the same time various other anatomical changes have to take place, the most important of which is the complete transformation of the tail region. The hind part of the body must have begun to twist on the fore part, and this twisting must have continued until the sideways movement of the tail developed into an up-and-down movement. While this twisting went on the hind limbs and pelvis must have diminished in size, until the former ceased to exist as external limbs in all, and completely disappeared in most, whales.[650]

Piece of cake. Happens every day.

What Can We Learn from This?

So even though no transitional fossils have ever been found, that doesn't mean that no *valid* fossils have been found. *Millions* of fossils have indeed been found, and are on display in museums the world over. But the aggravating fact to evolutionists is that *not a single one* has ever been found that is a transitional form between one kind and another.

It's very likely that some readers will think my previous statement is an exaggeration. So, to quell that suspicion, let me quote an evolutionist author. Dr. Colin Patterson, the senior paleontologist at the British Museum of Natural History, wrote a book entitled *Evolution*. Luther Sunderland, a creationist, wrote to Dr. Patterson and asked him why there were no photos of transitional fossils in his book.

Dr. Patterson wrote back:

> I fully agree with your comments on the lack of direct illustration of evolutionary transitions in my book. **If I knew of any, fossil or living, I would certainly have included them.**

---

[650] Michael Denton, *Evolution: A Theory in Crisis,* Adler & Adler Publishers, Bethesda, Maryland, 1986, pp. 217–218.

> You suggest that an artist should be used to visualise such transformations, but where would he get the information from? I could not, honestly, provide it, and if I were to leave it to artistic license, would that not mislead the reader?
>
> Yet Gould [Stephen J. Gould—a Harvard professor of paleontology, now deceased] and the American Museum people are hard to contradict when **they say there are no transitional fossils.** . . . You say that I should at least "show a photo of the fossil from which each type of organism was derived." I will lay it on the line—**there is not one such fossil for which one could make a watertight argument.**
>
> **It is easy enough to make up stories** of how one form gave rise to another, and to find reasons why the stages should be favoured by natural selection. **But such stories are not part of science, for there is no way of putting them to the test.**[651]

It's refreshing to see that there are some evolutionists who still acknowledge that the theory of evolution should be supportable by observations of the fossil record, and testable experiments.

You can see above that Dr. Patterson quotes Stephen J. Gould as stating transitional forms have not been found. Here is one such quote:

> **The extreme rarity of transitional forms in the fossil record persists as the trade secret of paleontology.** The evolutionary trees that adorn our textbooks have data only at the tips and nodes of their branches. . . in any local area, a species does not arise gradually by the gradual transformation of its ancestors; **it appears all at once and "fully formed."**[652]

It's interesting that Gould calls the absence of transitional fossils to be paleontology's "trade secret." But why is it a secret? Because it's just *embarrassing* that there's no transitional-fossil evidence for evolution, when it would be the most conclusive form of evidence, and it would be everywhere on the planet, were evolution actually a thing. Paleontologists have not found a single transitional fossil, in spite of an enormous political machine behind evolution,

---

[651] Sunderland, Luther, *Darwin's Enigma*, (Master Books, Arkansas, 1998), pp. 101–102. Patterson's letter was written in 1979.
[652] Stephen Jay Gould, "Evolution's erratic pace," *Natural History* 86(5):14, May 1977. Emphasis added.

and billions of dollars spent trying to find enough supportive data to at least keep the theory on life support. Because, just *think* of the alternative: what would happen if people discovered they are not alone, but are truly loved by their Creator, they have value and significance, and that they actually have a purpose for being alive! Horrors!

Herbert Nilsson of Sweden's Lund University concurs, on the topic of the fossil record's inadequacy:

> It is not even possible to make a caricature of evolution [Darwin's gradualism] out of paleobiological facts. The fossil material is now so complete that a lack of transitional series cannot be explained by the scarcity of material. **The deficiencies are real; they will never be filled.**[653]

What we have seen from the fossil record is this: evolutionists claim that life "arose" about 4.3 billion years ago, and what we see in the fossil record is about 4.3 billion years' worth of *missing* fossil evidence of any transitional forms whatsoever. (Again, assuming that the geologic column actually *does* indicate 4.3 billion years of history. This conjecture, as we've seen already, and will see again, is greatly contested by evidence.)

So if we have all these fossils, even though none of them is transitional, where did they come from? Glad you asked. We'll cover this in the next section.

---

[653] Geoffrey Simmons, M.D., *What Darwin Didn't Know*, 2004 (Harvest House Publishers, Eugene, Oregon), p. 304.

The Worldwide Flood

Few things are mocked by evolutionists more mercilessly than the idea of a worldwide flood on Earth, even though 71% of the Earth's surface is covered with liquid water. But curiously, as pointed out in the "Mars" section above, evolutionists *are* willing to entertain the idea of a planet-wide flood on Mars, even though it is not capable of supporting liquid water. Curious, isn't it?

Many people, scientists and general public alike, who were raised on a diet of evolutionism, dismiss the idea of Noah's worldwide flood as a fanciful myth at best and a preposterous hoax at worst, and have therefore never actually looked into the plausibility of the story. The whole idea is dismissed out of hand as just silliness that couldn't possibly be supported by actual evidence, so they don't feel a need to check it out. That, my friends, is bad science.

The Timeline of Noah's Flood

Since proponents of evolutionary theories want to steer clear of Noah's Flood, for fear of being branded "one of *them*," they have no theories about it, so we'll need to refer more to the Bible in this section than most other sections in this book.

Most people are familiar with the "forty days and forty nights" aspect of the Flood timeline, but there is much more to it than that, once we look into the actual chronology. From beginning to end, the Flood took a bit more than a year. In the following table is the timeline of events pertaining to Noah's Flood. All calculations are based on a 360-day calendar, which Noah and Moses probably used; 365-day calendars didn't appear until later.

Note that this table does not include the ark's construction, preparations, and getting the animals on board and settled; it only includes the timeline of the Flood itself.

Period Duration	Period Starting Point	Main Events During This Period
—	Day 0 (600th year of Noah's life, 2nd month, 17th day)	Genesis 7:11: In the six hundredth year of Noah's life, in the second month, the seventeenth day of the month, on that day **all the fountains of the great deep were broken up, and the windows of heaven were opened.**
40 days	Day 40 (3rd month, 27th day, after 40 days of rain and fountain eruptions)	Genesis 7:12, 17–20: And the rain was on the earth **forty days and forty nights.** [17]Now the flood was on the earth **forty days. The waters increased and lifted up the ark**, and it rose high above the earth. [18]The waters prevailed and greatly increased on the earth, and the ark moved about on the surface of the waters. [19]And the waters prevailed exceedingly on the earth, and **all the high hills under the whole heaven were covered.** [20]The waters prevailed fifteen cubits upward, and **the mountains were covered.**
150 days (including initial 40 days)	Day 150 (mo 7, day 17)	Genesis 7:24–8:5a: And the waters prevailed on the earth **one hundred and fifty days.** [8:1]Then God remembered Noah, and every living thing, and all the animals that were with him in the ark. And God made a wind to pass over the earth, and the waters subsided. [2]**The fountains of the deep and the windows of heaven were also stopped, and the rain from heaven was restrained.** [3]And the waters receded continually from the earth. **At the end of the hundred and fifty days the waters decreased.** [4]Then the ark rested in the seventh month, the seventeenth day of the month, on the mountains of Ararat. [5]And the waters decreased continually until the tenth month.
74 days	150 + 74 = Day 224 (mo 10, day 1)	Genesis 8:5b: In the tenth month, on the first day of the month, **the tops of the mountains were seen.**
40 days	224 + 40 = Day 264 (mo 11, day 11)	Genesis 8:6–7: So it came to pass, **at the end of forty days**, that Noah opened the window of the ark which he had made. [7]Then **he sent out a raven**, which kept going to and fro until the waters had dried up from the earth. [8]**He also sent out from himself a dove...**
7 days	264 + 7 = Day 271 (mo 11, day 18)	Genesis 8:8–9: He also sent out from himself a dove, to see if the waters had receded from the face of the ground. [9]But **the dove found no resting place for the sole of her foot, and she returned** into the ark to him, for the waters were on the face of the whole earth. So he put out his hand and took her, and drew her into the ark to himself.
7 days	271 + 7 = Day 278 (mo 11, day 25)	Genesis 8:10–11: And **he waited yet another seven days, and again he sent the dove out from the ark.** [11]Then the dove came to him in the evening, and behold, a freshly plucked olive leaf was in her mouth; and Noah knew that **the waters had receded from the earth.**

Period Duration	Period Starting Point	Main Events During This Period
7 days	278 + 7 = Day 285 (mo 12, day 2)	Genesis 8:12: So he waited yet another seven days and sent out the dove, which did not return again to him anymore.
29 days	285 + 29 = Day 314 (601st year of Noah's life: mo 1, day 1)	Genesis 8:13: And it came to pass in the six hundred and first year, in the first month, the first day of the month, that the waters were dried up from the earth; and Noah removed the covering of the ark and looked, and indeed the surface of the ground was dry.
56 days	314 + 56 = Day 370 (mo 2, day 27)	Genesis 8:14–17: And in the second month, on the twenty-seventh day of the month, the earth was dried. [15]Then God spoke to Noah, saying, [16]"Go out of the ark, you and your wife, and your sons and your sons' wives with you. [17]Bring out with you every living thing of all flesh that is with you: birds and cattle and every creeping thing that creeps on the earth, so that they may abound on the earth, and be fruitful and multiply on the earth."

(Bible quotations in the above table are from NKJV.)

There are some points referred to in the above table that bear further investigation; these are covered below.

The "Fountains of the Great Deep"

Genesis 7 refers to the "fountains of the great deep"—what could that mean? There are two major theories for what happened during the time of Noah's Flood, which endeavor to explain the flood processes and the results we see in the geologic record worldwide.

One theory is called "Catastrophic Plate Tectonics"[654] and the other is called the "Hydroplate Theory,"[655] and both are pretty good at explaining the geologic layers, the fossils, the mid-Atlantic ridge, continental drift, and other observable geologic phenomena of our planet. Both theories have landmasses and seafloors as tectonic plates that can move relative to each other, and this movement is the cause of the geological state of things today. In my opinion, the Catastrophic Plate Tectonics explains the details a bit better, although again, both explain a great deal of what we see in geological observations, and both are far superior to the uniformitarian speculations.

Let's take a look at some of the first verses in Genesis, which describe how God created the earth:

Genesis 1:2 (NKJV): The earth was without form, and void; and darkness was on the face of **the deep**. And the Spirit of God was hovering over **the face of the waters.**

The word translated "waters" above, unsurprisingly, means "waters." But the Hebrew word translated "deep" is תְּהוֹם (*thowm*, H8415). The Hebrew definition of *thowm* is "an abyss (as a surging mass of water), especially the deep (the main sea or the subterranean water-supply):—deep (place), depth." That is fascinating, because of the ideas of "*main* sea" (perhaps as opposed to a smaller, lesser sea above), as well as the idea of "*subterranean* water supply" (as opposed to the surface water supply).

Both of these phrases imply the earth's surface originally being entirely water. The rocky crust of the planet was not absent, but merely submerged, until God made it appear, as described in Genesis 1:9. And then below the rocky crust, was another body of water, perhaps larger than the surface one.

With that in mind, let's look at one of the Psalms of David:

Psalm 24:1–2 (NKJV): The earth is the Lord's, and all its fullness, the world and those who dwell therein. ²For He has founded it **upon the seas,** and established it **upon the waters.**

> v. 2, NIV: for he founded it **upon the seas** and established it **upon the waters.**
>
> GNT: He built it **on the deep waters beneath the earth** and laid its foundations **in the ocean depths.**
>
> NET: For he set its foundation **upon the seas,** and established it upon the **ocean currents.**
>
> NLT: For he laid the earth's foundation **on the seas** and built it **on the ocean depths.**
>
> WYC: For he founded it **on the seas;** and made it ready **on floods.** (For he founded it **upon the seas;** and established it **upon the depths below.**)

Fascinating: God founded the earth "*upon* the seas" and established it "*upon* the waters."

All theories make one or more assumptions as a starting point, and the two theories above make only one: subterranean water, as shown in Genesis 1:2 above. Then when the Flood actually starts:

> Genesis 7:11 (NKJV): In the six hundredth year of Noah's life, in the second month, the seventeenth day of the month, on that day all **the fountains of the great deep were broken up,** and the windows of heaven were opened.

You guessed it: the "great deep" spoken of here is the same Hebrew word *thowm* referred to above.

Any water beneath the rocky crust would be hot and under enormous pressure, of course. So if a rupture occurred in the rocky crust—if the "great deep" was broken up—supercritical water would explode skyward in supersonic jets of rapidly expanding steam—the "fountains"—carrying much ocean water and debris with it. This would add a great deal of moisture to the air, which would contribute to the torrential rain that had already started falling.

The above is a very superficial glance at the Hydroplate Theory, but if you are interested, I would highly recommend reading the detailed description found in *In the Beginning: Compelling Evidence for Creation and the Flood,* by Walt Brown.[656]

Starting with only the assumption of subterranean water, and then following where known laws of physics took him, Dr. Brown demonstrates how and why *so many* geological features could have formed: The Grand Canyon, the mid-Atlantic Ridge, continental shelves, ocean trenches, the "Ring of Fire," earthquakes, magnetic variations on the ocean floor, submarine canyons, coal and oil, methane hydrates, the Ice Age, mountain ranges, geothermal heat, volcanoes and lava, strata and layered fossils, metamorphic rock, salt domes, and so much more.

The *Mountains* were Covered?

One part of the Flood account that sometimes throws people for a loop is this part:

> Genesis 7:20 (NKJV): The waters prevailed fifteen cubits upward, and **the mountains were covered.**

People who have a hard time with the above verse often respond with something like this: "What? There's no way that the water could have covered all the mountains! Mount Everest is more than 29,000 feet high! There's not nearly enough water on earth to cover all the mountains!"

The resolution of this perceived problem is not at all difficult once we realize that the height of Earth's current mountains was caused by the collision of tectonic plates, whose movement was a *result* of the Flood. In other words, the mountains were not as high in the pre-Flood world as they are now. But even so, there is still an awful lot of water on earth: if the earth's surface was a perfectly smooth sphere, there is enough water in the oceans to cover the entire planet more than *two miles deep*.[657] Put another way, the volume of all the water on earth is *ten times greater* than the volume of all land masses above sea level.[658]

Not only that, but the tectonic activity of the Flood would have caused seafloors to rise—raising the water level—and the relatively heavy mountain ranges to settle—lowering the mountaintops. So once all the relevant cause-and-effect chains are taken into account, it's not difficult at all to see how a worldwide flood could cover the hills and mountains.

How Could the Ark Hold All Those Animals?

The answer to this question has several factors, each one of which, individually, is plausible and easily implemented, and which, collectively, accomplish the goal of accommodating all the animals, with plenty of room for food and other supplies.

Let's take a look at these factors.

The Ark Was Big

The ark was big, to be sure, but *how* big? The size of the ark was measured in cubits, and there are different definitions of "cubit," depending on which ancient culture it came from. The "royal cubit," the older unit of measure, was about 20.4 inches, and the newer "common cubit" was 18 inches.[659] Because the Bible apparently uses the older royal cubit,[660] it is likely that Noah's

Ark used the royal cubit, although for John Woodmorappe's book *Noah's Ark: A Feasibility Study*, he used the smaller common cubit, and there was still plenty of room.[661]

This illustration gives a size comparison between Noah's ark and some other well-known vessels: the Santa Maria (Christopher Columbus's largest ship, which sailed in 1492), the Titanic (launched in 1912), and the Queen Mary 2 (launched in 2004).

The ark is no small vessel:

Image by Dave Arns

A size comparison between Noah's Ark and other well-known ships.

> Genesis 6:15 (NKJV): And this is how you shall make it: The length of the ark shall be **three hundred cubits**, its width **fifty cubits**, and its **height thirty cubits**.

Since a royal cubit is 20.4 inches, or 1.7 feet, the ark's dimensions of 300×50×30 cubits is equal to 510×85×51 feet, which is 2,210,850 cubic feet of volume!

The ark had three decks:

> Genesis 6:16 (NKJV): You shall make a window for the ark, and you shall finish it to a cubit from above; and set the door of the ark in its side. **You shall make it with lower, second, and third decks.**

Since it had three decks, each deck would have been about 510 feet long and 85 feet wide, making the potential floor space more than 130,000 square feet. Assuming 20 feet or so of each deck was devoted to ramps for loading and unloading animals and supplies, that is still around 125,000 square feet of floor space that could be used for Noah's family, the animals, food, and other supplies.

One of the reasons that many modern-day people have trouble imagining Noah building an ark of this size and with this capacity is because of a subconscious worldview of people thousands of years ago being primitive brutes, the stereotypical "caveman"—*surely* not nearly as clever as we are today. That, in itself, indicates the influence of evolutionary thinking seeping into our minds and influencing them away from Biblical statements.

It Wasn't *That* Many Animals

A common misconception for people who are suspicious of the whole Noah's Ark story is the enormous number of animals that they imagine would have needed to be on the ark. But let's examine that idea in more detail.

Take a look at what God told Noah about which animals would be coming onto the ark:

> Genesis 6:19–20, 7:2–3 (NASB) And of every living thing of all flesh, you shall bring **two of every kind** into the ark, to keep them alive with you; they shall be male and female. ²⁰Of the birds **after their kind**, and of the animals **after their kind**, of every creeping thing of the ground **after its kind**, two **of every kind** will come to you to keep them alive. ²You shall take with you of every clean animal by sevens, **a male and his female**; and of the animals that are not clean two, **a male and his female**; ³also of the birds of the sky, by sevens, **male and female, to keep offspring alive** on the face of all the earth.

This tells us several very important things. Firstly, and most well-known, is that there were at least *two* animals of each kind—a breeding pair—for the purpose of repopulating the earth after the floodwaters abated. Clean animals came by sevens: three breeding pairs, plus one animal of each kind to be offered as a burnt offering in thanksgiving for preserving Noah's family and enough animals to repopulate earth after the Flood:

> Genesis 8:20 (TEV): Noah built an altar to the Lord; he took **one of each kind of ritually clean animal and bird, and burned them whole as a sacrifice** on the altar.

Secondly, only land animals and birds were on the ark. Even though many marine animals were killed in the violence of the Flood—almost 95% of the fossil record is composed of shallow-water marine organisms[662]—enough of them survived to continue the existence of most kinds of marine animals, although there were undoubtedly some extinctions during or shortly after the Flood.

Thirdly, the animals came into the ark "after their kind." What does that mean? It does *not* mean that every single species, as we understand the term today, needed to have a breeding pair on board the ark. In today's taxonomic progression of kingdom, phylum, class, order, family, genus, and species, the Biblical "kind" most closely corresponds to "order" or "family," which means that different genera and species did not need to be individually represented on the ark. In general, if a male animal and a female animal can mate and bear fertile offspring, they are of the same Biblical "kind;" dogs and cats are not interfertile, because they are of different kinds.

So, for example, in the dog department, Noah did not need to have a breeding pair for collies, dachshunds, spaniels, terriers, bulldogs, chihuahuas, poodles, huskies, pit bulls, greyhounds, dalmations, German shepherds, and every other breed recognized today—there are hundreds.[663] But since all of these breeds are of the "dog" kind, only one breeding pair would be needed on the ark.

All of the dog species we see today are results of various combinations of characteristics already latent in the genetic information of the original dogs' DNA. The potential for all dogs' characteristics—length of hair, color of hair, length of legs, body build, face shape, ear shape and size, muscularity, and so forth—was existent in the original DNA, and selective breeding eventually emphasizes some attributes at the expense of others. Such selective breeding may be intentional, as through a human-designed breeding program, or unintentional, as through a subset of the population being isolated from others of its kind.

This kind of selective breeding, whether intentional or not, is an example of the ill-named "microevolution" discussed earlier (it is ill-named because it is not actually evolution, as the theory defines it, but merely changes within a kind). Microevolution, as you remember, is only the expression of different combinations of attributes already defined in the organism's DNA, and no new genetic information is required. In fact, if selective breeding goes on too long, certain attributes are completely bred out of the gene pool and cannot be recovered; there is a *loss* of genetic information. In such a case, for example, a male and a female that each come from a long line of short-haired dogs *cannot* have a long-haired puppy.

Hybrid dogs, created by cross-breeding different species, are possible because the dogs are of the same Biblical kind—they can have fertile puppies. Examples include "huskamute" (husky and malamute), "labroxer" (Labrador retriever and boxer), "chihuachshund" (chihuahua and dachshund), "poogle" (beagle and poodle), and so forth.[664]

Because Noah only needed one or three breeding pairs of each *kind*, instead of each *species*, that reduces the number of required animals dramatically.

The Animals Used Little Floor Space

How much floor space did each animal need? Not as much as you might think.

All the animals were not simply running free on the ark; they had specific places to be. God told Noah to build enclosures for all the animals:

> Genesis 6:14 (AMP): Make yourself an ark of gopher or cypress wood; make in it **rooms (stalls, pens, coops, nests, cages, and compartments)** and cover it inside and out with pitch (bitumen).

Most children's Bible-story books emphasize large animals in their illustrations of the ark: elephants, giraffes, zebras, hippos, lions, and so forth, but most animals are much smaller than this. The ark had three decks that totaled 30 cubits, or 51 feet in height. Not counting the thickness of the hull, decks, and roof, each deck, floor to ceiling, was still around 16 feet high. Many of the smaller animals could be put into smaller cages stacked six or eight high.

And not only that, but it is unlikely that God brought full-grown animals to the ark. More likely, God brought younger ones, about a year younger than the beginning of their reproductive stage of life. This has many advantages: since they were smaller, the animals would have taken less floor space, they would have eaten less food, they would have created less waste, there would be less agitation from mating rituals, and it would have maximized the number of years of fertility once the Flood was over.

Some people have wondered what the carnivores ate during the Flood year; wouldn't Noah have needed some extra animals—a *lot* of them—just to be food for the carnivorous animals?

Actually not. There *were* no carnivores until after the Flood. Originally, both man and animals were vegetarians:

> Genesis 1:29–30 (NLT): Then God said, "Look! I have given you **every seed-bearing plant throughout the earth and all the fruit trees for your food.** ³⁰ And I have given **every green plant as food for all the wild animals, the birds in the sky, and the small animals** that scurry along the ground—everything that has life." And that is what happened.

The eating of meat, by humans or other animals, began *after* the animals left the ark:

> Genesis 9:2 (CEV): **All animals, birds, reptiles, and fish** will be afraid of you. I have placed them under your control, ³and **I have given them to you for food. From now on, you may eat them**, as well as the green plants that you have always eaten.

So yes, the ark had plenty of room for two (or seven) of each kind of animal and bird, plus room for food and other supplies, and living quarters for Noah and his family.

For More Detail. . .

It is beyond the scope of this book to get into greater detail of how Noah's flood is not only plausible as a historical event, but it explains observations better than a uniformitarian view or a punctuated-equilibrium view. Such observations include the thickness and expanse of sedimentary rock layers, the fossil record overall, polystrate fossils, fossilization of soft tissue, tectonic plate movement, genetic diversity, bone beds, irreducibly complex structures, and so forth.

I encourage you to research the whole idea of Noah's flood and Noah's ark; I think you might be surprised at how reasonable the idea is. I can recommend these books on the topic (see "Appendix A: Bibliography" for more complete information):

- *A Flood of Evidence: 40 Reasons Noah and the Flood Still Matter* (Ken Ham and Bodie Hodge)
- *The Genesis Flood: The Biblical Record and Its Scientific Implications* (John C. Whitcomb and Henry M. Morris)
- *Noah's Ark: A Feasibility Study* (John Woodmorappe)

The entire contents of the above books (and there are many more) are devoted to showing the plausibility of Noah's flood and the Ark. Other books that talk about a variety of creation-related topics have whole chapters dedicated to the topic of the Flood and the ark.[665]

The Lack of Transitional Fossils

The above information, and *so* much more I didn't have room to include, shows the absolute dearth of transitional fossils—those fossils that should outnumber non-transitional fossils many to one.

Even from earliest days of the promotion of evolution, it was known that a lack of fossils of transitional forms would be a deathblow to the theory. Even Charles Darwin acknowledged this fact:

> But just in proportion as this process of extermination has acted on an enormous scale, so must the number of intermediate varieties, which have formerly existed on the earth, be truly enormous. **Why then is not every geological formation and every stratum full of such intermediate links? Geology assuredly does not reveal any such finely graduated organic chain; and this, perhaps, is the most obvious and gravest objection which can be urged against my theory.** The explanation lies, as I believe, in the extreme imperfection of the geological record.[666]

So Darwin admitted that the lack of transitional fossils, when they should be everywhere, would be a showstopper to the credibility of his theory. In the last sentence of the above quote, he attributed the absence of transitional fossils

CHAPTER 7: THE FOSSIL RECORD

to the fact that the geological record had not yet been thoroughly explored. However, in the 160 years since Darwin's book was published, we have found millions more fossils all over the world, quite thoroughly exploring the fossil record, and still, *not one* transitional form has yet been discovered.

In the paragraph above, I make the claim that the fossil record has now been "quite thoroughly" explored. What exactly does that mean? *How* thoroughly? The table below shows the actual numbers, as of 1985, more than 120 years after Darwin's lament about the incompleteness of the fossil record.

Of the taxonomic levels of kingdom, phylum, class, order, family, genus, and species, the table below[667] describes terrestrial (non-ocean-dwelling) vertebrates at the level of Order and Family, and compares the number that have been found as fossils to the number that are still living today.

Groups Compared	Still Living	Found as Fossils	Percentage Found
Orders	43	42	97.7%
Families	329	261	79.3%
Non-bird Families	178	156	87.6%

So, of all the Families of terrestrial vertebrates alive today, almost 80% have also been found as fossils. If we don't include birds, the number rises to almost 90%. And at the level of Orders, the number rises to almost 98%. And still, *not a single example* of a transitional half-this-and-half-that form has been found. And with each new type of fossil discovered, the more statistically unlikely it is that they actually exist, but we've just "coincidentally" not discovered them yet. The more research is done, more far-fetched the whole idea of transitional forms becomes.

How does evolution—or, I should say, how do *evolutionists*—deal with that fact? It's been an embarrassment for more than a century and half. Here are a few more examples of just how bad a problem it is:

> Surely the lack of gradualism—**the lack of intermediates— is a major problem.**[668]

---

[667] Michael Denton, *Evolution: A Theory in Crisis* (Bethesda, MD: Adler and Adler, 1985), p. 190.

[668] Dr. David Raup, is taken from page 16 of an approved and verified transcript of a taped interview conducted by Luther D. Sutherland on July 27, 1979. Emphasis added.

> In fact, **the fossil record does not convincingly document a single transition** from one species to another.[669]
>
> . . .fossil species remain unchanged throughout most of their history and **the record fails to contain a single example of a significant transition.**[670]
>
> Well, we are now about 120 years after Darwin and the knowledge of the fossil record has greatly expanded. We now have a quarter of a million fossil species but the situation hasn't changed much. The record of evolution is still surprisingly jerky and, **ironically, we have even fewer examples of evolutionary transition than we had in Darwin's time.** By this I mean that some of the classic cases of Darwinian change in the fossil record, such as the evolution of the horse in North America, have had to be discarded or modified as a result of more detailed information—what appeared to be a nice simple progression when relatively few data were available now appears to be much more complex and much less gradualistic. So **Darwin's problem has not been alleviated in the last 120 years. . .**[671]
>
> It is not even possible to make a caricature of evolution *[Darwin's gradualism]* out of paleobiological facts. The fossil material is now so complete that a lack of transitional series cannot be explained by the scarcity of material. **The deficiencies are real; they will never be filled.**[672]

So the absence of evolution, as shown by the fossil record, is a big problem for evolution. But a new approach has now been proffered: the fossil record doesn't matter anymore.

---

[669] Steven M. Stanley, *The New Evolutionary Timetable* (New York: Basic Books, Inc., 1981), p. 73. Emphasis added.
[670] David S. Woodruff, "Evolution: The Paleobiological View," *Science*, Volume 208, May 16, 1980, p. 716. Emphasis added.
[671] David M. Raup, "Conflict Between Darwin and Paleontology," *Field Museum of Natural History Bulletin*, Volume 50, No. 1, January 1979, p. 25. Emphasis added.
[672] Geoffrey Simmons, M.D., *What Darwin Didn't Know*, 2004 (Harvest House Publishers, Eugene, Oregon), p. 304.

Seriously? Yes. Mark Ridley, from Oxford University, made the following statement:

> **No real evolutionist**, whether gradualist or punctuationist, **uses the fossil record as evidence** in favor of the theory of evolution as opposed to special creation. This does not mean that the theory of evolution is unproven.[673]

That sounds remarkably like "Since we got no evidence, let's pretend like we don't need it," doesn't it? *Of course* evolutionists don't use the fossil record as evidence for evolution; you'd need transitional forms to do that. The fossil record shows that life forms appear abruptly and fully formed; no transitional forms in sight. (Which, by the way, in case I hadn't mentioned it before, strongly supports the Creation model.) The above quote from Ridley was in a section whose heading announced, "The evidence for evolution simply does not depend upon the fossil record."

Ridley goes on to say that:

> **The gradual change of fossil species has never been part of the evidence for evolution.** In the chapters on the fossil record in *The Origin of Species* Darwin showed that the record was useless for testing between evolution and special creation because it has great gaps in it. The same argument still applies.

Um, no. That is some serious revisionist history. As we saw above, Darwin himself said that the absence of transitional forms was "the most obvious and gravest objection which can be urged against my theory." And *that* statement still applies.

It appears that Dr. R. B. Orr, in his report to the Ontario (Canada) Minister of Education, hit the proverbial nail on the head:

> When a writer has lost faith in the supernatural and then surrenders himself to skeptical theories, **there is no limit to which he will not go to support his own views.** No man may be called a scientist because he accepts as a certainty that which is but a theory—such, for example, as the evolution of man from the worm or the ascent of civilized man from a

---

[673] Mark Ridley, "Who Doubts Evolution?" *NewScientist* 25 (June 1981): p. 831.

savage. For **science, if it means anything, implies demonstration leading to stern truth.**[674]

Note: science "implies demonstration"—repeatable experiments that can demonstrate the truthfulness or falsity of a theory. Then British journalist Malcolm Muggeridge said much the same thing:

> One of the peculiar sins of the twentieth century which we've developed to a very high level is the sin of credulity. It has been said that when human beings stop believing in God they believe in nothing. The truth is much worse: **they believe in anything.**[675]

Few places is this credulity more evident than in the minds of evolutionists claiming that order is the result of disorderly processes.

Chapter 7 Summary

Here are the most important points in this chapter:

- **The fossil record contains no partly-this-and-partly-that intermediate forms.** This is exactly what the creation model predicted, but it is contrary to evolution's predictions.
- In evolutionary thinking, **rocks are often dated by the fossils in them, but fossils are dated by the rocks they're in.**
- Polystrate fossils are **individual fossils that extend through multiple sedimentary layers,** showing that those layers couldn't have taken "millions of years" to form. Polystrate fossils have been found of leaves, whales, jellyfish, tadpoles, dinosaur tracks, nautiloids, and more.
- Multilayer **slabs of sedimentary rock are *bent*, not broken, showing that they were all soft when the bending force was applied.** This too disproves the "millions of years" timeframe.
- Bioturbation, the process of animals and plants churning up soil, also shows that the sedimentary layers did not take long ages to form. **Bioturbation mixes layers within a few years, so if the sediment were laid

---

[674] Fourth session of the 14th legislature of the province of Ontario, Session 1918, Appendix to the Report of the Minister of Education, Ontario, 1917: by Dr. R. B. Orr., p. 65. Emphasis added.

[675] Malcolm Muggeridge, "An Eighth Deadly Sin," *Woman's Hour* radio broadcast, March 23, 1966. Quoted in Malcolm Muggeridge and Christopher Ralling, *Muggeridge Through the Microphone: B.B.C. Radio and Television* (London: British Broadcasting Corporation, 1967).

down over long ages, they would no longer have any layered appearance.
- **Soft tissue—skin, blood cells, intact DNA, and so forth—appears in many fossils,** even though it couldn't have survived for long ages.
- **Ecosystems can form** (as in Surtsey Island), or recover (as is the Mount St. Helens area), **much faster than uniformitarians believed possible.**
- Loose volcanic ash, thought by evolutionary geologists to require thousands of years to turn into the glassy palagonite tuff, **required only a few decades on Surtsey Island.**
- Evolutionary geologists thought that sedimentary layers took long ages to form even thin layers. But since Mount St. Helens erupted in 1980, up to **400 feet of strata have formed in less than 40 years. In some places, a 25-foot-thick series of thin sedimentary layers formed in just a few hours!**
- The Toutle River, blocked by debris from the Mount St. Helens, was forced to change course, and **since 1980, it has carved out a canyon 140 feet deep** in some places. Deep time unnecessary.
- Given the Grand Canyon's volume, the Colorado River's flow rate and sediment content, **it would only take a few thousand years to dig out the Grand Canyon.**
- The "Cambrian Explosion" is a burst of **complex new organisms appearing abruptly and fully formed in the fossil record, and with no ancestry,** exactly like evolution wouldn't be.
- The Cambrian Explosion was so creation-like that evolutionists invented **a rescue device called "punctuated equilibrium," which allows things to evolve almost instantaneously,** with vast amounts of genetic information appearing all at once, out of nowhere, exactly as if they'd been, well, created.
- The evolutionary pathway from invertebrates to fish: **No fossil evidence.**
- The evolutionary pathway from fish to amphibians: **No fossil evidence.**
- The evolutionary pathway from amphibians to reptiles: **No fossil evidence.**
- The evolutionary pathway from walking reptiles to flying reptiles: **No fossil evidence.**
- The evolutionary pathway from reptiles to mammals: **No fossil evidence.**

- The evolutionary pathway from walking mammals to swimming mammals: **No fossil evidence.**
- The evolutionary pathway from reptiles to birds: **No fossil evidence.**
- The **Genesis Flood explains** everything we find in the fossil record that uniformitarianism explains, plus **much that uniformitarianism** *cannot* **explain.**
- Since transitional fossils, essential to support evolutionism, are nowhere to be found, **a new approach of claiming that "fossils don't matter anyway" is being offered.**

Chapter 8:

The Next Step

So now you have read a smattering of the immense amount of data, experiments, and repeatable observations that militate against macroevolution. And "smattering" is a good word for it: one of the hardest things in writing this book was deciding which evidence for creation, and which evidence against evolution, *not* to include. There is *so* much of both, that this book easily could turn into a multi-volume set of thousand-page books.

Retaining Your Brain

There seems to be an impression among evolutionists that in order for anyone to actually embrace creationism, one would necessarily have to throw away your brain and commit intellectual suicide. As in, either you'd be an idiot in the first place, to accept creationism, or accepting creationism would turn you into an idiot.

Now again, there may be some readers who think I am overreacting, alarmist, or at least unnecessarily melodramatic. But considering the following quotes from evolutionists (some of which you've seen earlier) might adjust such a perception.

Richard Dawkins, the Charles Simonyi Professor of the Public Understanding of Science at Oxford University, quoted in Phillip Johnson's book:

> When he contemplates the perfidy of those who refuse to believe, Dawkins can scarcely restrain his fury. "It is ab-

solutely safe to say that, if you meet somebody who claims not to believe in evolution, **that person is ignorant, stupid or insane (or wicked,** but I'd rather not consider that)." Dawkins went on to explain, by the way, that **what he dislikes particularly about creationists is that they are intolerant.**[676]

When asked about his thoughts on school teachers who believe in creation, Dawkins replied:

We are failing in our duty to children, if we staff our schools with **teachers who are this ignorant—or this stupid.**[677]

We saw this quote from famous Russian-American evolutionist Theodosius Dobzhansky earlier, but it bears repeating. He wrote:

Evolution is a light which illuminates all facts, a trajectory which **all lines of thought must follow.**[678]

Michael Ruse, Professor of Philosophy at the Florida State University:

Scientific creationism is not just wrong; it is ludicrously implausible. It is a **grotesque parody of human thought and a downright misuse of human intelligence.** In short, to the Believer, it is an **insult to God.**[679]

This point of view is absolutely fascinating. Ruse is saying that it is "an insult to God" for a believer in God to actually believe what God says. One wonders how in the world he arrived at that conclusion. . . Ah: because his religion—and he freely admits that evolution is a religion—his religion doesn't have a God,[680] and that according to his religion, evolution is a "full-fledged alternative to Christianity." So here, Ruse is stating that, according to his religion, the sciences—the experiments and repeatable observations—that show his religion is mistaken are invalid.

---

[676] Phillip E. Johnson, *Darwin on Trial* (Regnery Publishing, Washington DC, 1991), p. 7. Emphasis added.

[677] Richard Dawkins and Steve Jones, "Richard Dawkins and Steve Jones give their views on creationism teaching poll," *The Guardian,* December 22, 2008. Emphasis added.

[678] Theodosius Dobzhansky, "Nothing in Biology Makes Sense Except in the Light of Evolution," *The American Biology Teacher* (March 1973): p. 129. Emphasis added.

[679] Michael Ruse, *Darwinism Defended: A Guide to the Evolution Controversies* (Menlo Park, CA: The Benjamin/Cummings Publishing Company, 1982), p. 303.

[680] Michael Ruse, "Saving Darwinism from the Darwinians," *National Post,* May 13, 2000, p. B-3.

CHAPTER 8: THE NEXT STEP

Stephen Law, senior lecturer in philosophy at the University of London:

> Would I include **young-earth creationism** in the school curriculum anywhere at all? I might put it next to, say, some conspiracy theories as examples, as illustrations of how people can be sucked into belief systems which are **utterly absurd**. And yet they themselves are convinced that everyone else is wrong and they are right. Once the person becomes sucked into that kind of belief system, they're never coming out. It makes them **intellectual prisoners**, and they become **intellectually unreachable**. It also encourages them to think in ways which are, which would normally be considered **symptomatic of mental illness**.[681]

As we can see, it is not uncommon for evolutionists to imply—or to flat-out state—that if you believe in creation, you are ignorant, stupid, or suffering from a mental illness. But let's think about it: if creationists are just a bunch of wackos, why do evolutionists get so enraged by our challenges to their beliefs? If creationists' beliefs are really that ludicrous, why do evolutionists get into such a frothing fury? If creationism really is a hopeless case, why do evolutionists bother with all the venomous hostility? Why give yourself an ulcer?

I'd like to propose that the blind rage expressed above, and other places on the internet (some far less tactfully), is so intense because this is not just an intellectual issue. There is a spiritual aspect of the creation/evolution question. When creationists present repeatable observations and experimentation that support the creation model and conflict with the evolutionary model, this is a very real threat to the naturalistic ("there is no God") worldview. And for those who have spent their lives promoting evolution, this is quite an emotional shock to the system.

It is a shock for at least two reasons. First, it would be terribly humiliating to acknowledge that what you've devoted your life to is demonstrably false and even destructive to people's lives. Second, it can be terrifying to acknowledge the God that you've hated your whole life is real and is exactly who He says He is.

But, I daresay, God is going to reveal himself to many evolutionists, and they will be confronted with His love and power in an undeniable way. Some will run into His arms, overwhelmed with joy and gratitude that they have fi-

---

[681] Stephen Law, "Should Creationism Be Taught in Schools?" 4thought.tv/themes/should-creationism-be-taught-in-schools/stephen-law. Emphasis added.

nally found what they've been searching for their entire lives. But some will not: some will harden their hearts even more, to the point that no amount of undeniable evidence will sway them.

So are real people actually hindered in the sciences because they believe in a literal six-day creation as outlined in Genesis? Not at all, in spite of evolutionists' statements (as in the title of the article by Dobzhansky, quoted above) that "nothing in biology can be understood outside the context of evolution."

Not even in biology does a rejection of evolution, or worse yet, a belief in creationism, hinder understanding. PhD cell biologist (and creationist) Dr. David Menton observed:

> The fact is that though widely believed, **evolution contributes nothing** to our understanding of empirical science and thus plays no essential role in biomedical research or education.[682]

Now some of my readers might be saying, "Well, of course *he* says that. He's a creationist." But it's not just creationists. Dr. Philip Skell, Emeritus Evan Pugh Professor of Chemistry at Penn State, researched it himself, and here's what he reported:

> I recently asked more than 70 eminent researchers if they would have done their work differently if they had thought Darwin's theory was wrong. **The responses were all the same: No.**
>
> I also examined the outstanding biodiscoveries of the past century: the discovery of the double helix; the characterization of the ribosome; the mapping of genomes; research on medications and drug reactions; improvements in food production and sanitation; the development of new surgeries; and others. I even queried biologists working in areas where one would expect the Darwinian paradigm to have most benefited research, such as the emergence of resistance to antibiotics and pesticides. Here, as elsewhere, I found that **Darwin's theory had provided no discernible guidance**, but was brought in, after the breakthroughs, as an **interesting**

---

[682] David Menton, "A Testimony to the Power of God's Word," Answers in Genesis, June 12, 2003. [Online at AnswersInGenesis.org/docs2003/0612menton_testimony.asp; accessed July 12, 2020.] Emphasis added.

narrative gloss. . . From my conversations with leading researchers it had became *[sic]* clear that modern experimental biology gains its strength from the availability of new instruments and methodologies, not from an immersion in historical biology.[683]

Fascinating. Evolution was brought in *after* the breakthroughs were made, almost as if obligatory. But it was of no benefit in actually *making* the breakthroughs.

Smartphones, smartwatches, computers, orbiting satellites, planetary probes, MRI machines, and every other bit of technology we use are made possible because of the laws of physics, which God created when He spoke our orderly, logical universe into existence. Then He created man with the ability to be creative and think logically, so that all the above technology is possible.

It's fascinating to see that all the above devices are created by intelligent people envisioning, designing, and manufacturing them. How then could it be helpful for them to have a belief that complex biological machines many orders of magnitude more complex do *not* need an intelligent designer?

The Parable of the Car Lot

Bringing the creation/evolution question into the automotive context may make certain concepts more understandable, so please indulge me as I tell you this parable. . .

A creationist and an evolutionist walk into a new-car lot at the same time. All the latest models had just been delivered a few days earlier, straight off the assembly line, and sit gleaming in the sunlight.

Having entered the dealership from opposite sides, the creationist and the evolutionist unintentionally meander toward each other. As they are walking, they are both marveling at the glistening examples of automotive engineering surrounding them; both are impressed at the attractiveness, power, and variety of the cars they see.

Approaching each other somewhere toward the middle of the car lot, they strike up a conversation. . .

Creationist: These cars sure are beauties, aren't they?

---

[683] Philip Skell, "Why Do We Invoke Darwin?" *The Scientist* 16:10. Emphasis added.

Oh, Evolve! (Good Luck With That...)

Evolutionist: They sure are!

Creationist: Look at that one! Such beautiful lines!

Evolutionist: And what an amazing paint job!

Creationist: Those car designers are amazing...

Evolutionist: Well, these cars weren't actually *designed,* of course. They evolved over long periods of time.

Creationist: What do you mean? These cars were just delivered from the factory about 6,000 minutes ago.

Evolutionist: Don't be silly. These cars have been right here, evolving little by little for the past 4.3 billion minutes.

Creationist: You're saying that these cars are more than four billion minutes old?

Evolutionist: Naturally; it takes a long time for cars to evolve out of dirt and rocks and water.

Creationist: They didn't evolve; they were manufactured just recently! See the logos on all the cars? Each one is like a fancy signature that shows who built them.

Evolutionist: Oh. Sorry; I wouldn't have taken you for the superstitious type. You believe these cars were *manufactured?* You believe they were intentionally *designed?*

Creationist: Well, of course! They were made for people to drive, for our benefit. Do you think it's a coincidence that they fit us so well? And that the transparent windshield is right where we need to look? And that the pedals are so conveniently positioned for our feet?

Evolutionist: Of course, they have the *appearance* of having been designed, just like this watch on my wrist. But all these cars—and my watch—are the result of random processes powered by sunlight and lightning strikes over long periods of time. Anything can happen, given enough time.

Creationist: Well, that doesn't make sense. Think about it: all these cars have tires, and all these tires have air in them. Air slowly leaks out of tires, even if they aren't damaged. We've measured how fast air pressure is lost, and if these tires were billions of minutes old, they would all be flat!

Chapter 8: The Next Step

Evolutionist: True, we know that air leaks out of tires at a known rate, but the air in the tires is being constantly replenished by air reservoirs inside the wheels.

Creationist: What? How did you come up with the idea of these air reservoirs? Do you have any actual evidence that they exist?

Evolutionist: Well, it's obvious that they exist. Like you say, air leaks out of tires at a known rate, so these reservoirs in the wheels must exist to replenish it.

Creationist: So, supposing that these reservoirs actually do exist, how could they have contained enough air to keep the tires inflated for all these billions of minutes?

Evolutionist: They just did, that's all. We all know that these cars are billions of minutes old.

Creationist: We do?

Evolutionist: Of course we do. It's completely unscientific to believe in a car manufacturer. No scientist worth his salt believes in a car manufacturer.

Creationist: Actually, there are a lot of us who do. . .

Evolutionist: Don't be silly. *(Whips out a chart from his pocket.)* See here? First, there was an original wheel. Over the eons, this wheel evolved by numerous tiny changes into the unicycle, which then evolved into the bicycle, then the tricycle, then the little red wagon—do you see how one wheel was added at each evolutionary stage? Then came the gas-powered go-cart, and finally the pinnacle of evolution: these beautiful cars all around us!

Creationist: You just arranged those into the order that would represent your assumptions!

Evolutionist: But my assumptions are clearly correct.

Creationist: So in the transition from the unpowered wagon to the gas-powered go-cart, the engine just "arose," tiny bit by tiny bit?

Evolutionist: Yes.

Creationist: So the gas tank, the carburetor, the flywheel, the spark plug, magneto, piston, crankshaft, valves, and all the rest, just "fell together" by accident, with no intention whatsoever? All these things, with just the right shape, size, composition, and each

	fulfilling a necessary function none of the others ones do, just *randomly* appeared because of lightning strikes and sunlight?
Evolutionist:	Correct.
Creationist:	But what good would it do to evolve a gas tank if there isn't a carburetor? What good would it do to have a piston without a crankshaft? Or a spark plug without a magneto? None of them would be of any benefit at all until *everything* was present! This is an irreducibly complex mechanism.
Evolutionist:	Ah, the mysteries of evolution. . .
Creationist:	So in order for the engine to "arise" through evolution, all the parts would have to have evolved at the same time, in a compatible way. And we haven't even talked about the engine block, that connects everything together in exactly the right way, with very precise tolerances.
Evolutionist:	It's remarkable that it happened that way, isn't it?
Creationist:	It's *impossible* that it happened that way.
Evolutionist:	I happen to believe differently. It's obvious to me that all these cars arose spontaneously. There's no other option, since these cars are 4.3 billion minutes old. And *(glaring at the creationist)* since there's no such thing as a car manufacturer.
Creationist:	*(Exasperated sigh.)* Here, check this out. *(Pops the hood of a car and tests its battery with a voltmeter he took from his pocket.)* Look there! It's still fully charged! If these cars were billions of minutes old, the batteries would all be dead! You know that batteries lose their charge over time, even when they're not used!
Evolutionist:	That is puzzling. But there must be some mechanism to maintain the charge, or replenish it, over the long ages. There has to be some such mechanism, because these cars clearly have been evolving for billions of minutes.
Creationist:	"Clearly?"
Evolutionist:	Yes! You can see that they exist! That *proves* they evolved!
Creationist:	So how did the first wheel in your little chart actually come to be?
Evolutionist:	It was the result of natural processes operating on rocks.
Creationist:	Oh, for Pete's sake! *(Opens a car door and grabs something from the glove compartment.)* Look at this! It's the Manufacturer's

CHAPTER 8: THE NEXT STEP

Handbook! It tells us how to operate these cars, so we avoid making stupid driving mistakes and end up having a wreck! Do you mean to tell me this Handbook just "evolved" out of ink molecules and really thin sheets of wood?

Evolutionist: Of course not. That handbook was written by people like yourself who need the psychological crutch of believing in some invisible "car manufacturer" in order to cope with the existence of cars...

A Prophetic Message

As you probably already know, Darwin made his famous voyage on the *HMS Beagle* from 1831–1836. One of the stops the ship made was at the Galapagos Islands, where Darwin observed the numerous unique species on the islands, and concluded that they had evolved from simpler species in that isolated environment. Darwin's mistake, similar to that of so many other evolutionists since, was a failure to distinguish between microevolution and macroevolution, as we discussed above, in the section "Microevolution vs. Macroevolution." Failing to distinguish between microevolution and macroevolution makes it sound reasonable to conclude that numerous small losses of genetic information can result in a net gain of genetic information.

In that section, we saw that microevolution, or change within a kind, is simply a reshuffling of attributes already described in the DNA, and requires no new genetic information. Macroevolution, on the other hand, is where one kind of organism gains entirely new functionality, described by new genetic information in the DNA. This is what the theory of evolution requires to have happened trillions of times, but not a single case of this actually occurring has ever been observed.

On one of the islands in the Galapagos archipelago is a natural rock formation that was named "Darwin's Arch," in honor of his groundbreaking work in discovering How Everything Came To Be; this arch is shown in the photo at right. But as scientific knowledge continues to grow, more and more of evolution's claims have been disproven, because its "evidence" was misinterpreted,

Creative Commons CC BY-SA 3.0
Darwin's Arch in the Galapagos Islands, in 2006.

manipulated, fabricated, or simply assumed. As this continues, more and more intellectually honest scientists the world over are becoming disenchanted with evolution as a plausible explanation for anything, and are cautiously entertaining the possibility that maybe, just maybe, creation actually happened. After all, the evolution model requires *so* many repeatable observations to be ignored, and *so* many laws of physics to be set aside, that the required suspension of disbelief is becoming ever more burdensome.

Almost as a divine affirmation of the accelerating collapse of the religion of evolutionism, on May 17, 2021, Darwin's Arch collapsed. This is no coincidence; rather, it shows all the earmarks of a prophetic sign, visible to those who have eyes to see, that Darwinism's days are numbered, and soon there will be a wholesale abandonment of the theory. When the blind devotion to evolutionism is dispensed with, scientists will be astonished that they ever believed in evolution in the first place, and they will be overwhelmed at how much better observation and experimentation support the creation model.

Image credit unknown—it appears on scores of websites around the world. *Darwin's Arch, after it collapsed on May 17, 2021.*

But what about people who choose to see the collapse of Darwin's Arch as simply a coincidence, and not a prophetic message from God? That's okay; it's still an excellent example of the Second Law of Thermodynamics, which consistently and repeatably shows that complexity does not arise by itself, and random processes cannot create order. Rather, the Law shows that, when left to themselves, things wear out, break, collapse, and fall apart. Like Darwin's Arch.

Requiem for Deep Time

Evolution, by its very definition, requires and assumes vast amounts of time, because the necessary processes, if they happened quickly, would too closely resemble a miracle, and that is strenuously resisted by evolutionists. Hence, the idea of "deep time."

Also, in an effort to avoid the Mind that would be required to create the information we see in every living thing, evolution requires and assumes that nothing operated over said deep time except natural processes operating in a uniform manner. Hence, the idea of "uniformitarianism."

So evolution absolutely depends on these two concepts, but, as shown in numerous places through this book, those very concepts are highly problematic because they repeatedly and insistently contradict repeatable observations and experimentation. Here's how it usually goes:

Evolutionist: "This process has been happening uniformly for millions of years."
Scientist: "Okay, let's measure how fast that process happens and see if your statement makes sense."
Evolutionist: "Oh, it will, because that is the only possible answer."
Scientist: *(after measuring the rate of the process)* "Wait a minute! We know how fast the process happens—we just measured it—and if that process has been happening uniformly for millions of years, that means that the situation would have been *(description)* back then."
Evolutionist: "That's ridiculous!"
Scientist: "I know! But do the math!"
Evolutionist: "I'd rather not. . ."

So one way that deep-time uniformitarianism get into trouble is that it can be repeatedly demonstrated that deep time simply is not necessary for many processes (geologic processes, for example) to happen, even though evolutionists vociferously claim otherwise.

Another way the problems with deep time and uniformitarianism show up is not at the *rate* of a certain process being refuted by observations, but simply that if a certain event had happened *at all*, it would have left evidence of a certain kind (like biological evolution, for example, in which transitional fossils would be everywhere). The absence of this "smoking gun" is also a recurring problem for evolution, because of its dependence on deep time and uniformitarianism.

And yet another way that deep time gets into trouble is that, in many cases, deep time is actually counterproductive to the given process, even though evolution claims that "millions" or "billions" of years were needed for certain processes to take place. If such processes actually did take that long, the results would be dramatically different than what we observe (for example, folded rock layers, continental erosion, radiocarbon dating, intact DNA in soft-tissue fossils, Arches National Park, and so forth).

OH, EVOLVE! (GOOD LUCK WITH THAT...)

The only two logical resolutions to these common problems for evolution is either 1) to acknowledge that uniformitarianism is not always the case, and therefore catastrophism must happen at least sometimes, or 2) the time component is much less "deep" than evolutionists would prefer. But since both of those options are usually deemed unacceptable, evolutionists come up with a third option whenever necessary: invent a rescue device. These are supplemental theories that are appended to the main theory, and invariably, there is no actual evidence to support these supplemental theories; their sole purpose for existing is to make the original theory's flaws less incriminating.

So, just to remind ourselves again how the combination of deep time and uniformitarianism, or simply the application of known laws of physics, repeatedly contradict science (using repeatable real-world observations and experimentation to determine the validity of theories), let's take another glance:

- The Big Bang is the deep-time assumption of How Everything Came To Be. But there are numerous scientific problems with the theory:
  - If the Big Bang happened, there would be magnetic monopoles still around. They're not.
  - If the Big Bang was really a random occurrence, we sure are "lucky" that the critical density of the universe was tuned to the right value, with a necessary precision of 1 part on 10^{62}. Rescue device: Inflation.
  - If the Big Bang happened, half the mass of the universe would be antimatter. It's not; if it were, life would not be possible.
  - If the alleged singularity from which the Big Bang supposedly came were real, all the mass of the universe would have been contained in it, giving it an escape velocity far above c. So how could it have exploded?
  - If the Big Bang happened, the expansion of the matter in the universe would be slowing down because of gravity. But observations show it to be *speeding up!*
  - If the Big Bang happened, Population III stars—stars composed of only hydrogen, helium, and lithium—should still exist. Not one has ever been found.
  - If the Big Bang happened, the universe should be homogeneous—the same average density throughout. It is definitely not; the universe is "clumpy," with galactic clusters and superclusters, separated by vast expanses of empty space. Such structures could not form naturally in less than 150 billion years, way more than

even deep time allows. Rescue device: Cold Dark Matter (CDM).
- If the Big Bang happened, the Cosmic Microwave Background (CMB) would be homogeneous to the same extent as the matter density. But it's not: even though the matter is clumpy, the CMB is homogeneous!
- If the Big Bang happened, how did the galaxies start spinning? Linear motion is easy to explain from an explosion, but how did galaxies start spinning? Where did the angular momentum come from?

- Blue stars use their nuclear fuel at a known rate, so they couldn't be more than a couple million years old, but evolutionary cosmology says they're billions of years old. Rescue devices: stellar nurseries, siphoning matter off other stars, slow consumption rate until just recently, collisions with other stars, blueness is a new thing.
- Supernovas have been observed to happen at a particular rate, but if they had been exploding at that rate for as long as uniformitarian deep time says they have, space should be littered with visible supernova remnants. It isn't.
- If spiral galaxies have been spinning as long as deep time says they have, the spiral arms would have blurred together into a featureless disk, because of the different rotation speeds as you go out from the center. They have not blurred together. Rescue devices: Density waves, recurrent arms.
- The uniformitarian deep-time Nebular Hypothesis of solar-system creation couldn't work because of death spirals (planets crashing into the sun), lack of an accretion process (to stick particles together), and turbulence (collisions knocking things apart).
- The Nebular Hypothesis of star formation is supposedly caused by a cloud of gas and dust contracting under its own gravity until fusion starts in its core. But this contraction would never happen because of thermal expansion and magnetic fields. Also, angular momentum would prevent this collapse, if the cloud were spinning, but evolutionary cosmologists can't explain why the cloud would start spinning. But it "must have," to make the planets orbit as they do.
- Exoplanets, which were supposed to show how our solar system "formed," via the Nebular Hypothesis, don't have nearly enough matter in their dust disks to make planets, even if they could contract. Many have 90%–99% too *little* mass.

- The Roche Limit is that distance, inside of which, the primary's gravity is stronger than that of the particles of the alleged dust clouds trying to turn into planets. So the Nebular Hypothesis can't explain how planets close to the sun, or moons close to their planets, would form.
- Because of the mechanisms of nuclear fusion, the sun would have been 30% fainter in the deep-time distant past, which means that when life was supposedly evolving on earth, all of earth's oceans would have been frozen solid. Rescue device: "Let's say there was 1000 times more carbon dioxide back then!"
- Mercury's magnetic field is weakening, and we have measured how fast it does so. It would have been gone eons ago if Mercury was really billions of years old. But it's still there, weakening by measurable amounts in only decades.
- Mercury is still geologically active, but it shouldn't be if it were really billions of years old. Rescue device: outgassing of volatile material. But those materials, too, would have been exhausted eons ago if Mercury were really billions of years old. Also, those volatile substances wouldn't have even survived planetary formation if the Nebular Hypothesis were how that planet came to be. But there they are.
- Venus rotates backwards, disqualifying the Nebular Hypothesis as the mechanism by which that planet came to be.
- Venus' crustal rocks, craters, mountains, and volcanoes all appear to be the same age, which is only about 10% of the deep-time presumptions about its age.
- Earth has a magnetic field that is weakening with a half-life of ≈ 1465 years. How could it still exist, if deep time were real? It couldn't; it would have dissipated long ago. The Earth's magnetic field cannot be more than 20,000 years old.
- Earth's continents are constantly eroding from wind and water; measured erosion rates are 40–650 feet per million years, varying by elevation and exposure to water. How long could the continents last before being eroded all the way down to sea level? At the slowest observed rate, the continents would be completely gone 70 times over, if deep time were a real thing. At the fastest rate, *1100 times over!* Rescue device: Continental tectonic uplift. Although this is plausible, if it had actually happened, *even once,* the process would have destroyed all the rock layers as well as the entire fossil record. Deep time counterproductive.

Chapter 8: The Next Step

- At the observed rates of river-borne sediment carried into the ocean, the current level of sediment would build up in only 12 million years, but evolutionists say the ocean has been here for 3 billion years. What happened to the other 99.6% of the time? Did no sediment at all go into the oceans?
- At the observed rates of river-borne sodium washing into the ocean, it would reach toxic levels in only 42 million years. Did no sodium at all wash into the oceans for the first 98.6% of their existence?
- At the observed rates of river-borne nickel washing into the ocean, it would reach current levels in only 25,000 years. Did no nickel at all wash into the oceans for the first 99.17% of their existence? And it would take barely more than a million years to reach toxic levels!
- And the same observations, with the same deep-time-defying answers, for *dozens* of other minerals.
- Earth's moon has a magnetic field. Like the other planets and moons discussed above, it wouldn't, if deep time were a real thing.
- The gravity of Earth's moon causes tides to happen on earth. The water's friction on the ocean floor slows the Earth's rotation, but since the angular momentum of the system must be preserved, the Moon edges farther away from Earth by 38mm every year. Backing up this non-linear process, the Moon would have been *touching* the Earth only 1.4 billion years ago, more than 3 billion years after the Nebular Hypothesis supposedly formed it! But long before it could have touched the Earth, the Roche Limit would have prevented the moon from forming in the first place.
- Mars, at some point in the past, had enough flowing water to form riverbeds and deltas, which we have observed, and evolutionists posit a planetwide flood may have caused these water features, even though no liquid water is there now. These are the same people who dismiss a planetwide flood on Earth as preposterous, even though Earth surface is 70% covered with water.
- Mars' atmosphere has too little $^{13}CO_2$, as compared to $^{12}CO_2$, to be billions of years old. Rescue device: the planet is replenishing the $^{12}CO_2$ from subsurface reservoirs. (Never mind that they, too, would have run out eons ago if Mars were really billions of years old.)
- Jupiter has too much argon, krypton, and xenon in its atmosphere to have "formed" where it is, but if it formed farther away from the sun (where those gases could have condensed), how could it have moved to where it is now? And even if those gases condensed farther out, they

would have evaporated eons ago with Jupiter at its present distance from the sun, if deep time were a thing.
- Jupiter has several retrograde moons; that is, they orbit the opposite direction of Jupiter's rotation. This disqualifies the Nebular Hypothesis as the cause of Jupiter's moons.
- Jupiter has several moons in orbits that are significantly off the axis of Jupiter's rotation; that is, their orbital planes are not straight above Jupiter's equator. This too disqualifies the Nebular Hypothesis as the cause of Jupiter's moons.
- Jupiter's moon Io has 400 active volcanoes, which radiate much heat away into space. If Io were as old as deep-time enthusiasts claim, it would have run out of heat eons ago, and there would no longer be any active volcanoes. Rescue device: tidal flexing (even though this produces only 10% of the required heat, and the volcanoes would be in different places).
- If Io had been erupting at only 10% of its current rate, it would have erupted its *entire mass 40 times over,* if the deep-time guesses of the solar system's age were accurate.
- Io's magma shows dark-colored minerals erupting from the surface, but such elements would have sunk to the core during moon formation, if Nebular Hypothesis had actually been the mechanism of solar system formation.
- Jupiter's moon Europa is too smooth to be as old as deep-time enthusiasts estimate; it has only 5% of the craters that would be expected if deep time were a real thing.
- Jupiter's moon Ganymede still has a magnetic field, and it wouldn't if deep time were real.
- Saturn's rings are too bright and shiny to be as old as evolutionary planetary scientists say: spacecrafts have measured the influx of micrometeorite dust, and at the observed rate, the rings would be sooty and dingy, if deep time were a real thing.
- Saturn's rings have tiny moons in them, but gravitational interactions would have flung them away in only a few million years at most. So these moons would no longer be there, if the rings had been there as long as deep time claims.
- Saturn's rings are spreading out, and the observed rate of spread doesn't fit deep time.

- Some of Saturn's moons are so close to the ring system that they are constantly being chipped away from collisions with ring particles. If deep time were real, those moons wouldn't even be there anymore.
- Saturn's rings are being pulled into the planet at the rate of six Olympic-sized swimming pools *per hour*. At that rate, Saturn's rings wouldn't exist at all anymore, if deep time were a real thing.
- Because of all the above evidence that Saturn and its rings are recent creations, another rescue device was absolutely essential: Maybe Saturn captured a comet and tore it apart, so it became the rings. Never mind that the comet would have to be larger than Earth's moon (194,000 times larger than the biggest comet ever seen), just to create the B Ring, let alone the others.
- Saturn's moon Enceladus constantly has more than 100 active geysers erupting, shooting liquid water into space. It quickly freezes and some of it falls back to the surface as snow, the rest escaping the moon's gravity altogether and becoming the E Ring. All the heat energy powering these geysers would have long since been gone, and its core of liquid water would have frozen solid long ago, if deep time were a real thing.
- Saturn's moon Titan has lakes of liquid methane along its equator, dozens of miles long and wide, but they should have evaporated long ago, if deep time were a thing. In fact, they should be gone after only a few thousand years. How are they still there, if deep time is real? Rescue device: Replenishment by subsurface reservoirs. But they too would have run out eons ago if deep time were real.
- Saturn's moon Titan should have large amounts of ethane, because ethane is chemically very stable, and its ratio, compared to methane, should be constantly increasing. There should be much more ethane than methane, if deep time were an actual thing. But there isn't.
- Saturn's moon Iapetus has large amounts of frozen carbon dioxide (dry ice) on its surface. But it is sublimating at 2.2 billion tons per orbit! How long would it last, at that rate? Certainly not the 4.5 billion years that deep-time enthusiasts claim to be the age of the solar system. Rescue device: Comets crashing into Iapetus replenish the carbon dioxide. But that makes one wonder why comets don't crash into the other moons and replenish them too.
- Uranus' axis of rotation is almost 90° off the plane of its orbit around the sun! The Nebular Hypothesis, which is the standard go-to assumption for what created the solar system, cannot explain how that happened.

- Uranus' magnetic field is 60° off-axis from the planet's axis rotation, and it doesn't even go through the center of the planet! Again, the Nebular Hypothesis can't explain this, and if deep time were valid, the magnetic field wouldn't even exist anymore. Rescue device for magnetic field's existence: Dynamo Model. Except the off-axis and off-center magnetic field is incompatible with dynamo models.
- Neptune radiates more than twice as much heat as it receives from the sun. How could it sill be doing that, after 4.5 billion years? It couldn't; it would have run out of heat long ago.
- Neptune's magnetic field (which wouldn't even be there anymore if deep time were real), is not only off-axis from the planet's axis of rotation, but it doesn't go through the planet's center. And again, both of those feature defy the Nebular Hypothesis.
- Neptune's moon Triton has a retrograde orbit, and its orbital plane is off Neptune's equator by 23°, both of which disqualify the Nebular Hypothesis. Rescue device: Neptune gravitationally captured Triton as it was passing by. (However, the chances of falling into the almost perfectly circular orbit it has, are infinitesimally small.)
- Pluto has a huge area, called Tombaugh Regio, that has very few impact craters, but there would be innumerable craters by now if deep time were real. And on Tombaugh Regio's west edge is an area named Sputnik Planitia, which has *no* craters at all. This is impossible to explain if you assume deep time.
- Pluto is geologically active, and even has convection cells on its surface, where warm fluid rises and cold fluid sinks. This convection process is constantly losing heat, and there is *no way* convection could continue after the 4.5 billion years that evolutionists claim is the age of the solar system; it would have solidified ages ago.
- Pluto's moon Charon is huge in comparison to Pluto, and astronomers are trying to figure out how it "formed." (Remember the Roche Limit?) Rescue device: Charon was caused by a glancing collision between the two. But that would have caused Pluto to spin so fast that it would have an equatorial bulge. It doesn't.
- Comets (the short-period ones, with less-than-200-year orbits) evaporate so fast during the parts of their orbits that are close to the sun, that there's no way they could be left over from the solar system's formation. Rescue device: the Kuiper Belt—an imagined belt of trillions of comet bodies out beyond the orbit on Pluto, from which gravitational perturbations dislodge one every so often, forming the short-period comets

CHAPTER 8: THE NEXT STEP

we see. Problem: Voyager 2 flew past in 1989, and it saw no comet bodies waiting to be launched.
- Comets (the long-period ones, with more-than-200-year orbits) have the same fast-evaporation problem, so they couldn't be left over from the alleged 4.5-billion-year-ago formation of the solar system either. Rescue Device: The Oort Cloud, a postulated shell of trillions of comet bodies far outside the solar system, and these too are occasionally nudged into cometary orbits so we can see them. But the Oort Cloud is too far away to see, even with our best instruments. That's convenient. . .
- Radiocarbon dating is unreliable for any estimates greater than 50,000 years, because of Carbon-14's short half-life. Yet it is routinely used to "prove" dates in the range of tens of millions, or even billions of years. Even though a sample of Carbon-14 as massive as the entire universe would be entirely decayed in only 1½ million years! (Do the math.) Deep time counterproductive.
- Using a random selection process at a thousand attempts per second, even the *simplest* protein (a paltry 50 amino acids) would take quintillions of times longer than even the deep-time age of the universe. But some proteins have 30,000 amino acids! Conclusion: Deep time is utterly insufficient to assemble even simple things, so proteins (and amino acids, DNA, etc.) must have been assembled intentionally.
- At the measured rate of mutations in humans—more than 63 new mutations per generation—the human race would have gone extinct long ago if deep time were real. State-of-the-art genetics-modeling simulations, with only 60 new mutations per generation (fewer than what was measured), predicts the extinction of humanity after only 350 generations—about 7,000 years.
- Numerous polystrate (multi-layer) fossils have been found worldwide: individual fossils that extend through multiple rock layers, thus showing that not only is uniformitarian deep time *unnecessary* to cause what we observe, but uniformitarian deep time would actually *prevent* what we observe, because the fossilized organisms would have rotted completely away long before uniformitarian processes could fossilize them.
- Folded rock layers—multiple layers folded as a unit—are also found worldwide. These too overturn the deep-time article of faith, because if rock layers were actually formed like evolutionary geologists say, the bottom layers would have been solidified when the force was applied,

and the whole structure would have shattered, not bent. Deep time is unnecessary, and even counterproductive to cause what we see.
- If new sedimentary layers are not buried quickly and deeply, small animals, plant roots, and weather will soon obliterate any layered appearance by churning the layers together, if uniformitarianism were the rule. The fact that rock layers are seen all over the world attests to the fact that these sediment layers were quickly and deeply buried—as, for example, by a worldwide flood. Deep time unnecessary.
- DNA quickly decays when an organism dies; its half-life has been measured as 541 years, so it will not last any more than about 10,000 years in a natural environment. But intact DNA has been found in dinosaur fossils supposedly 65 million years old, and bacteria 250 million years old. If deep time were real, all traces of DNA would be long gone. Deep time counterproductive.
- Surtsey, a volcanic island that formed near Iceland in 1963, has a thriving ecosystem, even though it was red-hot magma little more than 60 years ago. About 20 new species are discovered there every year! Deep time not needed.
- Mount St. Helens erupted in 1980, and numerous processes, thought by evolutionists to take millions of years, happened in a very short time:
    - Deposition of sedimentary layers was thought (by evolutionists) to take thousands or millions of years, but on Mount St. Helens, 400 feet have built up since 1980, and a 25-foot-thick series of thin layers built up in only a few hours. No deep time necessary.
    - Erosion of rock was thought to take millions of years, but when a volcanically formed lake burst its banks, the rushing water dug a canyon 140 feet deep, 100 feet wide, and 2,000 feet long, in only *nine hours!* No deep time necessary.
    - Ecosystems were thought to take many thousands of years to spread across barren landscape. After Mount St. Helens sterilized 250 square miles of forest with 1000° gas and ash in 1980, it only took 20 years for most of the animals to return, and now, only 40 years after the eruption, lush trees cover many areas in that recently sterile environment. No deep time necessary.
    - Gems were thought to take thousands of years to form, but the eruption of Mount St. Helens showed otherwise. Numerous gems were produced by the eruption, and are on sale at the gift shop. No deep time necessary.

- Arches National Park in Utah has over 2000 sandstone arches. But through normal weathering and erosion, they are collapsing at the rate of almost one per year! If deep time were a thing—the Park Service says they're five million years old—the arches would all have collapsed eons ago. Deep time counterproductive.
- Many processes, commonly thought to take millions of years, all over the world, are being discovered to require far less than what evolutionists say:
  - Rock formation: Things that have been found inside solid rock include a clock, a spark plug, a ship's bells, and even teddy bears! No deep time necessary.
  - Coal formation: Wood has been turned into coal in only nine months. No deep time necessary.
  - Oil formation: Slaughterhouse waste can be transformed into fuel oil *better than crude oil* in less than a day, using conditions common in nature. Algae can be transformed into oil in less than an hour. No deep time necessary.
  - Speleothem formation (cave formations): Growth rates of stalagmites and stalactites have been measured at more than a meter in less than a century, inches in a matter of days, and three feet within a year. The basement of a pub in England has thin stalactites *twelve feet long!* In a man-made structure! No deep time necessary.
  - Gem formation: Diamonds are routinely made by numerous companies; you can even turn your cremated loved ones into diamonds. Opals, rubies, garnet, topaz, emeralds, and more, are made all over the world. No deep time necessary.
  - Petrification: Objects such as a bowler hat, a ham, candles, pieces of wood, and more have been petrified in less than a century. The chapel of Santa Maria de Salute in Venice, Italy, was built on wooden pilings in 1630. They petrified in the years since, and the chapel now rests on stone pilings! No deep time necessary.
  - Fossilization: Scientists have fossilized soft animal tissues in the lab, and it is now so easy that faked fossils are a problem, as the famous "feathered dinosaur" debacle attests. A study in 2010 estimated that up to 80% of marine reptile fossils on display in museums in China have been manufactured or manipulated in some way. No deep time necessary.

- Glacier formation: Once thought to take hundreds of thousands of years to form, a glacier in Greenland built up and buried a squadron of World War II planes that crash-landed in 1942 under 250 feet of ice—in only 46 years! No deep time necessary.
- Coral reef formation: Corals have been observed growing more than 17 inches a year, and some coral colonies have reached more than 30 inches in diameter in less than five years. No deep time necessary.

- Canyon formation: Again, no deep time necessary.
  - Using published values for the Colorado River flow rate and sediment content, plus the volume of the Grand Canyon, the whole canyon could have formed in only a few thousand years! Even without catastrophism! Do the math.
  - Canyon Lake Gorge in Texas is more than a mile long and 20–50 feet deep, and it was formed in only three days during a period of heavy rain in 2002.
  - Mount St. Helens "Little Grand Canyon" is 1,000 feet wide, 2,000 feet long, and up to 140 feet deep, and it was formed in only nine hours when a dam of volcanic debris broke, emptying the whole lake behind it.
  - Burlingame Canyon in Washington state formed from a plant-choked irrigation ditch when the obstructions finally gave way in 1926. The six-foot-wide irrigation ditch became a canyon 1500 feet long, 120 feet deep, and 120 feet wide in only six days.
  - Providence Canyon in Georgia became 160 feet deep, 600 feet wide, and 1300 feet long, simply from erosion caused by poor farming practices in the 1800s. This didn't even require catastrophism!

So in *all* of the above points, deep time fails in one or more of three areas. First, deep time is completely unnecessary to get a given process done, and we have experimental and observational evidence confirming that deep time is unnecessary. Or, second, deep time would actually be detrimental to the process; so much time would allow uniformitarian processes to actually damage or destroy the thing, and this too is confirmed by measuring the rate at which good things are running down and problematic things are accumulating. Or, deep time is utterly insufficient to accomplish even the smallest of design features via random means.

So it is perfectly understandable why evolutionists want to hang onto the ideas of uniformitarianism and deep time, because if the stunning complexity we see everywhere came into existence too quickly, that would seem alarmingly similar to a creative miracle done by a supernatural being. And such an admission is just too frightening to those who want to maintain the illusion they are in control of their own lives.

Unreal Scientists

In this section I'll list a few scientists (in order of birth year) who apparently weren't "real" scientists, and whose contributions to science are apparently irrelevant and of no value, because they were all creationists. They actually believed in God and believed that God created the heavens and the earth. How quaint and rustic, wouldn't you say?

Leonardo da Vinci (1452–1519)

Image in the Public Domain

Leonardo da Vinci

Leonardo da Vinci was an Italian polymath of the High Renaissance who is widely considered one of the greatest painters of all time, despite less than 25 of his paintings having survived. The Mona Lisa is the most famous of Leonardo's works and the most famous portrait ever made. The Last Supper is the most reproduced religious painting of all time, and his Vitruvian Man drawing is also regarded as a cultural icon.

He is also known for his notebooks, in which he made drawings and notes on science and invention; these involve a variety of subjects including anatomy, cartography, painting, and paleontology. Leonardo's collective works compose a contribution to later generations of artists rivalled only by that of his contemporary Michelangelo.

Leonardo was born in Vinci, in the region of Florence, Italy. He was educated in the studio of the renowned Italian painter Andrea del Verrocchio. Although he had no formal academic training, many historians and scholars regard Leonardo as the prime exemplar of the "Universal Genius" or "Renaissance Man," an individual of "unquenchable curiosity" and "feverishly inventive imagination." He is widely considered one of the most diversely talented individuals ever to have lived. According to art historian Helen Gardner, the scope and depth of his interests were without precedent in recorded history, and "his mind and personality seem to us superhuman, while the man himself mysterious and remote." Scholars interpret his view of the world as being based in logic, though the empirical methods he used were unorthodox for his time.

Leonardo is revered for his technological ingenuity. He conceptualized flying machines, a type of armoured fighting vehicle, concentrated solar power, an adding machine, and the double hull. Relatively few of his designs were constructed or even feasible during his lifetime, as the modern scientific approaches to metallurgy and engineering were only in their infancy during the Renaissance.[684]

What is often overlooked, when admiring da Vinci's artistic and scientific prowess, is that his works and his words strongly imply he was a Christian and a creationist. This is attested by not only his paintings, which often portrayed Biblical people and scenes, but by his own notebooks:

> Good Report soars and rises to heaven, for virtuous things find favor with God. Evil Report should be shown inverted, for all her works are contrary to God and tend toward hell.[685]

Referring to the fearfully-and-wonderfully-made human body, da Vinci wrote:

> O you who look on this our machine, do not be sad that with others you are fated to die, but rejoice that our Creator has endowed us with such an excellent instrument as the intellect.

Marco Rosci, who wrote a biography of da Vinci, described his views as follows:

> Man is the handiwork of a God who retains few links with traditional orthodoxy. But man is emphatically no mere 'instrument' of his Creator. He is himself a 'machine' of extraordinary quality and proficiency and thus proof of nature's rationality.[686]

Although da Vinci understood we were created by the Creator, he didn't accept the historicity of Noah's Flood, because he couldn't understand how

---

[684] en.wikipedia.org/wiki/Leonardo_da_Vinci; accessed July 30, 2020.
[685] da Vinci, Leonardo, Manuscript H.
[686] "Was Leonardo da Vinci religious?" October 25, 1999. [Online at StraightDope.com/columns/read/1697/was-leonardo-da-vinci-religious; accessed July 30, 2020.]

fragile seashells could be swept so far inland without breaking. The misunderstanding was caused by an unwarranted assumption that the mountains had their current height *during* the Flood, instead of forming *after* the Flood, which, because of the discovery of tectonic plate movement, is much more plausible.[687] (See the books mentioned in the section "The Worldwide Flood" for details on flood geology and hydrology.)

Sir Francis Bacon (1561–1626)

Image in the Public Domain

Sir Francis Bacon

Sir Francis Bacon was an English philosopher and statesman who served as Attorney General and as Lord Chancellor of England. His works are credited with developing the scientific method and remained influential through the scientific revolution.

Bacon has been called the father of empiricism. His works argued for the possibility of scientific knowledge based only upon inductive reasoning and careful observation of events in nature. Most importantly, he argued science could be achieved by use of a sceptical and methodical approach whereby scientists aim to avoid misleading themselves. His ideas of the importance and possibility of a sceptical methodology makes Bacon the father of the scientific method. This method was a new rhetorical and theoretical framework for science, the practical details of which are still central in debates about science and methodology.

Francis Bacon was a patron of libraries and developed a functional system for the cataloging of books by dividing them into three categories—history, poetry, and philosophy—which could further be divided into more specific subjects and subheadings.[688]

Francis Bacon wrote in his 1605 book *The Advancement of Learning*:

> To conclude, therefore, let no man. . . think or maintain that a man can search too far or be too well studied in the book of God's word, or in the book of God's works; divinity or philosophy; but rather let men endeavor an endless progress or proficience in both.

---

[687] Verderame, John "The 'evidence' for a biblical worldview," November 13, 2000. [Online at Creation.com/the-evidence-for-a-biblical-worldview; accessed July 30, 2020.]

[688] en.wikipedia.org/wiki/Francis_Bacon; accessed July 12, 2020.

Bacon's statement is a classic acknowledgement of the truth that there are two ways to understand God better: through His word, the Bible, and through His works, the universe.

Galileo Galilei (1564–1642)

Image in the Public Domain

Galileo Galilei

Galileo Galilei was an Italian astronomer, physicist, and engineer, sometimes described as a polymath, from Pisa. Galileo has been called the "father of observational astronomy," the "father of modern physics," the "father of the scientific method," and the "father of modern science."

Galileo studied speed and velocity, gravity and free fall, the principle of relativity, inertia, projectile motion and also worked in applied science and technology, describing the properties of pendulums and "hydrostatic balances," inventing the thermoscope and various military compasses, and using the telescope for scientific observations of celestial objects. His contributions to observational astronomy include the telescopic confirmation of the phases of Venus, the observation of the four largest satellites of Jupiter (now dubbed the "Galilean" moons; see the "Jupiter" section in "Chapter 4: A Young Solar System" for details), the observation of Saturn's rings, and the analysis of sunspots.

Galileo's championing of heliocentrism (the sun as the center of the solar system) and Copernicanism was controversial during his lifetime, when most subscribed to geocentric models (the Earth as the center of the solar system) such as the Tychonic system. He met with opposition from Aristotelian astronomers, who doubted heliocentrism because of the absence of an observed stellar parallax.[689] Contrary to popular opinion, the greatest resistance to Galileo's heliocentric views was not the church, but Aristotelian astronomers.[690]

The geocentric Aristotelians were the status quo, and they felt Galileo's theories preposterous and impossible, much like today's evolutionists regard creationism. Galileo eventually got in trouble with the church also, because of unwise and abrasive communication practices, but his greatest conflict was

---

[689] en.wikipedia.org/wiki/Galileo_Galilei; accessed July 13, 2020.

[690] Thomas Schirrmacher, "The Galileo affair: history or heroic hagiography?", *Journal of Creation* 14(1):91–100, April 2000. [Online at Creation.com/the-galileo-affair-history-or-heroic-hagiography; accessed July 14, 2020.]

with scientists who were still hanging on to widespread, but wrong, theories. The battle was first and foremost science vs. science, not science vs. religion.

In spite of the grief that Galileo's unfortunate pride and harsh words caused him, he never abandoned his faith in the Bible. Galileo's Catholic faith was completely unshaken by his discovery, and besides, Galileo was just continuing the work of a Canon in the church, whose name was Nicolaus Copernicus—the mathematician and astronomer who developed the heliocentric, or sun-centered, model of the solar system.[691] Galileo remained a young-earth creationist throughout his life.[692]

Johannes Kepler (1571–1630)

Image in the Public Domain

Johannes Kepler

Johannes Kepler was a German astronomer and mathematician, and a key figure in the 17th-century scientific revolution, best known for his laws of planetary motion, and his books *Astronomia nova, Harmonices Mundi*, and *Epitome Astronomiae Copernicanae*. These works also provided one of the foundations for Newton's theory of universal gravitation.

Kepler was a mathematics teacher at a seminary school in Graz, where he became an associate of Prince Hans Ulrich von Eggenberg. Later he became an assistant to the astronomer Tycho Brahe in Prague, and eventually the imperial mathematician to Emperor Rudolf II and his two successors Matthias and Ferdinand II. He also taught mathematics in Linz, and was an adviser to General Wallenstein. Additionally, he did fundamental work in the field of optics, invented an improved version of the refracting (or Keplerian) telescope, and was mentioned in the telescopic discoveries of his contemporary Galileo Galilei. He was a corresponding member of the Accademia dei Lincei in Rome.[693]

Kepler plainly stated that he was a Christian, and he acknowledged that God was "the kind Creator who brought forth nature out of nothing." In spite of his brilliance and great success in the sciences, he remained humble,

---

[691] en.wikipedia.org/wiki/Nicolaus_Copernicus; accessed September 21, 2020.

[692] Jonathan Sarfati, "Galileo's Quadricentennial," Creation Ministries International, July 9, 2009. [Online at Creation.com/galileo-quadricentennial; accessed July 14, 2020.]

[693] en.wikipedia.org/wiki/Johannes_Kepler; accessed July 12, 2020.

saying, "Let my name perish if only the name of God the Father is thereby elevated."[694]

Toward the end of his life, Kepler noted,

> I had the intention of becoming a theologian. . . but now I see how God is, by my endeavours, also glorified in astronomy, for 'the heavens declare the glory of God.'[695]

Blaise Pascal (1623–1662)

Blaise Pascal

Blaise Pascal was a French mathematician, physicist, inventor, writer, and Catholic theologian. He was a child prodigy who was educated by his father, a tax collector in Rouen. Pascal's earliest work was in the natural and applied sciences, where he made important contributions to the study of fluids, and clarified the concepts of pressure and vacuum by generalising the work of Evangelista Torricelli. Pascal also wrote in defence of the scientific method.

In 1642, while still a teenager, he started some pioneering work on calculating machines. After three years of effort and 50 prototypes, he built 20 finished machines (called Pascal's calculators and later Pascalines) over the following 10 years, establishing him as one of the first two inventors of the mechanical calculator.

Pascal was an important mathematician, helping create two major new areas of research: he wrote a significant treatise on the subject of projective geometry at the age of 16, and later corresponded with Pierre de Fermat on probability theory, strongly influencing the development of modern economics and social science. Following Galileo Galilei and Torricelli, in 1647, he rebutted Aristotle's followers who insisted that nature abhors a vacuum.[696]

---

[694] Tiner, J. H., Johannes Kepler-Giant of Faith and Science, Mott Media, Milford, Michigan (USA), p. 193–197, 1977.

[695] Psalm 19:1–3 (NKJV): The heavens declare the glory of God; and the firmament shows His handiwork. ²Day unto day utters speech, and night unto night reveals knowledge. ³There is no speech nor language where their voice is not heard.

[696] en.wikipedia.org/wiki/Blaise_Pascal; accessed July 13, 2020.

On the topic of why God doesn't reveal Himself so openly that His reality is inescapable, Pascal wrote:

> It is not in this manner that he chose to appear in the gentleness of his coming; because since so many men had become unworthy of his clemency, he wished them to suffer the privation of the good that they did not want. It would not have been right therefore for him to appear in a way that was plainly divine and absolutely bound to convince all mankind; but it was not right either that he should come in a manner so hidden that he could not be recognized by those who sought him sincerely. He chose to make himself perfectly knowable to them; and thus, wishing to appear openly to those who seek him with all their heart, he tempered the knowledge of himself, with the result that he has given signs of himself which are visible to those who seek him, and not to those who do not seek him.[697]

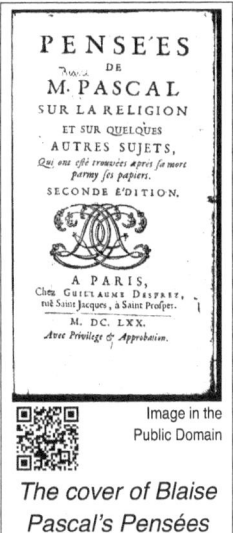

The cover of Blaise Pascal's Pensées

Today, this thought is often expressed as, "God always speaks loudly enough to hear, but quietly enough to ignore."

Robert Boyle (1627–1691)

Robert Boyle

Robert Boyle was an Anglo-Irish natural philosopher, chemist, physicist, and inventor. Boyle is largely regarded today as the first modern chemist, and therefore one of the founders of modern chemistry, and one of the pioneers of modern experimental scientific method. He is best known for Boyle's law, which describes the inversely proportional relationship between the absolute pressure and volume of a gas, if the temperature is kept constant within a closed system. With all the important work he accomplished in physics—the discovery of the part taken by air in the propagation of sound, and investigations on

---

[697] Blaise Pascal, *Pensées,* #309, 1670.

the expansive force of freezing water, on specific gravities and refractive powers, on crystals, on electricity, on color, on hydrostatics, and so forth—chemistry was his favorite topic of study.[698]

In 1661, at 34 years of age, Boyle published *The Skeptical Chymist*. In this book, he replaced Aristotle's idea of only four elements (earth, water, air, and fire) and with the modern understanding of an element—specifically, an element is a substance that cannot be separated into simpler components by chemical methods. *The Skeptical Chymist* is recognized as the foundation-stone of modern chemistry. Boyle is also known for his pioneering experiments on the physical properties of gases, his role in creating the Royal Society of London, and his philanthropy in the American colonies.

Boyle was a solid Christian and a diligent student of the Bible—he studied the Scriptures in their original languages (Hebrew for the Old Testament and Greek for the New Testament) to glean more understanding of them. Because of his Gaelic roots, Boyle even funded and supervised the translation of the Bible into Gaelic.[699]

Boyle died in 1691, but the year before that, he published *The Christian Virtuoso*. In this book, he expounded upon the fact that the study of nature and responsible management of nature are God-given duties of man. He based his thesis on this passage:

> Genesis 1:28 (NKJV): Then God blessed them, and God said to them, "Be fruitful and multiply; fill the earth and subdue it; **have dominion over the fish of the sea, over the birds of the air, and over every living thing that moves on the earth.**"

Boyle was a prolific lecturer and writer, and he showed that science and faith in God can peacefully coexist. He praised his Creator for all the scientific discoveries he had made, and urged others to do the same. He recognized that the universe works in accordance with well-defined and finely tuned natural laws that God set up for its operation. As a powerful apologist for the Christian faith, he made provision in his last will and testament for the Boyle Lectures for the defence of Christianity. He strongly supported missionary work to other nations, and gave great support to organizations that furthered the Gospel.

---

[698] en.wikipedia.org/wiki/Robert_Boyle; accessed August 17, 2020.

[699] Robert Doolan, "The man who turned chemistry into a science" December 1989. [Online at Creation.com/the-man-who-turned-chemistry-into-a-science; accessed August 17, 2020.]

Modern chemistry owes an enormous debt of gratitude to Robert Boyle for his work and his writings. He was a creation scientist with a passionate love for God's truth, as expressed both in the Bible and in nature. His belief in an orderly God led him to discover and reject the errors of the theory of alchemy, which had been for centuries a roadblock to the development of actual scientific chemistry.[700]

Isaac Newton (1642–1727)

Sir Isaac Newton

Isaac Newton was an English mathematician, physicist, astronomer, theologian, and author (described in his own day as a "natural philosopher") who is widely recognised as one of the most influential scientists of all time and as a key figure in the scientific revolution. His book *Philosophiæ Naturalis Principia Mathematica* (Mathematical Principles of Natural Philosophy), first published in 1687, laid the foundations of classical mechanics. Newton also made seminal contributions to optics, and shares credit with Gottfried Wilhelm Leibniz for developing the infinitesimal calculus.

In *Principia*, Newton formulated the laws of motion and universal gravitation that formed the dominant scientific viewpoint until it was superseded by the theory of relativity. Newton used his mathematical description of gravity to prove Kepler's laws of planetary motion, account for tides, the trajectories of comets, the precession of the equinoxes and other phenomena, eradicating doubt about the Solar System's heliocentricity.

He demonstrated that the motion of objects on Earth and celestial bodies could be accounted for by the same principles. Newton's inference that the Earth is an oblate spheroid was later confirmed by the geodetic measurements of Maupertuis, La Condamine, and others, convincing most European scientists of the superiority of Newtonian mechanics over earlier systems.

Newton built the first practical reflecting telescope and developed a sophisticated theory of colour based on the observation that a prism separates white light into the colours of the visible spectrum. His work on light was collected in his highly influential book *Opticks*, published in 1704. He also formulated an empirical law of cooling, made the first theoretical calculation of the speed of sound, and introduced the notion of a Newtonian fluid. In

---

[700] Doolan, Robert "The man who turned chemistry into a science" December 1989. [Online at Creation.com/the-man-who-turned-chemistry-into-a-science; accessed July 30, 2020.]

addition to his work on calculus, as a mathematician Newton contributed to the study of power series, generalised the binomial theorem to non-integer exponents, developed a method for approximating the roots of a function, and classified most of the cubic plane curves.[701]

Newton was a godly man who believed the Bible was God's word, and thus, understood that God created all things. In his words:

> I have a fundamental belief in the Bible as the Word of God, written by men who were inspired. I study the Bible daily.[702]

When studying the orbital characteristics of the planets, Newton was so awestruck at what he saw that he wrote:

> This most beautiful system of the sun, planets, and comets, could only proceed from the counsel and dominion of an intelligent Being. . . This Being governs all things, not as the soul of the world, but as Lord over all; and on account of his dominion he is wont to be called "Lord God," or "Universal Ruler." . . . The Supreme God is a Being eternal, infinite, absolutely perfect.[703]

Leonhard Euler (1707–1783)

Image in the Public Domain
Leonhard Euler

Leonhard Euler was a Swiss mathematician, physicist, astronomer, geographer, logician, and engineer who made important and influential discoveries in many branches of mathematics, such as infinitesimal calculus and graph theory, while also making pioneering contributions to several branches such as topology and analytic number theory. He also introduced much of the modern mathematical terminology and notation, particularly for mathematical analysis, such as the notion of a mathematical function. He is also known for his work in mechanics, fluid dynamics, optics, astronomy and music theory.

---

[701] en.wikipedia.org/wiki/Isaac_Newton; accessed July 30, 2020.

[702] J. H. Tiner, *Isaac Newton—Inventor, Scientist and Teacher*, Mott Media, Milford (Michigan), 1975.

[703] *Principia*, Book III; cited in; *Newton's Philosophy of Nature: Selections from his writings*, p. 42, ed. H. S. Thayer, Hafner Library of Classics, NY, 1953.

CHAPTER 8: THE NEXT STEP

Euler was one of the most eminent mathematicians of the 18th century and is held to be one of the greatest in history. He is also widely considered to be the most prolific, as his collected works fill 92 volumes, more than anyone else in the field. A statement attributed to another world-class scientist and mathematician, Pierre-Simon Laplace, expresses Euler's influence on mathematics: "Read Euler, read Euler, he is the master of us all."[704]

In the badly misnamed era called "the Enlightenment," skeptical philosophers who called themselves "Freethinkers"—Voltaire, Hume, Kant, and others—ridiculed the Biblical concept of God, rejected Christian faith, and claimed that man could improve himself solely through intellectual processes. In this era, Euler shone as one of the beacons of light, both scientific and spiritual. In one of his works, entitled *Defence of the Revelation Against the Objections of Freethinkers* (where "the Revelation" refers to the whole Bible, not just its final book), Euler pointed out that the Freethinkers do not believe that the world had a beginning or will have an end, because this would constitute and acknowledgement of the direct action of the God they had rejected (paragraph #47).

In 1759, Friedrich Heinrich, Euler's close friend, asked him to tutor his 14-year-old daughter, Friederike Charlotte Leopoldine Louise. Friederike was a second cousin of King Frederick II and became known as the Princess of Prussia. Over the next two years, Euler wrote 234 letters to her, using layman's terms, and omitting technical equations or formulae. These letters included all manner of scientific topics, and included many scriptural references, including the story of Adam and Eve—showing that Euler was a solid creationist.

In several of the letters, Euler included discussions about God, prayer, eternal life, evil and sin, divine justice, the usefulness of adversity, and the conversion of sinners. In Letter 41, he wrote about the intricacy of how the eye was designed, and commented:

> Though we are very far short of a perfect knowledge of the subject, the little we do know of it is more than sufficient to convince us of the power and wisdom of the Creator. We discover in the structure of the eye perfections which the most exalted genius could never have imagined.[705]

---

[704] en.wikipedia.org/wiki/Leonhard_Euler; accessed July 12, 2020.
[705] *Letters of Euler to a German Princess,* translated by Henry Hunter, Vol. 1, No. 41, p. 165. Note: all quotations and hence page numbers are from Hunter's 1802 English Edition).

Oh, Evolve! (Good Luck With That. . .)

In Letter 43, p. 174, in the second French Edition, the Freethinker editor, Condorcet, deliberately left out Euler's final paragraph, which referred to God as the Creator of the eye. During Henry Hunter's translation of Euler's work into English, he restored the omitted content as a footnote. Here is what Condorcet didn't want people to consider:

> But the eye which the Creator has formed is subject to not one of all the imperfections under which the imaginary construction of the freethinker labours. In this we discover the true reason why infinite wisdom has employed several transparent substances in the formation of the eye: it is thereby secured against all the defects which characterize every work of man. What a noble subject of contemplation! How pertinent the question of the Psalmist! *He who formed the eye, shall he not see? And he who planted the ear, shall He not hear?*[706] The eye alone being a master-piece that far transcends the human understanding, what an exalted idea must we form of Him, who has bestowed this wonderful gift, and that in the highest perfection, not only on man, but on the brute creation, nay, on the vilest of insects!

Princess Friederike encouraged the publication of the letters, making science as taught by Euler accessible to the masses. By 1800, they had gone through thirty editions and had been translated into Danish, Dutch, English, German, Italian, Russian, Spanish, and Swedish.[707] Biographer Ronald Calinger describes these letters as "The most exhaustive and authoritative science popularization written during the 18th century."

---

[706] Psalm 94:9.

[707] Ronald S. Calinger, *Leonard Euler: Mathematical Genius in the Enlightenment*, Princeton University Press, USA, 2016.

Carl Linnaeus (1707–1778)

Carl Linnaeus

Carl Linnaeus was a Swedish botanist, zoologist, and physician who formalised "binomial nomenclature," the modern system of naming organisms. For example, the fruit fly is *Drosophila melanogaster*. He is known as the "father of modern taxonomy."

Linnaeus was born in the countryside of Småland in southern Sweden. He received most of his higher education at Uppsala University and began giving lectures in botany there in 1730. He lived abroad between 1735 and 1738, where he studied and also published the first edition of his *Systema Naturae* in the Netherlands. He then returned to Sweden where he became professor of medicine and botany at Uppsala. In the 1740s, he was sent on several journeys through Sweden to find and classify plants and animals. In the 1750s and 1760s, he continued to collect and classify animals, plants, and minerals, while publishing several volumes. He was one of the most acclaimed scientists in Europe at the time of his death.[708]

Linnaeus was a firm creationist, and he believed that God had chosen him to reveal the logical and orderly character of everything in creation. In the preface to his book *Systema Naturae,* he stated:

> The end [goal] of the Earth's creation is the glory of God, as seen from the works of Nature by man alone.[709]

Linnaeus' worldview is summed up well in this statement (in which "Nemesis" refers to God's inescapable and just retribution for wrongdoing):

> Theologically, man is to be understood as the final purpose of the creation; placed on the globe as the masterpiece of the works of Omnipotence, contemplating the world by virtue of sapient reason, forming conclusions by means of his senses, it is in His works that man recognizes the almighty Creator, the all-knowing, immeasurable and eternal God, learning to live morally under His rule, convinced of the complete justice of His Nemesis.[710]

---

[708] en.wikipedia.org/wiki/Carl_Linnaeus; accessed July 12, 2020.

[709] Russell Grigg, "Carl Linnaeus: the scientist who saw evidence for God in everything in nature" *Creation* 37(4):52–55, October 2015. [Online at Creation.com/carl-linnaeus; accessed July 14, 2020.]

[710] Linnaeus, C., *Nemesis Divina,* 1834, as trans. by Michael John Petry, in *Nemesis Divina,*

Louis Pasteur (1822–1895)

Image in the Public Domain
Louis Pasteur

Louis Pasteur was a French biologist, microbiologist and chemist renowned for his discoveries of the principles of vaccination, microbial fermentation and pasteurization. He is remembered for his remarkable breakthroughs in the causes and prevention of diseases, and his discoveries have saved many lives ever since. He reduced mortality from puerperal fever, and created the first vaccines for rabies and anthrax.

His medical discoveries provided direct support for the germ theory of disease and its application in clinical medicine. He is best known to the general public for his invention of the technique of treating milk and wine to stop bacterial contamination, a process now called pasteurization. He is regarded as one of the three main founders of bacteriology, together with Ferdinand Cohn and Robert Koch, and is popularly known as the "father of microbiology."

Pasteur was responsible for disproving the doctrine of spontaneous generation. Today, he is often regarded as one of the fathers of germ theory. Pasteur made significant discoveries in chemistry, most notably on the molecular basis for the asymmetry of certain crystals and racemization. Early in his career, his investigation of tartaric acid resulted in the first resolution of what is now called optical isomers. His work led the way to the current understanding of a fundamental principle in the structure of organic compounds.[711]

In one intriguing puzzle Pasteur took upon himself to solve, he investigated tartrate and paratartrate crystals. In looking into this question, which had baffled even the greatest scientists of the day, he discovered that there were two types of paratartrate crystals: right-handed and left-handed, like the amino acids and sugars mentioned earlier in this book. His description of his microscopic examination of the crystals was that he "looked upon them as direct evidence of the artistic expression of God the Creator."

As mentioned, one of Pasteur's major contributions was to disprove the theory of spontaneous generation—life arising from non-life. It had already been disproven in 1668, by Italian biologist Francesco Redi, for larger organisms like flies. But scientists still were hanging on the idea of spontaneous generation of microbes, and Pasteur devised an experiment to disprove that

---

p. 21, Springer, 2001; *Encyclopedia of Quotations*, Linnaeus.
[711] en.wikipedia.org/wiki/Louis_Pasteur; accessed July 13, 2020.

as well. Evolutionists *still* are hanging on to spontaneous generation—life from non-life (although they call it **abiogenesis** now), because it's the only option if you've already rejected the idea of a Creator. But, "spontaneous generation by any other name. . ." For this and other reasons, Louis Pasteur was a strong opponent of Darwin's theory.

Far from seeing conflict between science and Christianity, Pasteur saw harmony. He stated that "science brings men nearer to God." In his scientific work, he perceived wisdom and design, not randomness and chaos. Pasteur commented: "The more I study nature, the more I stand amazed at the work of the Creator."[712]

Louis Pasteur died on September 28, 1895, after a long and fruitful life. His contributions to science were truly outstanding, and his Christian faith sustained him through many trials. He firmly believed in creation, and strongly opposed Darwin's theory of evolution because it did not fit well with scientific evidence.[713]

Lord Kelvin, William Thomson (1824–1907)

Image in the Public Domain

Lord Kelvin: William Thomson

William Thomson was a British mathematical physicist and engineer who was born in Belfast. At the University of Glasgow he did important work in the mathematical analysis of electricity and formulation of the first and second laws of thermodynamics, and did much to unify the emerging discipline of physics in its modern form.

He also had a career as an electric telegraph engineer and inventor, which propelled him into the public eye and ensured his wealth, fame, and honor. For his work on the transatlantic telegraph project he was knighted in 1866 by Queen Victoria, becoming Sir William Thomson. He had extensive maritime interests and was most noted for his work on the mariner's compass, which previously had limited reliability.

Absolute temperatures are stated in units of Kelvin in his honour. While the existence of a lower limit to temperature (absolute zero) was known prior

---

[712] J. H. Tiner, *Louis Pasteur—Founder of Modern Medicine*, Mott Media, Milford, Michigan, USA, 1990, pp. 75, 90

[713] Ann Lamont, "Louis Pasteur (1822–1895) Outstanding scientist and opponent of evolution," *Creation* 14(1):16–19, December 1991. [Online at Creation.com/louis-pasteur; accessed July 14, 2020.]

to his work, Kelvin is known for determining its correct value as approximately −273.15°C or −459.67°F.[714]

Lord Kelvin was also a staunch creationist. Concerning the origin of life on the Earth, he said:

> Mathematics and dynamics fail us when we contemplate the earth, fitted for life but lifeless, and try to imagine the commencement of life upon it. This certainly did not take place by any action of chemistry, or electricity, or crystalline grouping of molecules under the influence of force, or by any possible kind of fortuitous concourse of atoms. **We must pause, face to face with the mystery and miracle of creation of living creatures.**[715]

James Clerk Maxwell (1831–1879)

Image in the Public Domain

James Clerk Maxwell

James Clerk Maxwell was a Scottish scientist in the field of mathematical physics. His most notable achievement was to formulate the classical theory of electromagnetic radiation, bringing together for the first time electricity, magnetism, and light as different manifestations of the same phenomenon, thus paving the way for radio, television, radar, wi-fi, bluetooth, and more. Maxwell's equations for electromagnetism have been called the "second great unification in physics" after the first one realized by Isaac Newton.

With the publication of *A Dynamical Theory of the Electromagnetic Field* in 1865, Maxwell demonstrated that electric and magnetic fields travel through space as waves moving at the speed of light. He proposed that light is an undulation in the same medium that is the cause of electric and magnetic phenomena. The unification of light and electrical phenomena led his prediction of the existence of radio waves. Maxwell is also regarded as a founder of the modern field of electrical engineering.

He helped develop the Maxwell-Boltzmann distribution, a statistical means of describing aspects of the kinetic theory of gases. He is also known

---

[714] en.wikipedia.org/wiki/William_Thomson,_1st_Baron_Kelvin; accessed July 28, 2020.

[715] Barnes, Thomas. "Physics: A Challenge to 'Geological Time'" *Acts & Facts,* July 1, 1974. [Online at ICR.org/article/physics-challenge-geological-time; accessed July 28, 2020.] Emphasis added.

for presenting the first durable color photograph in 1861 and for his foundational work on analysing the rigidity of rod-and-joint frameworks (trusses) like those in many bridges.

His discoveries helped usher in the era of modern physics, laying the foundation for such fields as special relativity and quantum mechanics. Many physicists regard Maxwell as the 19th-century scientist having the greatest influence on 20th-century physics. His contributions to the science are considered by many to be of the same magnitude as those of Isaac Newton and Albert Einstein.[716]

In 1796, Pierre-Simon LaPlace, a French atheist, proposed the Nebular Hypothesis of solar system formation: a giant cloud of dust and gas that spontaneously collapsed over millions of years to form the sun and planets; this was covered extensively in "Chapter 4: A Young Solar System." This hypothesis was quite popular among atheists, because it did away with the need for God as the Creator. One of Maxwell's contributions to science was that he demonstrated two fatal flaws in LaPlace's theory, and proved mathematically that such a process could not occur.[717] Yes, it was disproven more than *two centuries ago*, but it is still being taught in schools today.

Maxwell was a committed creationist, as shown by this prayer that he wrote and kept with his notes:

> Almighty God, Who hast created man in Thine own image, and made him a living soul that he might seek after Thee, and have dominion over Thy creatures, teach us to study the works of Thy hands, that we may subdue the earth to our use, and strengthen the reason for Thy service; so to receive Thy blessed Word, that we may believe on Him Whom Thou hast sent, to give us the knowledge of salvation and the remission of our sins. All of which we ask in the name of the same Jesus Christ, our Lord.[718]

---

[716] en.wikipedia.org/wiki/James_Clerk_Maxwell; accessed July 12, 2020.

[717] Ann Lamont, "Great creation scientists: James Clerk Maxwell," *Creation* 15(3):45–47, June 1993. [Online at Creation.com/great-creation-scientists-james-clerk-maxwell; accessed July 14, 2020.]

[718] Maxwell, J. C., "Discourse on Molecules," a paper presented to the British Association at Bradford in 1873, as cited in: E. L. Williams and G. Mulfinger, *Physical Science for Christian Schools*, Bob Jones University Press, Greenville, South Carolina, 1974, p. 487.

...And So Many Others!

`<irony mode=on>`

It's really too bad that Leonardo da Vinci, Francis Bacon, Galileo Galilei, Johannes Kepler, Blaise Pascal, Robert Boyle, Isaac Newton, Leonhard Euler, Carl Linnaeus, Louis Pasteur, William Thomson, James Clerk Maxwell, and so many more not mentioned here, were Bible-believing creationists. If they had only believed in evolution, or that everything came about with no need for God, they really could have accomplished something with their lives...

`<irony mode=off>`

So no, you don't have to throw away your brain in order to embrace creationism. You can not only retain your brain, but you will be amazed at what God shows you, once you are willing to follow His lead:

> Proverbs 25:2 (ESV): It is the glory of God to conceal things, but the glory of kings is to search things out.

But How Important Is It, Really?

There are some Jesus-followers today who seem to think that it's not really that important whether or not they believe the Genesis accounts of Creation and the Flood are actually true. It seems to them like an inconsequential detail whether they believe things came about as Genesis describes it, or by evolution, as long as we believe the New Testament.

Such an opinion reveals a serious absence of having done one's homework. The reason is simple: one can't really believe the contents of the New Testament without also believing that the Creation account (the beginning of the universe, Adam and Eve, the Garden of Eden, etc.) and the Flood account (Noah, the ark, the worldwide flood, etc.) in Genesis. Attempting to do so requires either a perilous lack of knowledge or an acceptance of intellectual dishonesty that results in confusion, self-contradictory thinking, and self-deception.

This is why I make such a bold statement: In every category of the New Testament—the Gospels, Acts, the Epistles, and Revelation—the writers matter-of-factly talk about Creation, Adam and Eve, Noah, the Ark, and the Flood, and so on. If we consider the New Testament to be reliable, when we realize that those New Testament writers considered Genesis to be factual, we are logically obligated to consider Genesis factual also.

Consider the following New Testament passages:

Matthew 19:4–5 (NLT): "Haven't you read the Scriptures?" Jesus replied. "They record that **from the beginning 'God made them male and female.'**" ⁵And he said, "This explains why a man leaves his father and mother and is joined to his wife, and the two are united into one."

Matthew 24:38–39 (NKJV): For as in **the days before the flood,** they were eating and drinking, marrying and giving in marriage, **until the day that Noah entered the ark,** ³⁹and did not know until **the flood came and took them all away,** so also will the coming of the Son of Man be.

Mark 10:6 (AMP): But **from the beginning of creation God made them male and female.**

Mark 13:19 (ASV): For those days shall be tribulation, such as there hath not been the like from **the beginning of the creation which God created** until now, and never shall be.

Luke 3:38 (ESV): . . .the son of Enos, the son of Seth, the son of **Adam, the son of God.**

Luke 17:26–27 (AMP): And [just] as **it was in the days of Noah,** so will it be in the time of the Son of Man. ²⁷[People] ate, they drank, they married, they were given in marriage, right up to **the day when Noah went into the ark, and the flood came** and destroyed them all.

John 1:3 (MSG): **Everything was created through him *[Jesus]*; nothing—not one thing!—came into being without him.**

John 1:10 (CSB): He *[Jesus]* was in the world, and **the world was created through him,** and yet the world did not recognize him.

John 5:46–47 (ERV): **If you really believed Moses, you would believe me,** because he wrote about me. ⁴⁷**But you don't believe what he wrote, so you can't believe what I say.**

Acts 4:24 (CEV): When the rest of the Lord's followers heard this, they prayed together and said: **Master, you created heaven and earth, the sea, and everything in them.**

Acts 7:49–50 (ESV): "'Heaven is my throne, and the earth is my footstool. What kind of house will you build for me, says the Lord, or what is the place of my rest? ⁵⁰**Did not my hand make all these things?**'"

Acts 14:15 (DARBY): and saying, Men, why do ye these things? We also are men of like passions with you, preaching to you to turn from these vanities to the living **God, who made the heaven, and the earth, and the sea, and all things in them...**

Acts 17:24, 26 (ERV): He is the **God who made the whole world and everything in it.** He is the Lord of the land and the sky. He does not live in temples built by human hands. [26]**God began by making one man, and from him he made all the different people who live everywhere in the world.** He decided exactly when and where they would live.

Acts 17:28 (GWORD): **Certainly, we** live, move, and **exist because of him.** As some of your poets have said, 'We are God's children.'

Romans 1:20 (GWORD): **From the creation of the world,** God's invisible qualities, his eternal power and divine nature, have been **clearly observed in what he made.** As a result, people have no excuse. *[Note that if evolution could explain the universe in a scientifically plausible way, people would have an excuse for not believing in God. Since people have no excuse for not believing in God, evolution must fail in its attempt to explain the universe. And as we've seen throughout this book, it does fail, in spades. Therefore, a Creator is the only alternative.]*

Romans 1:25 (HCSB): They exchanged the truth of God for a lie, and worshiped and served something created instead of **the Creator,** who is praised forever. Amen.

Romans 5:14 (NIV): Nevertheless, death reigned from **the time of Adam** to the time of Moses, even over those who did not sin by breaking a command, as did **Adam,** who was a pattern of the one to come.

Romans 8:19–22 (MSG): **The created world** itself can hardly wait for what's coming next. [20]Everything in **creation** is being more or less held back. God reins it in [21]until **both creation and all the creatures** are ready and can be released at the same moment into the glorious times ahead. Meanwhile, the joyful anticipation deepens. [22]All around us **we observe a pregnant creation.** The difficult times of pain throughout the world are simply birth pangs. But it's not only around us; it's within us. The Spirit of God is arousing us within. We're also feeling the birth pangs.

Romans 11:36 (TEV): For **all things were created by him, and all things exist through him** and for him. To God be the glory forever! Amen.

I Corinthians 8:6 (NLT): But for us, There is one **God, the Father, by whom all things were created,** and for whom we live. And there is one

Lord, **Jesus Christ, through whom all things were created,** and through whom we live.

I Corinthians 11:12 (CEV): It is true that the first woman came from a man, but all other men have been given birth by women. Yet **God is the one who created everything.**

I Corinthians 12:18 (AMP): But as it is, **God has placed and arranged the limbs and organs in the body, each [particular one] of them, just as He wished and saw fit and with the best adaptation.**

I Corinthians 12:24 (GWORD): However, our presentable parts don't need this kind of treatment. **God has put the body together** and given special honor to the part that doesn't have it.

I Corinthians 15:22 (NASB): For as in **Adam** all die, so also in Christ all will be made alive.

I Corinthians 15:37 (GWORD): What you plant, whether it's wheat or something else, is only a seed. It doesn't have the form that the plant will have. [38]**God gives the plant the form he wants it to have. Each kind of seed grows into its own form.**

I Corinthians 15:45 (NABRE): So, too, it is written, **"The first man, Adam,** became a living being," the last Adam a life-giving spirit.

I Corinthians 11:3 (NLT): But I fear that somehow your pure and undivided devotion to Christ will be corrupted, just as **Eve was deceived by the cunning ways of the serpent.**

II Corinthians 4:6 (NLT): For **God, who said, "Let there be light in the darkness,"** has made this light shine in our hearts so we could know the glory of God that is seen in the face of Jesus Christ.

Ephesians 2:10 (NASB): For **we are His workmanship, created in Christ Jesus** for good works, which God prepared beforehand so that we would walk in them.

Ephesians 3:8–9 (NET) To me—less than the least of all the saints —this grace was given, to proclaim to the Gentiles the unfathomable riches of Christ [9]and to enlighten everyone about God's secret plan —the mystery that has been hidden for ages in **God who has created all things.**

Colossians 1:16–17 (NIV): For **by him** *[Jesus]* **all things were created:** things in heaven and on earth, visible and invisible, whether thrones or powers or rulers or authorities; **all things were created by him and for him.** [17]He is before all things, and **in him all things hold together.**

Colossians 3:10 (NLT): Put on your new nature, and be renewed as you learn to know **your Creator** and become like him.

I Timothy 2:13–14 (NLT): For **God made Adam** first, and afterward **he made Eve.** ¹⁴And it was not **Adam** who was deceived by Satan. **The woman** was deceived, and sin was the result.

I Timothy 4:3–4 (TEV): Such people teach that it is wrong to marry and to eat certain foods. But **God created those foods** to be eaten, after a prayer of thanks, by those who are believers and have come to know the truth. ⁴**Everything that God has created is good;** nothing is to be rejected, but everything is to be received with a prayer of thanks. . .

I Timothy 6:13 (NASB): I charge you in the presence of **God, who gives life to all things,** and of Christ Jesus, who testified the good confession before Pontius Pilate. . .

Hebrews 1:2 (TLV): In these last days He has spoken to us through a **Son,** whom He appointed heir of all things and **through whom He created the universe.**

Hebrews 1:10 (AMP): And [further], **You, Lord, did lay the foundation of the earth in the beginning, and the heavens are the works of Your hands.**

Hebrews 2:10 (CEV): **Everything belongs to God, and all things were created by his power.** So God did the right thing when he made Jesus perfect by suffering, as Jesus led many of God's children to be saved and to share in his glory.

Hebrews 3:4 (NIV): For every house is built by someone, but **God is the builder of everything.**

Hebrews 11:3 (WEYMTH): Through faith we understand that **the worlds came into being, and still exist, at the command of God,** so that what is seen does not owe its existence to that which is visible.

Hebrews 12:27 (AMP): Now this expression, Yet once more, indicates the final removal and transformation of all [that can be] shaken—that is, of **that which has been created**—in order that what cannot be shaken may remain and continue.

James 3:9–10 (CEV): My dear friends, with our tongues we speak both praises and curses. We praise our Lord and Father, and we curse **people who were created** to be like God, and this isn't right.

I Peter 4:19 (BBE): For this reason let those who by the purpose of God undergo punishment, keep on in well-doing and put their souls into **the safe hands of their Maker.**

II Peter 2:5 (NIV): . . .if he *[God]* did not spare the ancient world when **he brought the flood on its ungodly people, but protected Noah, a preacher of righteousness, and seven others.** . .

II Peter 3:3–6 (CEB): Most important, know this: in the last days scoffers will come, jeering, living by their own cravings, ⁴and saying, "Where is the promise of his coming? After all, nothing has changed—not since **the beginning of creation,** nor even since the ancestors died." ⁵But **they fail to notice that, by God's word, heaven and earth were formed long ago out of water and by means of water.** ⁶And it was through these that **the world of that time was flooded and destroyed.**

Jude 14 (GWORD): Furthermore, Enoch, from the seventh generation after **Adam,** prophesied about them. He said, "The Lord has come with countless thousands of his holy angels."

Revelation 3:14 (TEV): To the angel of the church in Laodicea write: "This is the message from the Amen, **the faithful and true witness, who is the origin of all that God has created."**

Revelation 4:11 (CSB): **Our Lord and God,** you are worthy to receive glory and honor and power, because **you have created all things, and by your will they exist and were created.**

Revelation 10:6 (ESV): . . .and swore by him who lives forever and ever, **who created heaven and what is in it, the earth and what is in it, and the sea and what is in it,** that there would be no more delay,

Revelation 14:7 (DARBY): . . .saying with a loud voice, Fear God and give him glory, for the hour of his judgment has come; and do homage to **him who has made the heaven and the earth and the sea and fountains of waters.**

Okay. So what we've seen in the above list of Scriptures is that the New Testament Gospels, Acts, Epistles, and Revelation *all* casually talk about the Creation account and/or Noah's Flood as self-evident truths, indisputable facts.

If a person believes that God inspired the Scriptures, and if he also desires any rational understanding of reality, he would have to acknowledge as factual the Creation and Flood accounts as well. And if a person *doesn't* believe that God inspired the Scriptures, he would at least have to acknowledge that his conclusion implies that Matthew, Mark, Luke, John, Paul, Peter, James, Jude, and Jesus Himself were all laboring under misconceptions when they spoke or wrote their respective words. And if that's the case, how can such a person trust anything they said? Indeed, how could such a person seriously think he is a Jesus-follower?

Especially if we actually believe what Jesus said (mentioned above):

> John 5:46–47 (CEV): If you believed Moses, you would believe me, because Moses wrote about me. **⁴⁷But if you don't believe what Moses wrote, how can you believe what I say?**
>
> ERV: If you really believed Moses, you would believe me, because he wrote about me. **⁴⁷But you don't believe what he wrote, so you can't believe what I say.**
>
> EXB: [For] If you really believed Moses, you would believe me, because Moses wrote about me [in the Torah, the first five books of the OT; for example, Deut. 18:15 quoted in Acts 3:22]. **⁴⁷But if you don't believe what Moses wrote, how can you believe what I say?**

Here is Jesus saying that it's at least questionable, if not impossible, to believe in Him if we don't even believe what Moses wrote. For Christians who claim Jesus as Savior but deny Biblical creation of the universe in six literal days, Adam and Eve, the Fall of Man, Noah's Flood, and the Tower of Babel, the above statements from Jesus should prompt some sober soul-searching.

According to the Biblical chronology, there are about 2000 years from Creation to the birth of Abraham (described in Genesis 1–11), and about 2000 more years from Abraham to Jesus. If someone dismisses the Creation-related content in Genesis 1 through 11 as merely metaphorical, he is stating that *fully 50% of the Bible's history is wrong.* Do we really want to do that? And if we believe that the first 50% is wrong, why would we believe that the second 50% is reliable?

And another thing: The Bible states that Jesus came to earth, endured humiliation and torture, offered Himself as an atoning sacrifice, dying on the cross to redeem mankind from the sin nature that began with Adam and Eve's rebellion against God. If we accept this Biblical doctrine, but simultaneously claim that Genesis 1–11 are metaphorical and therefore Adam and Eve weren't actually historical characters, this means that Jesus allowed Himself to be tortured and killed to fix a problem that didn't even exist! The problem of the fall of man would have been relegated to an imaginary state, brought about by the metaphorical actions of fictional characters, so why would Jesus bother?

Why would Jesus go through all that to redeem us from a nonexistent state supposedly brought about by metaphorical actions done by fictional characters? If we claim that Genesis 1–11 are not historical, we are necessarily also claiming that Jesus' entire ministry is meaningless. Are we sure we want to go there? If we conclude Jesus' ministry was meaningless, is it even possible to *be* a Christian?

But it shouldn't be surprising that evolution exerts such a powerful draw on the hearts of those who deny God's existence. The Bible very clearly said this would be the case: that people would mock Biblical ideas ("creationism is a pseudoscience"), that people would believe everything continues as it always has (uniformitarianism), that people would believe that creation needs no Creator ("Nothing exploded in the Big Bang, and it became everything") and therefore that people would deliberately choose to believe a lie ("special creation is unthinkable"), people would believe that the earth wasn't originally water (Nebular Hypothesis says it was a ball of magma), and that people would believe Noah's Flood was not an actual historical event.

Really? The Bible predicts all of these things? Yes:

> II Peter 3:3–6 (NLT): Most importantly, I want to remind you that **in the last days scoffers will come, mocking the truth** and following their own desires. ⁴They will say, "What happened to the promise that Jesus is coming again? From before the times of our ancestors, **everything has remained the same since the world was first created.**" ⁵**They deliberately forget** that God made the heavens long ago by the word of his command, and **he brought the earth out from the water** and surrounded it with water. ⁶Then **he used the water to destroy the ancient world with a mighty flood.**

Here's one more thought worth considering:

> Isaiah 29:16 (AMP): [Oh, your perversity!] You turn things upside down! Shall the potter be considered of no more account than the clay? **Shall the thing that is made say of its maker, He did not make me; or the thing that is formed say of him who formed it, He has no understanding?**
>
> CEB: You have everything backward! Should the potter be thought of as clay? **Should what is made say of its maker, "He didn't make me"? Should what is shaped say of the one who shaped it, "He doesn't understand"?**

> MSG: You have everything backward! You treat the potter as a lump of clay. **Does a book say to its author, "He didn't write a word of me"? Does a meal say to the woman who cooked it, "She had nothing to do with this"?**
>
> NLT: How foolish can you be? He is the Potter, and he is certainly greater than you, the clay! **Should the created thing say of the one who made it, "He didn't make me"? Does a jar ever say, "The potter who made me is stupid"?**

So we have seen in this book that the theory of evolution is based on numerous faulty assumptions—that random processes can create information, that the Second Law of Thermodynamics doesn't apply to life, that deep time is a real thing, and so forth. It's understandable, because some people are desperate to keep God out of their lives at any cost, but that doesn't change the fact that evolution is built on the very untrustworthy foundation of wishful thinking, fantasy, and unfounded presuppositions.

Does the Bible say anything about such a situation? Oh, look; it does:

> Matthew 7:26 (NASB): Everyone who hears these words of Mine and does not act on them, will be like **a foolish man who built his house on the sand.** ²⁷The rain fell, and the floods came, and the winds blew and slammed against that house; and **it fell—and great was its fall.**

Maybe a solid understanding of Genesis—*including* the Creation and Flood accounts—is more important than we thought. . .

Appendix A:

Bibliography

As mentioned earlier, one of the most difficult things about writing this book was deciding which information *not* to include. There is *so much* evidence for a supernatural creation of the universe by a Being who is outside of the realm of our physical laws, that much had to be omitted to prevent the book from being ponderously long.

One of the ways I dealt with this situation was to have many footnotes, directing the curious reader to pursue the topic at hand in more detail. Another way to address the situation is to have a bibliography of information sources so the inquisitive reader can delve into as much detail as he wants. Hence, this appendix.

Websites

I can recommend all of these websites; they are listed in alphabetic order of site name.

- Answers in Genesis (AnswersInGenesis.org)
- Apologetics Press (ApologeticsPress.org)
- Ark Encounter (ArkEncounter.com)
- Awesome Science Media (www.AwesomeScienceMedia.com)
- Center for Origins Research and Education (OriginsResearch.org)
- Center for Scientific Creation (CreationScience.com)
- The Creation Club (TheCreationClub.com)

- Creation Evolution Headlines (CREV.info)
- Creation Evolution Science Ministries (CreationMinistries.org)
- Creationism (Creationism.org)
- Creation Ministries International (Creation.com)
- Creation Moments (CreationMoments.com)
- Creation Museum (CreationMuseum.org)
- Creation Research Society (CreationResearch.org)
- Creation Science (CreationScience.net)
- Creation Science (CreationScience.co.uk)
- Creation Science 4 Kids (CreationScience4Kids.com)
- Creation Science Evangelism (DrDino.com)
- Creation Science Today (CreationScienceToday.com)
- Creation Studies Institute (CreationStudies.org)
- Creation Summit (CreationSummit.com)
- Evolution News (EvolutionNews.org)
- Genesis Park (GenesisPark.com)
- Genesis Science (GenesisScience.org)
- Geoscience Research Institute (GRISDA.org)
- Institute for Creation Research (ICR.org)
- Intelligent Design (IntelligentDesign.org)
- Is Genesis History? (IsGenesisHistory.com)
- Northwest Creation Network (NWCreation.net)

Books

There are many books that have been written by solid scientists with advanced degrees, and who have concluded, after looking at the evidence, either or both of two things. First, that the Biblical description of beginnings matches observations much better than we've been led to believe in school and media, and second, that the theory of evolution has a lot of smoke and mirrors, but not much substantive evidence.

I recommend the list of books below. They are listed in alphabetic order of the primary author's last name.

- Axe, Douglas: *Undeniable: How Biology Confirms our Intuition that Life is Designed* (HarperOne, San Francisco, 2017)
- Behe, Michael J.: *Darwin Devolves: The New Science about DNA that Challenges Evolution* (HarperOne, San Francisco, 2019)
- Behe, Michael J.: *Darwin's Black Box: The Biochemical Challenge to Evolution* (Free Press/Simon and Schuster, New York, 2006)

- Brown, Walt: *In the Beginning: Compelling Evidence for Creation and the Flood* (Center for Scientific Creation, Phoenix, Arizona, 1980, 2008)
- Clarey, Tim: *Dinosaurs: Marvels of God's Design* (Master Books, Green Forest, Arkansas, 2015)
- Dembski, William A.: *Intelligent Design: The Bridge Between Science and Theology* (Intervarsity Press, Downers Grove, Illinois, 1999)
- Dembski, William A.: *No Free Lunch: Why Specified Complexity Cannot Be Purchased Without Intelligence* (Rowman and Littlefield Publishers, Inc., Oxford, England, 2002)
- Denton, Michael: *Evolution: A Theory in Crisis* (Adler & Adler Publishers, Bethesda, Maryland, 1985)
- Gish, Duane T: *Dinosaurs by Design* (Master Books, Green Forest, Arkansas, 1999)
- Gish, Duane T: *Evolution: The Fossils Still Say No!* (Institute for Creation Research, El Cajon, California, 1985)
- Ham, Ken: *The Lie: Evolution/Millions of Years* (Master Books, Green Forest, Arkansas, 1987, 2012)
- Ham, Ken and Hodge, Bodie: *The New Answers Book 1* (Master Books, Green Forest, Arkansas, 2006)
- Ham, Ken and Hodge, Bodie: *The New Answers Book 2* (Master Books, Green Forest, Arkansas, 2008)
- Ham, Ken: *The New Answers Book 3* (Master Books, Green Forest, Arkansas, 2009)
- Ham, Ken and Hodge, Bodie: *The New Answers Book 4* (Master Books, Green Forest, Arkansas, 2013)
- Ham, Ken and Hodge, Bodie: *A Flood of Evidence: 40 Reasons Noah and the Flood Still Matter* (Master Books, Green Forest, Arkansas, 2016)
- Johnson, Phillip E.: *Darwin on Trial* (Regnery Publishing, Washington, DC, 1991)
- Lubenow, Marvin L.: *Bones of Contention: A Creationist Assessment of Human Fossils* (Baker Books, Grand Rapids, Michigan, 1992, 2004)
- Morrison, John: *Evolution's Final Days: The Mounting Evidence Disproving the Theory of Evolution* (ZML Corp, 2020)
- Simmons, Geoffrey, M.D.: *What Darwin Didn't Know,* (Harvest House Publishers, Eugene, Oregon, 2004)

- Strobel, Lee: *The Case for a Creator* (Zondervan, Grand Rapids, Michigan, 2004)
- Werner, Carl: *Evolution: The Grand Experiment, Volume 1* (New Leaf Press, Green Forest, Arkansas, 2007, 2014)
- Werner, Carl: *Living Fossils— Evolution: The Grand Experiment, Volume 2* (New Leaf Press, Green Forest, Arkansas, 2009)
- Whitcomb, John C. and Morris, Henry M.: *The Genesis Flood: The Biblical Record and Its Scientific Implications* (P & R Publishing, Phillipsburg, New Jersey, 2011)
- John Woodmorappe: *Noah's Ark: A Feasibility Study* (Institute for Creation Research, Dallas, Texas, 1996)

Glossary

Some technical jargon is unavoidable when discussing the topics like creation and evolution, so this glossary may help clarify some concepts.

Amino Acid: An organic compound composed primarily of carbon, oxygen, hydrogen, and nitrogen. Of the roughly 500 known amino acids, only 20 of them are used in the **genetic information** that controls the processes of life. Amino acids are assembled into **proteins**, and are thus often called the "building blocks" of proteins.

Biological Evolution: the process of simple life forms developing into more complex life forms through the processes of random **mutations** and **natural selection** (survival of the fittest).

Catastrophism: The theory that what we see on earth and in the universe today was caused at least in part by rapid, non-repeating, non-uniform events that made great changes in a short amount of time. Examples include creation itself and Noah's flood. Compare **uniformitarianism**.

Chemical Evolution: the process of random atoms spontaneously assembling themselves into the chemicals required to support life. These chemicals include **amino acids**, **proteins**, **DNA**, and others.

Chromatid: Half of a replicating **chromosome**: one "upright" of the twisted ladder of **DNA**.

Chromosome: A single strand of **DNA**. Humans have 22 pairs of chromosomes that define most of the human characteristics, plus one more pair that determines gender: XX for females, and XY for males. Each upright of the twisted ladder of a chromosome is called a **chromatid**; two chromatids stuck together by their "rungs" make up a chromosome.

Cosmological Evolution: the theory that the current state of the universe—its galaxies, stars, nebulae, its very existence—came from random and undirected natural laws operating over long periods of time.

Creation: The theory that the universe was created by a supernatural God—a transcendent being outside of the constraints of our laws of physics. Compare **evolution**.

Darwinism: The theory of **biological evolution** developed by the English naturalist Charles Darwin and others. The theory states that all species of organisms arise and develop through the **natural selection** of numerous, small, inherited **mutations** that add beneficial genetic information, thereby increasing the individual's ability to compete, survive, and reproduce.

Deep Time: The hypothesized "millions and millions of years" or "billions and billions of years" needed for **evolution** to take place. Also called "long ages."

DNA: Also known as deoxyribonucleic acid, it is the long molecule, in the shape of a double helix (like a twisted ladder), which contains all the **genetic information** required to define the organism of which it is a part.

Entropy: Technically, the degree of unavailability of heat energy available to do work. In layman's terms, the amount of disorder or randomness in a system. Barring direct intervention from God, the Second Law of Thermodynamics guarantees that entropy always increases. **Evolution** disregards this principle entirely by claiming that for evolving things (which supposedly includes the entire universe), entropy *decreases* all by itself.

Enzyme: A **protein** that acts as a catalyst in biological processes; enzymes make other biological processes happen faster. Enzymes are known to catalyze (accelerate) more than 5,000 biochemical processes, and without the increase in speed enabled by enzymes, the chemical reactions necessary for the processes of life would happen too slowly for the organism to survive.

Evolution: The theory that everything we see in the cosmos, the solar system, on Earth, and in life, both animate and inanimate, developed without intent, having been assembled by undirected, random inputs of energy. Depending on the context, it may refer to a belief in **cosmological evolution**, **geological evolution**, **chemical evolution**, or **biological evolution**. Evolution of all kinds presupposes **naturalism**, a religious belief that there is no God. Compare **creation**.

Geological Evolution: the belief that the current state of the earth—the landforms, oceans, tectonic plates, and so forth—is the result of long ages of gradual, constant, uniform processes.

Gene: A "unit of heredity" that determines some characteristic in the offspring: different combinations of the information in genes from the parents dictate genetic traits in the offspring. Genes can be damaged by **mutations**, invariably leading to a decrease in viability.

Genetic Information: Information, coded in an organism's **DNA**, containing instructions defining the characteristics of the life form in which it resides, as well as how to build **proteins** (molecular machines) required for all the processes of life. Information in DNA is another showstopper for the theory of **evolution**, because random processes cannot create information.

For more complex organisms (not counting viruses and bacteria), genetic information is hierarchically organized as follows:

- **Genome:** The complete set of genetic material, in groups called **chromosomes**, that describes an organism.
- **Chromosome:** A single strand of **DNA**. Humans have 46 (23 pairs of) chromosomes.
- **DNA:** Deoxyribonucleic acid, the double-helix molecule that contains all the information required to describes the organism of which it is a part. DNA is composed of **genes**.
- **Gene:** A "unit of heredity" that determines some characteristic in the offspring. Genes are specific sequences of **nucleotides**.
- **Nucleotide:** The basic structural unit of genes, nucleotides contain the instructions for how to build **proteins**.
- **Protein:** Large biomolecules that are the "molecular machines" required to execute the processes necessary for biological life. Proteins are composed of specific sequences of **amino acids**.
- **Amino Acid:** Simple organic compounds, the kind and sequence of which comprise programming instructions that determine the function of the protein of which they are a part.

Genome: The entire set of genetic information for a given organism. The Human Genome Project,[719] completed in 2003, mapped all—more than three billion—nucleotides of the generic human genome (not for a specific individual, but the human species in general).

---

[719] en.wikipedia.org/wiki/Human_Genome_Project; accessed November 8, 2020.

Intelligent Design: The "intelligent design" approach is, in general, an acknowledgment that the universe we live in could *not* have arisen by the random, undirected processes that naturalistic evolutionary theories promote. As such, Intelligent Design proponents differ from Creationists in that the Intelligent Design movement realizes that *some* superintelligence must have created the universe, but they stop short of identifying *Who* that intelligent designer is. Essentially, Creationists acknowledge the God of the Bible as the intelligent designer, but the Intelligent Design movement avoids that final acknowledgment, stopping at the point of understanding that *some* unknown entity had to architect the universe, because randomness is not capable of accomplishing it. Both approaches are good, in that they both expose the numerous shortcomings of evolutionary theories, but only Creationism identifies the Designer. That is, both acknowledge "Evolution couldn't make the universe," but only Creationism completes the thought by adding ". . .but Jesus could."

Irreducible Complexity: An attribute of numerous systems we see all over the world, in which *every* part of a multi-part system is essential for the operation of the system, and the absence, removal, or failure of any one part causes the entire system to fail. Thus, having only a subset of the parts would be detrimental to the organism, and **natural selection** would eliminate that variation. Irreducibly complex systems are yet another showstopper for **evolution**, since they cannot evolve by numerous small changes, because none of the parts would be beneficial until they are *all* present.

Long Ages: See **deep time**.

Macroevolution: The kind of **evolution** in which random **mutations** and **natural selection** result in increased complexity and "higher level" organisms. This requires the enormous amounts of **genetic information** in **DNA** to "arise" spontaneously and with no intelligent input, plus trillions of beneficial **mutations**, even though not even one beneficial mutation has ever been seen. **Macroevolution** is another name for **biological evolution**.

Microevolution: A misleading name for the process of change within a kind of organism. This word does not really describe **evolution**, in the sense that the theory requires; these minor changes are simply new expressions of existing **genetic information**. Microevolution does not require new genetic information, nor does it result in new kinds of organisms.

Mutation: A modification of the **genetic information** in **DNA**, which is passed on to offspring. Mutations are usually caused by DNA-copying errors or radiation. Most, if not all, mutations are harmful; the jury's still out on whether there is such a thing as a neutral mutation. Beneficial mutations, which add new functional genetic information, and which **macroevolution** requires in astronomical numbers, have never been seen.

Naturalism: A philosophical/religious belief that there is no supernatural God. Naturalism is an essential starting point of evolution, because if there *is* a supernatural God, He certainly could have created the universe, so there would be no need for the theory of **evolution**. But if there *isn't* a supernatural God, He certainly couldn't have created anything, so we "need" the theory of evolution to explain the universe.

Natural Selection: The process of "weeding out" organisms whose **DNA** has been mutated, because mutations damage DNA, which results in reduced fitness for survival: greater susceptibility to disease, reduced mental capacity, less strength, and so forth. Because of the decline in fitness, damaged organisms left to themselves are less likely to compete successfully for resources. Natural selection is a real thing, but it can only *preserve* existing **genetic information**; it cannot *create* new genetic information.

Neo-Darwinism: Any integration of Darwin's theory of **evolution** with genetics. There are many variations on this theme.

Nucleotide: The basic structural unit of **genes**, nucleotides contain the programming instructions for how to build **proteins**, which are the molecular machines that carry out the biochemical processes required for life.

Protein: A large molecule composed of chains of twenty different kinds of **amino acids**. The sequence of amino acids determines the function of the resulting protein, and proteins are assembled according to instructions in the **DNA**. About 200,000 different proteins are needed in human cells in order to maintain life processes.

Punctuated Equilibrium: A variation of **evolution**, Punctuated Equilibrium (also called "Punk Eek") allows evolution to behave like **Creation**, but still deny the existence or involvement of God. Firstly, because **uniformitarianism** is unable to produce what we see all over the planet (since it is limited to the laws of physics) Punk Eek allows **catastrophism** in certain, otherwise unexplainable circumstances. Secondly, for some unknown reason and using some unknown process, Punk Eek allows for the instantaneous, quite magical, appearance of enormous amounts of spectacularly intricate **genetic information**, so that parents of one species can produce functional, healthy offspring of a different species. The existence of the Punk Eek hypothesis allows processes that are alarmingly Creation-like to still be explained away as "evolution."

Ribosome: A macromolecular machine that performs **protein** synthesis, part of the process of **DNA** replication.

Saltation: A naturalistic theory that attempts to explain the gaps in the fossil record: it claims that huge increases in complexity, with its required huge increases of **genetic information**, can occur in a single generation. No explanation is offered as to where the information comes from, or how it could come about via random processes.

Second Law of Thermodynamics: The law of physics that states everything in the universe above the atomic level runs down and wears out; that is, entropy is always increasing. Without God's direct intervention, even intentional, directed, intelligent efforts only delay the inevitable increase in disorder and decay. This is significant for the topic of this book because **evolution** supposedly is unaffected by the Second Law of Thermodynamics.

Taxonomy: In general, the science of classification, noting and categorizing the similarities and differences between things. In biology, it is the science of classifying and naming organisms, usually with the method developed by Carl Linnaeus. The various levels or organization in the taxonomy of living creatures are usually defines as kingdom, phylum, class, order, family, genus, and species. Taxonomic names that are similar are often based on the assumption that Darwin's tree of life is a valid concept, but this has been thoroughly disproven (see "The Phylogenetic Tree").

Uniformitarianism: The theory that "the present is the key to the past;" in other words, things have always been the way they are today. Thus, the universe as we see it today came about through nothing more than unchanging, **naturalistic** processes acting over **deep time** (enormous periods of time). Compare **catastrophism**.

About the Author

David Arns was raised in church, but didn't start actually serving the Lord until his sophomore year of high school, in 1972. Being of a rather analytical turn of mind, he was delighted to see that there is a Biblical mandate for all Christians to be analytical: I Thessalonians 5:21 (NIV) says "Test everything. Hold on to the good." That, coupled with Paul's exhortation to teach what "the Holy Ghost teaches," not depending on man's wisdom (I Corinthians 2:11–14), and with the commendation of the Bereans, who "searched the Scriptures daily, whether those things were so" (Acts 17:11), pretty much define Dave's life, in the spiritual realm, as well as the natural realm. In the mid-1970s, Dave heard a sermon in which he was exhorted to "know *what* you believe and *why* you believe it," and he has been trying to put that into practice ever since. He has been known to abandon long-held beliefs when someone showed him that they were incompatible with Scripture; that attitude seems to be necessary if we want to continue to grow in the Lord.

Books in the "Thoughts On" Series

This book is a member of the "THOUGHTS ON" series of books. The phrase "Thoughts On" is deliberately ambiguous, because it is meaningful and accurate either way you interpret it. First, it indicates where the seeds of the whole series came from: they were from a large list of informal Bible studies Dave had put together for his own interest and edification as a result of his "thoughts on" various topics that occurred to him during his quiet times with the Lord. And second, it indicates one of Dave's goals as a teacher: to persuade people to turn their "thoughts on" and consider logically what God has said in His word, and how it is very much to our benefit to take heed to what He says.

When reading *The Chronicles of Narnia* to his son Matthew when he was little, Dave came across the Professor's exasperated musing: "'Logic!' said the Professor half to himself. 'Why don't they teach logic at these schools?'" Oh, did that ring true with him! Many are the times Dave has heard a preacher or Bible teacher make a statement from the pulpit, and the crowd responds with a hearty "Amen!" Dave looks around astonished, thinking, "That statement's not true! I can think of three Scriptures off the top of my head that refute it!" And he just grieves for the complacency evident in most Christians; there is *so* much God wants to bless them with, and they miss out because they don't check the Bible to verify statements they hear.

So, Dear Reader, please turn your Thoughts On. . .

OH, EVOLVE! (GOOD LUCK WITH THAT. . .)

To see the names and descriptions of the other books in the "THOUGHTS ON" series, see the list below. To see the sources from which they are available, or to contact the author, see the website BibleAuthor.DaveArns.com. Books are available both in electronic form and in paperback. Note that the numbers of these books within the THOUGHTS ON series are merely the order in which they were written; they do not need to be read in sequential order. All of them are stand-alone books, so Book 1, for example, does not need to be read before Book 2, and so on.

BOOK 1: *Prophets vs. Seers: Is There a Difference?* This book looks at that question from a Biblical viewpoint. There are Bible teachers teaching that prophets and seers are fundamentally different, and they offer some supporting evidence, while others say they are merely variations in the manifestation of fundamentally the same gift and calling. Is there enough Scriptural evidence to conclude that they are the same kind of person, or the same kind of calling, or are they indeed different? An in-depth analysis of related Scriptures leads the author to a solid conclusion.

BOOK 2: *Is It Possible to Stop Sinning?* There are a couple common beliefs in Christianity today: one holds that Christians living on earth will inevitably continue to sin until they graduate to heaven, and the other holds that it is possible for Christians to be without sin even while living on earth. Of course, the major factor in this discussion is what the Bible says. For example, What is sin? What does God say about it? What does God tell us to do about it? What did Jesus provide in the atonement? This book delves into great detail on the subject and includes Biblical support from many relevant Scriptures, showing God's heart on the matter, in a way that is both theologically relevant and practical in everyday life.

BOOK 3: *Extra-Biblical Truth: A Valid Concept?* There is a theory that says that God will not do nor say anything for which there is not a Biblical precedent, nor would He reveal a doctrine that was hitherto unheard of. Is this theory reasonable? Does the Bible itself address the question of God doing or saying things that are not already exemplified in the Bible itself? Actually, the Bible does address this question very clearly, and in several different ways. This book

OH, EVOLVE! (GOOD LUCK WITH THAT. . .)

illustrates how to analyze and discern, from a Scriptural point of view, events and practices for which the Bible doesn't have specific examples.

BOOK 4: *Gold Dust, Jewels, and More: Manifestations of God?* For the last couple of decades, there have been more and more reports of "unusual" occurrences taking place at meetings in which the Holy Spirit is allowed to move freely. These occurrences include gold dust appearing on people and things, jewels suddenly popping into existence, people "falling under the power" (a.k.a., being "slain in the Spirit"), glory clouds hovering or swirling, oil coming from people's hands, and more. Are these real manifestations of God, or just the result of overzealous but unethical leaders? Is there a Biblical basis for any of these? This book delves into the Scriptures and analyzes passages that are often overlooked, to give a thoughtful and Biblically sound response to these reports of unusual manifestations.

BOOK 5: *If It Be Thy Will* Many people in the Body of Christ, when they pray for physical healing, end their prayers with ". . .if it be Thy will." That brings up a very important point: *Is it* God's will to heal us? Never, sometimes, or always? How do we know? What does the Bible say? So often, Jesus said to the people He just healed, "Your faith has made you well." Where did they get that faith, and can we learn from them? This book goes into great detail about what the Bible says—and does *not* say—about physical healing, and whether praying for it is something we are forbidden, discouraged, permitted, encouraged, or commanded to do. The Bible has much to say on this subject, and we can learn a great deal by just looking at what it says, and noting the obvious implications.

BOOK 6: *Free to Choose?* One of the most hotly debated concepts in the last 500 years or so has been that of whether or not people actually have a free will to choose their eternal destiny. People debate each other with—shall we say, *religious* fervor—and people on both sides of the debate offer Scriptures to support their viewpoints. On the one hand, we have people who believe that God offers us a choice to voluntarily repent and turn to Him. On the other hand, we have people who believe that God is sovereign, and that sovereignty necessarily means that God determines the eternal destination of

everyone, with no regard to our choices. These two viewpoints can't both be correct, because they say mutually exclusive things. But fortunately, the Bible is remarkably unambiguous in its teachings: reading Scriptures in context and thinking about how various passages relate to each other make it abundantly clear which one of these viewpoints is actually the Biblical position.

BOOK 7: ***Be Filled With the Spirit*** In the last fifty years or so, there has been a tremendous resurgence of interest in the baptism of the Holy Spirit and the accompanying gifts of the Spirit. In some, the interest is entirely academic; in others, it is a passionate hunger to experience it firsthand. But there are people who claim that such things faded away around the end of the first century, and are therefore no longer available. Did they really fade away? We need to know because other people claim to have been baptized in the Holy Spirit and use the gifts of the Spirit every day, as a normal part of Christian life. As always, the Bible is the normative standard for living the Christian life, so what does the Bible say on this topic? Quite a lot, and if we follow what the Bible says, our Christian lives will become much more exciting and fruitful in the things of the Kingdom.

BOOK 8: ***Going Beyond Christianity 101*** What would be the content of a "Christianity 101" class? In other words, what is "elementary" Christianity? To avoid pet doctrines and the inevitable differences of opinion, we should see what the Bible itself describes as the "elementary doctrines" or the "foundational principles" of the faith. These are enumerated in Hebrews 6:1–2 as: repentance from sin, faith toward God, baptisms (plural), the laying on of hands, the resurrection of the dead, and eternal judgment. Listening to the amount of heated discussion in the body of Christ on these topics, we soon realize that as a whole, the body of Christ doesn't even have a good handle on the *elementary* doctrines yet. The Bible says much on these doctrines that is often overlooked by those doing only a casual study. This book looks at the Scriptures pertaining to these six topics in great detail, and then speculates on what it might mean to "go beyond" these foundational teachings, as Hebrews 6:1 encourages us to do.

BOOK 9: *Searching the Word: Bible Word-Search Puzzles on Steroids* What do you get when you cross a word-lover with a Word-lover? In other words, what do you get when you cross a person who enjoys word games with a student of the Bible? And then, for good measure, throw in a teacher and a writer. What do you get? This book. Much more than just a book of word-search puzzles, and much more than just a book of Bible lists, this book combines the fun of solving word problems with a fascinating way to study the Bible. Words or phrases from the Bible, and which fit into the same category, are used as the word lists for the puzzles. While you're looking for words, sooner or later you're bound to think, "What does *that* mean?" and when you check the info section for that puzzle, you'll learn something and realize you've discovered a delightful new way to study the Bible!

BOOK 10: *Hearing from God: A Daily Devotional* Many daily devotionals are in very small, bite-sized installments that you can read in three minutes or less. This may be very appropriate for people who are always on the go, and are doing so at God's leading. But such tiny tidbits, while they may be very good and very true, are still pretty small, and as such, have insufficient room to get very deep. As such, they are barely spiritual *hors d'oeuvres*, let alone a hearty spiritual meal of "strong meat." If you have a bit more time, this devotional is a good alternative. It goes into greater depth and breadth in the Scriptural support and elaboration. You may notice that the list of Scripture references at the bottom of each day's entry is longer than you have seen in other daily devotionals. This is deliberate: You'll be blessed if you read all the Scriptures for each day's devotional, even if two or three passages seem to say the same thing—when the Bible makes similar statements but expresses them slightly differently, the various nuances of meaning are significant and enlightening; they are not merely accidental. There is amazing depth in the Scriptures. . .

BOOK 11: *Lord of the Dance* The Bible talks about dancing in many places, both as an act of worship, and as a normal expression of joy. The church, after a lengthy period of thunder-fisted condemnation of all dance, as if it could not possibly occur without wallowing in sin, is recognizing their previous overreaction and seeing dance in many positive aspects: as an expression of worship, an enjoyable social activity, and a way to improve bodily fitness and mental acuity, to name but a few. Having been a dance instructor since 1999, and a student of the Word for even longer, the author could not help noticing that there are a great many correlations between a man and a woman dancing, and a husband and wife in a marriage. These correlations were vividly brought into focus while teaching engaged couples how to dance for their upcoming weddings—it's remarkable how often dance lessons included, almost unavoidably, significant premarital counseling. And those same correlations apply with even more eternal import in our relationship with Christ our Redeemer. This book explores many of those correlations and similarities in a way that presents concepts of dance almost as parables whose meanings, for those who have ears to hear, are nothing less than profound in the marital and spiritual realms.

BOOK 12: *Prophetic Ministry: A Biblical Look at Seeing* Scripture tells us to "eagerly desire spiritual gifts, especially the gift of prophecy" (I Corinthians 14:1, NIV). Why "especially" the gift of prophecy? The Bible seems to emphasize prophecy as the highest gift, so there must be a good reason. And indeed, there is; in fact, there are many. This book examines Scriptures that tell us about how prophecy works: Who is authorized to pursue this gift, how people can perceive messages from God, what forms they can take, how to deliver them to the intended recipients, the necessary attitude and demeanor when doing so, common pitfalls, and more. If you have been hungering to hear the voice of God, rest assured that you can, because Jesus said, "My sheep hear my voice" (John 10:27). You *do* hear His voice. That is wonderful in itself. But when you have the privilege of speaking into someone else's life God's own words *for that specific person and moment and situation*, that is even more wonderful. Yes, eagerly desire spiritual gifts, *especially* the gift of prophecy. You'll be glad you did.

BOOK 13: *Arise, My Beloved Daughter* Recently, an increasing number of prophetic words from established, world-class prophets—of both genders—are calling for women to arise and fulfill the callings and destinies that God ordained for them before the world began. And women are rising to the call: thoughtful, godly, competent women, with compassion for the lost, deep intimacy with God, and a passion to see the mercy and blessings of Jesus poured out onto a seriously damaged world. Also, there is a dawning awareness on the part of males in church leadership that they have been missing out on much of what God wants to pour out because highly gifted women have been disregarded, ignored, passed over, and even actively suppressed in their attempts at ministry. God is opening up revelation about things that have been in the Word all along, but about which we have long had a flawed understanding. Why is God revealing it now? Because with the glory that God is intending to pour out in the Third Great Awakening, the Church no longer has the "luxury" of limping along with half of its soldiers in the brig because the other half thinks they're incapable.

BOOK 14: *Oh, Evolve! (Good Luck With That. . .)* When the question of Creation vs. Evolution comes up, many people immediately assume it is a question of science vs. religion. But is it really? There are many scientists with impressive credentials in a variety of fields—many of them clearly *not* creationists—who are becoming more vocal all the time about the problems with the whole Darwinian idea of how everything came to be. And it's true: there are more discoveries every year that militate against the ideas of the Big Bang, deep time, the Nebular Hypothesis of how the solar system was formed, uniformitarianism, life "arising" by random and undirected processes, and more. This book examines the problems with a variety of evolutionary assumptions, many of which are expressed by evolutionists themselves, and shows, in laymen's terms, why the theory of evolution is collapsing under the weight of its own presuppositions. Evidence from cosmology, geology, chemistry, genetics, biology, and more, is becoming increasingly hostile to evolutionary notions. Because of this, more and more "rescue devices" (supplementary theories intended to explain why observations don't match evolutionary predictions) are needed each year, to prop up the teetering theory. Not only will you see that evolution is no less a religion than Christianity, but you'll see that the Creation vs. Evolution debate is science vs. science. Check out the evidence, and see which model is more supported by real-world observations!

BOOK 15: *One Nation Under God . . .Again!* One of the recent discussions that has been generating more heat than light lately pertains to the spiritual underpinnings of the Founding Fathers of these United States: whether or not they intended to include Biblical/Christian principles in the founding documents, and therefore the entire fabric of our American society. There are some modern scholars who say the Founders were godless and secular, and other modern scholars who say they were solid Biblical Christians. Who is right? Rather than simply quoting recent writings concerning what the Founders "must have" meant, it is much more reliable to look at the writings of the Founders themselves, in context, compare their content to the Bible, and see how well they match. Unlike some modern scholars who "omit the scholarly practice" of including citations, expecting their readers to simply trust their conclusions, this book includes hundreds of footnotes containing citations, so you can go to the original documents themselves and verify the statements herein. When you do, you will see that our Declaration of Independence, Constitution, and Bill of Rights are not at all "godless" documents written from a secular mindset, but are filled with Biblical references, concepts, and wisdom. Armed with that knowledge and understanding, you will be able to confidently promote, as did the Founders, the strength of character and solid societal foundations that originally formed the basis of this country. If the Body of Christ rises to the challenge, we will indeed be one nation under God . . .*again!*

Music in the "Worship On" Series

Dave's current music project is the "Worship On" series of albums. The phrase "Worship On" not only parallels Dave's "Thoughts On" series of books, but it also points out a very significant truth about the destiny of those who choose to make Jesus Christ the Lord of their lives: Though many other aspects of normal Christian life on earth—evangelism, healing the sick, casting out demons, raising the dead, suffering persecution, and so forth—will go away once we're in heaven, worship will not. Throughout all eternity, we will worship Jesus, the King of Kings. Far from being an arduous chore we will be required to do, we will spontaneously burst out into joyous praise and worship every time we see another aspect of God's goodness and love and holiness. As we discover more of God's marvelousness moment by moment, it will be more clear than ever that He is the only One worthy of our worship—no one and nothing else even comes close. Indeed, the word "worship" comes from the Old English phrase "worth-ship," and He is certainly worth all of our worship.

Oh, Evolve! (Good Luck With That. . .)

So, Dear Listener, when listening to this music, feel free to Worship On. And on and on. . . :)

The music below is available both in downloadable electronic form and as CDs, and is available from the sources mentioned on the website Music.DaveArns.com.

CD 1: *Songs of the Tribe of Judah* In the early 1980s, Dave was a member of the worship team at the church he attended. In addition to that, a subset of that worship team formed a band that sang on other occasions and in other, more public venues. This smaller group called themselves the Tribe of Judah, after the name referring to Jesus in Revelation 5:5. Dave and one other member of the group wrote most of the songs they performed, and in this album are the songs that Dave wrote, along with improved orchestration. The reason for the name of this album is twofold: first, these songs were written when Dave was writing songs to be performed by the band called the Tribe of Judah, and second, because Judah means "praise and worship," which is what Dave prays this music will inspire in you.

CD 2: *Worship the King* The second album in the "Worship On" series, *Worship the King* is intended to draw the listener from a passive "listening" mode and into a more active "worshiping" mode. As you listen to the words of these songs, you'll notice than many of them are taken straight from the Bible, and as such, are excellent tools with which to learn Scripture. Even the ones that are not taken directly from the Bible are laden with Scriptural concepts, whether their context is worshiping Him in the beauty of holiness, the story of an Appalachian moonshiner who encounters the living God, a description of every believer's job on earth, a joyous proclamation of God's glorious traits, or a simple acknowledgement of the most basic understanding of every believer: that the Lord is good.

CD 3: *Go Into All the World* This album, the third in the "Worship On" series, acknowledges the importance of Jesus' exhortation to "Go into all the world" and preach the gospel to everyone (Mark 16:15–20). The wheat field image recalls Jesus' commands to pray that laborers will go into the fields, because the harvest is plentiful (Matthew 9:37–38). Because of that emphasis, this album contains songs echoing Isaiah's cry "Send me!", marveling at God's mercy, showing how a Caribbean man sees Jesus gloriously working among his people, expressing the hunger that God's children feel to get into His presence, recalling a portion of one of David's psalms that he gave to Asaph and the other worshippers to sing, and more. My hope is that your heart will be touched with compassion for those who don't yet know the inexpressible joy of being a child of God.

CD 4: *I Have Not Forgotten You* This album, the fourth in the "Worship On" series, endeavors to respond to those in the body of Christ who have heard God's promises, both those in the Bible and those He has spoken to them personally, who remember His prophetic words, and who feel like it is taking for*ever* for those promises to come to pass. To such people, as well as to those who have experienced great hardship in their lives, God's unchanging faithfulness comes through in *I Have Not Forgotten You*, and His love in a conversation between the heavenly Father and one of His beloved children in *That Will I Seek After*. In *Hear and Do*, a believer discovers the simple but profound secret to living in God's presence, and in *Truckin'*, a truck driver has a divine appointment with a couple of the Lord's servants. Other songs include the word of the Lord coming to a cattle driver crossing Death Valley, a believer echoing Moses' heartfelt cry to see God's glory, and an expression of intense spiritual hunger when such a large outpouring of that glory—a "glory storm"—is seen building on the horizon.

CD 5: *Dry Bones to Living Stones*

This album, the fifth in the "Worship On" series, describes several different aspects of God's process of building His people—His "living stones—into a holy and glorious temple He can inhabit. One song tells of the Father's desire to give us the Kingdom; another tells about a surfer hearing the voice of God promising a wave of the Holy Spirit; another portrays the hunger to drink deeply of God's Spirit—a hunger we should all have. Yet another describes the realization that the long-awaited revival of societal transformation into wholeness and health has finally arrived; another tells the story of a bored and lukewarm Christian discovering that there is more! Another relates the little-known key to Jesus' success in ministry, and another tells in a new way the story of Shadrach, Meshach, and Abednego being thrown into the fiery furnace. And finally, a song that expresses the passion of a believer who doesn't want to miss out on what God is doing in these days.

Books: BibleAuthor.DaveArns.com

Music: Music.DaveArns.com

www.ingramcontent.com/pod-product-compliance
Lightning Source LLC
Chambersburg PA
CBHW071347210526
45465CB00001B/6